THE THEORY OF
ADSORPTION
AND CATALYSIS

PHYSICAL CHEMISTRY
A Series of Monographs

ERNEST M. LOEBL, *Editor*

Department of Chemistry, Polytechnic Institute of Brooklyn
Brooklyn, New York

THE THEORY OF
ADSORPTION
AND CATALYSIS

Alfred Clark

PHILLIPS PETROLEUM COMPANY
RESEARCH AND DEVELOPMENT CENTER
BARTLESVILLE, OKLAHOMA

ACADEMIC PRESS   New York and London   1970

ACADEMIC PRESS, INC.
111 Fifth Avenue, New York, New York 10003

United Kingdom Edition published by
ACADEMIC PRESS, INC. (LONDON) LTD.
Berkeley Square House, London W1X 6BA

LIBRARY OF CONGRESS CATALOG CARD NUMBER: 78-107553

PRINTED IN THE UNITED STATES OF AMERICA

CONTENTS

III. Localized Adsorption—Dependent Systems

IV. Localized Adsorption—Dependent Systems (Continued)

V. Nonlocalized Adsorption

VI. Physical Forces of Adsorption

VII. The Chemical Forces of Adsorption—Metals

VIII. The Chemical Forces of Adsorption—Semiconductors

IX. The Kinetics of Chemisorption

Part II. CATALYSIS

X. Adsorption and Catalysis

XI. Kinetics of Heterogeneous Catalysis—Diffusion Steps Neglected

PREFACE

One of the main functions of science is to develop theories that explain and predict. In adsorption and catalysis, explanations are still tentative and predictions precarious, for surface phenomena are unusually difficult to observe and measure. So, unavoidably, the following pages are full of fledgling theories, some of which may never pass the test of sustained flight. However, such theories are worth reviewing because they help to polarize thinking and to pose sharp questions.

A quantitative and nonempirical approach has been adopted so far as possible. Although empirical equations often can be made to fit experimental data better, they explain less about mechanisms. The approach works best for physical adsorption, where surprisingly simple models frequently turn out to be useful caricatures of reality. The treatment grows increasingly qualitative and empirical, for it becomes increasingly difficult to devise suitable models as the complexity mounts from physical adsorption to chemisorption to catalysis. Tacitly indicated is the enormous amount of research, theoretical and experimental, yet to be done.

The author wishes to thank Mrs. Mary Townsend Crow and Mrs. Pat McGlasson for skillfully and patiently typing the difficult manuscript; Professors M. Boudart and J. J. Carberry for reading the manuscript and making valuable suggestions; his colleagues, too numerous to mention individually, for taking the time for many clarifying discussions; Phillips Petroleum Company for providing the facilities used in preparing the manuscript and figures; and Academic Press for their excellent editorial and stylistic suggestions.

THE THEORY OF
ADSORPTION
AND CATALYSIS

ADSORPTION

CHAPTER

I

THERMODYNAMICS OF ADSORPTION

1.1 Introduction

Solid catalysts work by adsorbing at least one reactant. Therefore, a knowledge of adsorption is necessary for a fundamental understanding of catalysis.

Adsorption on solid surfaces is a complex phenomenon. Most catalytic surfaces are not uniform. They consist of sites with a broad distribution of adsorption energies and irregular surface patterns. The density of sites may be extremely low or so high as to form a continuum. The adsorbed molecules may be mobile or immobile. They may be independent of each other or they may interact with their nearest neighbors or even with more distant ones, and, as a result, undergo phase transformations. The forces binding foreign molecules to a surface may be physical, including van der Waals' attraction forces and the forces developed by dipoles and ions as they approach surfaces; or they may be equivalent to the forces in chemical bonds involving molecular orbital overlap. In theory, physical adsorption and chemical adsorption are easily distinguishable, but no experimental criteria are available yet to make completely reliable distinctions in practice. Perturbations of surface atoms or ions may occur as a result of adsorption. It is usually assumed that perturbations intensify with increasing strength of the adsorption bond. Since the forces of physical adsorption are often much weaker than chemical forces, the assumption is frequently made that perturbations can be neglected in the case of physical adsorption. In any event, no satisfactory methods are available for dealing with these perturbations.

Obviously, specific models must be used in the attempt to bring order into the complexity of the adsorption picture. We propose to treat adsorption models in the order of their increasing specificity. The classical thermodynamic approach of this chapter introduces the least physical restrictions into the treatment of adsorption phenomena. The relationships developed are more generally true than those of subsequent chapters. But the price we pay for not being more specific in our model is the loss of detailed information about the adsorption process. When, as in subsequent chapters, we specify that the adsorbed molecules are mobile (or immobile), our model loses generality, but gains much in detailed information about those real systems for which it is an approximation. Even more detailed information accrues from models in which the nature of the adsorption forces are stipulated.

We shall now derive the most important results of the application of classical thermodynamics to surfaces. We shall assume that the adsorbent is unperturbed, that it may be considered as an inert material supplying a potential field for the adsorbate. First the differential energy equation will be derived in the form applicable to adsorption, and from it the other thermodynamic functions. Next, these functions will be used to derive expressions for four important isothermal heats of adsorption: the isosteric heat, the differential heat, the equilibrium heat, and the equilibrium energy of adsorption. The relationships between these heats will then be derived. Finally, we shall develop an expression for the adiabatic heat of adsorption. A knowledge of elementary classical thermodynamics is assumed.

1.2   Fundamental Equations

In ordinary three-dimensional thermodynamics, the total energy E is a function of three independent variables which are most conveniently selected to be S, the total entropy; V, the total volume; and n_i, the number of moles of each species present in the system. The differential energy is written

$$dE = T\,dS - P\,dV + \sum \mu_i\,dn_i, \qquad (1)$$

which may be integrated holding intensive variables temperature T, pressure P, and chemical potentials μ_i, constant to give

$$E = TS - PV + \sum \mu_i n_i. \qquad (2)$$

In two-dimensional, or adsorption thermodynamics, we will show that an additional work term is required. Following Hill (*1–3*), we start from the thermodynamics of solutions and, with appropriate assumptions, arrive at equations applicable to adsorption. Consider a two-component condensed phase containing n_A moles of a nonvolatile component and n_s moles of a volatile component in equilibrium with the gas phase. The differential energy of the *condensed* phase is

$$dE = T \, dS - P \, dV + \mu_A \, dn_A + \mu_s \, dn_s, \tag{3}$$

where P is the *hydrostatic* pressure exerted by a hypothetical piston or (in part) by a hypothetical inert additional gas on the volume of the condensed phase.

The equation applies to such diverse systems as argon–graphite, hydrogen–tungsten, hydrogen–charcoal, hydrogen–palladium, benzene–rubber, water–sodium chloride, and water–sulfuric acid. If the nonvolatile component is a solid, it is understood that a change dn_A in n_A refers to solid of the same state of subdivision, specific surface, etc. For the pure substance, we write

$$dE_{0A} = T \, dS_{0A} - P \, dV_{0A} + \mu_{0A} \, dn_A. \tag{4}$$

The following quantities are now defined:

$$E_s \equiv E - E_{0A}, \qquad V_s \equiv V - V_{0A}, \qquad S_s \equiv S - S_{0A}, \qquad \Phi \equiv \mu_{0A} - \mu_A. \tag{5}$$

So far these quantities have no special significance physically. For example, E_s is just the difference between the total energy of the condensed phase and the energy of n_A moles of pure substance. By subtracting Eq. (4) from Eq. (3), we obtain the differential energy

$$dE_s = T \, dS_s - P \, dV_s - \Phi \, dn_A + \mu_s \, dn_s. \tag{6}$$

In order to make the transition to adsorption thermodynamics, we stipulate that the n_A moles of nonvolatile substance are inert. For this special case, E_s, for example, becomes just the energy of n_s moles of adsorbed molecules in the potential field of the surface of the inert adsorbent (the energy of the adsorbent subtracts out except for the interaction energy between adsorbent and adsorbed molecules, which is left in E_s). Similar meanings apply to V_s and S_s. In the identity $\Phi \equiv \mu_{0A} - \mu_A$, μ_{0A} is the chemical potential of pure adsorbent with clean surface and μ_A is the chemical potential of pure adsorbent with a surface layer of adsorbate. We have

$$(\partial E_{0A}/\partial n_A)_{S_{0A}, V_{0A}} = \mu_{0A},$$

$$(\partial E/\partial n_A)_{S, V, n_s} = \mu_A, \tag{7}$$

$$\Phi \equiv \mu_{0A} - \mu_A = -(\partial E_s/\partial n_A)_{S_s, V_s, n_s}.$$

Thus, the difference $\Phi \equiv \mu_{0A} - \mu_A$ represents the energy change per unit of adsorbent in the surface spreading of adsorbate. We assume that, for an inert adsorbent, n_A is proportional to surface area α, so that

$$\Phi\, dn_A = \Phi c\, d\alpha \equiv \varphi\, d\alpha, \tag{8}$$

and

$$\varphi = -(\partial E_s/\partial \alpha)_{S_s, V_s, n_s}. \tag{9}$$

The quantity $\varphi\, d\alpha$ is the two-dimensional equivalent of the three-dimensional work term $P\, dV$, and φ is often called the "spreading pressure." It is well known that $\varphi = \gamma_0 - \gamma$, where γ_0 is the surface tension of the clean surface and γ is the surface tension of the surface with adsorbate.

We may now write

$$dE_s = T\, dS_s - P\, dV_s - \varphi\, d\alpha + \mu_s\, dn_s \tag{10}$$

for the energy differential of a one-component system of n_s moles of adsorbed gas.

So long as the adsorbent is truly inert, Eq. (10) is useful. From here on we shall assume an inert adsorbent. Whether or not this is a completely valid assumption is a difficult question. For weak adsorption, the assumption is usually considered justifiable, but not for strong adsorption. The question will be discussed further in Chapter VI.

The fundamental equations of adsorption thermodynamics can now be set up, based on the approximation of an inert adsorbent and a one-component system of n_s moles of adsorbed gas. A complete set of thermodynamic functions will be provided from which the various heats of adsorption may then be conveniently derived (see Section 1.3). We use Eq. (10) and the following definitions of enthalpy H_s, Gibbs free energy F_s, and the Helmholtz free energy A_s:

$$H_s \equiv E_s + PV_s,$$
$$F_s \equiv H_s - TS_s, \tag{11}$$
$$A_s \equiv E_s - TS_s.$$

From these, the following fundamental equations for thermodynamic differentials are obtained:

$$dE_s = T\, dS_s - P\, dV_s - \varphi\, d\alpha + \mu_s\, dn_s, \tag{12}$$

$$dH_s = T\, dS_s + V_s\, dP - \varphi\, d\alpha + \mu_s\, dn_s, \tag{13}$$

$$dA_s = -S_s\, dT - P\, dV_s - \varphi\, d\alpha + \mu_s\, dn_s, \tag{14}$$

$$dF_s = -S_s\, dT + V_s\, dP - \varphi\, d\alpha + \mu_s\, dn_s. \tag{15}$$

By integrating with all intensive variables constant:

$$E_s = E_s(S_s, V_s, \alpha, n_s) = TS_s - PV_s - \varphi\alpha + \mu_s n_s, \tag{16}$$

$$H_s = H_s(S_s, P, \alpha, n_s) = TS_s - \varphi\alpha + \mu_s n_s, \tag{17}$$

$$A_s = A_s(T, V_s, \alpha, n_s) = -PV_s - \varphi\alpha + \mu_s n_s, \tag{18}$$

$$F_s = F_s(T, P, \alpha, n_s) = -\varphi\alpha + \mu_s n_s. \tag{19}$$

In these equations, we see that four independent variables are required to describe completely each thermodynamic variable, whereas in ordinary three-dimensional thermodynamics, only three variables are needed for a one-component system. In each equation, the area α is taken as an independent variable.

From the fundamental equations, φ and μ_s are defined by

$$\varphi \equiv -(\partial E_s/\partial \alpha)_{S_s, V_s, n_s} = -(\partial H_s/\partial \alpha)_{S_s, P, n_s}$$
$$= -(\partial A_s/\partial \alpha)_{T, V_s, n_s} = -(\partial F_s/\partial \alpha)_{T, P, n_s}, \tag{20}$$

$$\mu_s \equiv (\partial E_s/\partial n_s)_{S_s, V_s, \alpha} = (\partial H_s/\partial n_s)_{S_s, P, \alpha}$$
$$= (\partial A_s/\partial n_s)_{T, V_s, \alpha} = (\partial F_s/\partial n_s)_{T, P, \alpha}. \tag{21}$$

1.3 Isothermal Heats of Adsorption

First, we shall derive expressions for two *differential* heats of adsorption, making use of the thermodynamic differentials, Eqs. (14) and (15). Then we shall derive expressions for two *integral* heats of adsorption from the integrated thermodynamic formulas, Eqs. (18) and (19). The definitions of these four isothermal heats of adsorption follow from their derivations. We begin by deriving an expression for the differential heat known as the isosteric heat of adsorption, q_{st}, using Eq. (15) in conjunction with the equation

$$d\mu_s = (\partial\mu_s/\partial T)_{P, n_s, \alpha}\, dT + (\partial\mu_s/\partial P)_{T, n_s, \alpha}\, dP$$
$$+ (\partial\mu_s/\partial n_s)_{T, P, \alpha}\, dn_s + (\partial\mu_s/\partial\alpha)_{T, P, n_s}\, d\alpha. \tag{22}$$

Substituting in this equation, $(\partial F_s/\partial n_s)_{T, P, \alpha} = \mu_s$, and noting that

$$\left[\frac{\partial(\partial F_s/\partial n_s)_{T, P, \alpha}}{\partial T}\right]_{P, n_s, \alpha} = \left[\frac{\partial(\partial F_s/\partial T)_{P, n_s, \alpha}}{\partial n_s}\right]_{T, P, \alpha}$$
$$= -(\partial S_s/\partial n_s)_{T, P, \alpha} \equiv -\bar{s}_s, \tag{23}$$

and similarly that

$$(\partial\mu_s/\partial P)_{T, n_s, \alpha} = (\partial V_s/\partial n_s)_{T, P, \alpha} \equiv \bar{v}_s, \tag{24}$$

$$(\partial\mu_s/\partial\alpha)_{T, P, n_s} = -(\partial\varphi/\partial n_s)_{T, P, \alpha}, \tag{25}$$

we obtain

$$d\mu_s = -\bar{s}_s\, dT + \bar{v}_s\, dP - (\partial\varphi/\partial n_s)_{T,P,\alpha}\, d\alpha + (\partial\mu_s/\partial n_s)_{T,P,\alpha}\, dn_s. \quad (26)$$

A change in μ_s at constant amount adsorbed ($dn_s = 0$) and at constant adsorbent area ($d\alpha = 0$) gives

$$d\mu_s = -\bar{s}_s\, dT + \bar{v}_s\, dP. \quad (27)$$

At equilibrium with the gas $d\mu_s = d\mu_G$, or

$$-\bar{s}_s\, dT + \bar{v}_s\, dP = -s_G\, dT + v_G\, dp, \quad (28)$$

where

$$s_G \equiv S_G/n_G = (\partial S_G/\partial n_G)_{T,p}, \qquad v_G \equiv V_G/n_G = (\partial V_G/\partial n_G)_{T,p},$$

and p is the equilibrium gas pressure. Note that $\bar{s}_s$ and $\bar{v}_s$ are differential molar quantities at constant temperature, pressure and surface area, and in general vary with n_s in contrast to s_G and v_G which are integral molar quantities and independent of n_G at constant T and p. We shall always express a molar quantity by a lower-case letter, and shall place a bar above it for the differential molar quantity at constant T, p, and α, for example,

$$(\partial F_G/\partial n_G)_{T,p} = \mu_G = F_G/n_G \equiv f_G;$$

$$(\partial F_s/\partial n_s)_{T,p,\alpha} = \mu_s = (\partial f_s/\partial n_s)_{T,p,\alpha}n_s + f_s \equiv (\partial f_s/\partial n_s)_{T,p,\alpha}n_s + F_s/n_s) \equiv \bar{f}_s.$$

Thus $\bar{f}_s = f_s$ only if $(\partial f_s/\partial n_s)_{T,p,\alpha} = 0$, which is not true in general.

Under usual conditions, the hydrostatic pressure P on the adsorbed layer consists only of the equilibrium gas pressure p and therefore $P = p$. Alternatively the hydrostatic pressure P may be assumed constant ($dP = 0$) while p changes with temperature. However, variations in P have little effect on the adsorbed volume V_s which is small in comparison to V_G and virtually incompressible. Thus

$$(\partial \ln p/\partial T)_{P,n_s,\alpha} \cong (\partial \ln p/\partial T)_{n_s,\alpha};$$

also

$$(\partial S_s/\partial n_s)_{P,T,\alpha} \cong (\partial S_s/\partial n_s)_{T,\alpha} \cong (\partial S_s/\partial n_s)_{V_s,T,\alpha}, \quad \text{etc.,}$$

and furthermore,

$$(\partial H_s/\partial n_s)_{T,\alpha} \cong (\partial E_s/\partial n_s)_{T,\alpha}.$$

From Eq. (28) we obtain

$$(\partial p/\partial T)_{n_s,\alpha} = (s_G - \bar{s}_s)/(v_G - \bar{v}_s). \quad (29)$$

In the usual approximation, $v_G \gg \bar{v}_s$, and assuming a perfect gas,

$$(\partial \ln p/\partial T)_{n_s, \alpha} = (s_G - \bar{s}_s)/RT. \tag{30}$$

At equilibrium, $\mu_G = \mu_s$, i.e., $f_G = \bar{f}_s = h_G - Ts_G = \bar{h}_s - T\bar{s}_s$, and therefore,

$$T(s_G - \bar{s}_s) = h_G - \bar{h}_s \equiv h_G - (\partial H_s/\partial n_s)_{T, P, \alpha}. \tag{31}$$

Also, n_s and α may be replaced in Eq. (30) by $\Gamma \equiv n_s/\alpha$, a surface concentration, since only the ratio, n_s/α, and not the total quantity of each is significant. Eq. (30) may, therefore, be written as

$$\left(\frac{\partial \ln p}{\partial T}\right)_\Gamma = \frac{h_G - \bar{h}_s}{RT^2} \equiv \frac{q_{st}}{RT^2}, \tag{32}$$

in which $h_G - \bar{h}_s \equiv q_{st}$ defines the isosteric heat of adsorption. It should be noted that, by convention, $\Delta h \equiv \bar{h}_s - h_G \equiv -q_{st}$.

Since $(\partial H_G/\partial p)_T = 0$ for a perfect gas, $h_G - \bar{h}_s$ corresponds to the enthalpy difference between one mole of perfect gas at any pressure and one mole of adsorbate in equilibrium with gas at pressure p. On the other hand, s_G is not independent of pressure and therefore $s_G - \bar{s}_s$ corresponds to the entropy difference between one mole of gas at equilibrium pressure p and one mole of adsorbate in equilibrium with that pressure.

There is nothing in Eq. (32) that restricts its application to systems with inert adsorbent. One can always use the equation to obtain heats of adsorption from experimental isosteres or isotherms, whether perturbations of the adsorbent exist or not. However, only in the absence of perturbations can significant interpretations of the data be made. Otherwise, the heat of adsorption is distributed in some unknown and complex manner between the adsorbate and adsorbent. Halsey (7) has discussed the formal thermodynamic equations in which a work term for perturbations is explicitly shown.

In order to show that q_{st} is actually the heat transferred to the constant temperature bath in an isothermal, isobaric process, we use

$$dQ = dE + p \, dV, \tag{33}$$

where $E = E_G + E_s$ and $V = V_G + V_s$. Setting $E_G = n_G \, e_G$, $E_s = n_s \, e_s$ and $dn_G = -dn_s$, while neglecting V_s, and remembering that e_G and v_G are constant, we find

$$\begin{aligned} dQ &= -e_G \, dn_s + n_s(\partial e_s/\partial n_s)_{\alpha, T} \, dn_s + e_s \, dn_s - pv_G \, dn_s \\ &= [(\partial E_s/\partial n_s)_{\alpha, T} - h_G] \, dn_s \cong [(\partial H_s/\partial n_s)_{\alpha, T} - h_G] \, dn_s \\ &\equiv -q_{st} \, dn_s. \end{aligned} \tag{34}$$

If there is more than one species adsorbing on the inert adsorbent, the energy differential becomes

$$dE_s = T \, dS_s - P \, dV_s - \varphi \, d\alpha + \sum_{i=1}^{m} \mu_{si} \, dn_{si}, \tag{35}$$

where μ_{si} and n_{si} are the chemical potential and number of moles of the ith adsorbed species. At equilibrium, $d\mu_{si} = d\mu_{Gi}$. The procedure used above will now yield an equation for each species, and for a perfect gas we find

$$(\partial \ln p_i/\partial T)_{\Gamma_1, \Gamma_2, ..., \Gamma_m} = (h_{Gi} - \bar{h}_{si})/RT^2 = q_{sti}/RT^2. \tag{36}$$

Proceeding in a similar manner, we now derive an expression for the *differential heat of adsorption* q_d. First an expression for $d\mu_s$ in terms of the variables T, V_s, α, n_s is obtained, using Eq. (14). Then, setting $d\mu_s = d\mu_G$ gives

$$(\partial S_s/\partial n_s)_{V_s, \alpha, T}\, dT - (\partial P/\partial n_s)_{V_s, \alpha, T}\, dV_s - (\partial \varphi/\partial n_s)_{V_s, \alpha, T}\, d\alpha$$
$$+ (\partial \mu_s/\partial n_s)_{V_s, \alpha, T}\, dn_s = -s_G\, dT + v_G\, dp. \tag{37}$$

Substituting RT/p for v_G and $d\ln(n_G RT/V_G)$ for $d\ln p$, and holding V_s, α, n_s, V_G constant gives

$$(\partial S_s/\partial n_s)_{V_s, \alpha, T}\, dT = -s_G\, dT + RT\, d\ln n_G + R\, dT, \tag{38}$$

or

$$\left(\frac{\partial \ln n_G}{\partial T}\right)_{V_s, V_G, \Gamma} = \frac{s_G - (\partial S_s/\partial n_s)_{V_s, \alpha, T}}{RT} - \frac{1}{T},$$

where n_G is the total number of moles of gas in a system of constant volume V_G. At equilibrium,

$$\mu_s = (\partial A/\partial n_s)_{V_s, \alpha, T} = (\partial E_s/\partial n_s)_{V_s, \alpha, T} - T(\partial S_s/\partial n_s)_{V_s, \alpha, T}$$
$$= \mu_G = h_G - Ts_G, \tag{39}$$

or

$$T[s_G - (\partial S_s/\partial n_s)_{V_s, \alpha, T}] = h_G - (\partial E_s/\partial n_s)_{V_s, \alpha, T},$$

so that we may write

$$\left(\frac{\partial \ln n_G}{\partial T}\right)_{V_s, V_G, \Gamma} = \frac{h_G - (\partial E_s/\partial n_s)_{V_s, \alpha, T} - RT}{RT^2}$$
$$= \frac{e_G - (\partial E_s/\partial n_s)_{V_s, \alpha, T}}{RT^2}, \tag{40}$$

in which $q_d \equiv e_G - (\partial E_s/\partial n_s)_{V_s, \alpha, T} \cong e_G - \bar{e}_s$ defines the differential heat of adsorption, and where $\bar{e}_s = (\partial E_s/\partial n_s)_{T, P, \alpha} \cong (\partial E_s/\partial n_s)_{V_s, \alpha, T}$.

For this isothermal, constant volume process,

$$dQ = dE = dE_G + dE_s, \tag{41}$$

and it is readily shown by the procedure used to derive Eq. (34) that

$$dQ = [(\partial E_s/\partial n_s)_{\alpha,\,T} - e_G]\,dn_s = -q_d\,dn_s. \tag{42}$$

We now calculate what Everett (4) calls the *equilibrium heat of adsorption*. From Eq. (19), we obtain by taking the differential,

$$dF_s = -\varphi\,d\alpha - \alpha\,d\varphi + \mu_s\,dn_s + n_s\,d\mu_s. \tag{43}$$

Setting this equal to Eq. (15) gives

$$n_s\,d\mu_s = -S_s\,dT + V_s\,dP + \alpha\,d\varphi,$$

or   (44)

$$d\mu_s = -s_s\,dT + v_s\,dP + (1/\Gamma)\,d\varphi.$$

At equilibrium with the gas

$$-s_s\,dT + v_s\,dP + (1/\Gamma)\,d\varphi = -s_G\,dT + v_G\,dp. \tag{45}$$

Putting $P = p$, as before, and considering the case of constant φ,

$$-s_s\,dT + v_s\,dp = -s_G\,dT + v_G\,dp, \tag{46}$$

or, dropping the insignificant term v_s,

$$(\partial \ln p/\partial T)_\varphi = (s_G - s_s)/RT. \tag{47}$$

We note that the surface thermodynamic functions obtained in this case are integral molar quantities. They are equal to partial (differential) molar quantities with all intensive properties held constant, as follows:

$$(\partial V_s/\partial n_s)_{T,\,P,\,\varphi} = V_s/n_s \equiv v_s,$$

$$(\partial \alpha/\partial n_s)_{T,\,P,\,\varphi} = \alpha/n_s \equiv 1/\Gamma, \tag{48}$$

$$(\partial S_s/\partial n_s)_{T,\,P,\,\varphi} = S_s/n_s \equiv s_s.$$

In contrast, the partial molar quantities developed previously in connection with the isosteric and differential heats of adsorption, namely, $\bar{s}_s = (\partial S_s/\partial n_s)_{T,\,P,\,\alpha}$, $\bar{v}_s = (\partial V_s/\partial n_s)_{T,\,P,\,\alpha}$, etc., are not taken with all intensive variables constant (φ varies while α is constant). Therefore, these quantities are not necessarily equal to integral molar quantities, but, in general, vary with n_s.

The question arises as to what is the enthalpy difference that corresponds to $s_G - s_s$ at equilibrium. To determine this, we note from Eq. (19) that

$$\mu_s = (F_s + \varphi\alpha)/n_s = (H_s - TS_s + \varphi\alpha)/n_s \equiv (\mathbf{H}_s - T\,S_s)/n_s, \tag{49}$$

where we have defined $\mathbf{H}_s \equiv H_s + \varphi\alpha$. At equilibrium, $\mu_s = \mu_G = f_G$ or

$$(\mathbf{H}_s - TS_s)/n_s \equiv \mathbf{h}_s - Ts_s = h_G - Ts_G, \tag{50}$$

which leads to

$$h_G - \mathbf{h}_s = T(s_G - s_s). \tag{51}$$

We may now write

$$(\partial \ln p/\partial T)_\varphi = (h_G - \mathbf{h}_s)/RT^2, \tag{52}$$

where $h_G - \mathbf{h}_s$ defines the equilibrium heat of adsorption.

This process is not only isothermal and isobaric, as in the production of the isosteric heat, q_{st}, but it also operates at constant surface pressure φ and thus requires a "surface piston" which delivers an additional amount of heat, $\varphi \, d\alpha$, to the surrounding bath during the adsorption of dn_s moles. We write for this process, neglecting V_s,

$$dQ = dE_G + dE_s + p \, dV_G + \varphi \, d\alpha, \tag{53}$$

which is readily shown to be equal to $(\mathbf{h}_s - h_G) \, dn_s$. The introduction of a "surface piston" poses experimental problems so that equilibrium heats cannot be measured calorimetrically like q_{st} and q_d.

It is interesting to note that starting with the equilibrium condition Eq. (45) and letting T be constant, the Gibbs equation is obtained (8):

$$d\varphi = \Gamma[RT/p - v_s] \, dp, \tag{54}$$

or, neglecting v_s, and integrating,

$$\varphi = RT \int_0^p \Gamma \, d \ln p \qquad (T \text{ constant}). \tag{55}$$

It was pointed out by Hill (2), that the best way to calculate the equilibrium heat of adsorption is to use this equation. The procedure is:

(1)   Use adsorption isotherm data in the calculation of φ from

$$\varphi = RT \int_0^p \Gamma \, d \ln p \qquad (T \text{ constant}).$$

(2)   Then use

$$(\partial \ln p/\partial T)_\varphi = (s_G - s_s)/RT = (h_G - \mathbf{h}_s)/RT^2$$

to determine the heat of adsorption.

Another integral quantity, called by Everett (4) *the equilibrium energy of adsorption*, is easily obtained using previous methods:

$$(\partial \ln n_G/\partial T)_{V_G, \varphi} = (e_G - e_s)/RT^2, \tag{56}$$

where

$$\mathbf{e}_s = \mathbf{E}_s/n_s \equiv (E_s + \varphi\alpha)/n_s.$$

There is no limit to the number of different heats of adsorption which can be expressed in terms of equations of the Clausius–Clapeyron type. To show this, take $p = p(\Gamma, T)$ and

$$d \ln p = (\partial \ln p/\partial T)_\Gamma \, dT + (\partial \ln p/\partial \Gamma)_T \, d\Gamma. \tag{57}$$

Let us keep some function $\chi = \chi(\Gamma, T)$ constant. Then

$$(\partial \ln p/\partial T)_\chi = (\partial \ln p/\partial T)_\Gamma + (\partial \ln p/\partial \Gamma)_T(\partial \Gamma/\partial T)_\chi$$
$$= (q_{st}/RT^2) + (\partial \ln p/\partial \Gamma)_T(\partial \Gamma/\partial T)_\chi. \tag{58}$$

Similar results are obtained by keeping any function $f(\chi) = f[\chi(\Gamma, T)]$ constant. It is clear from Eq. (58) that for each choice of χ, a different heat of adsorption is obtained. Only a few are useful, including the case where $\chi \equiv \Gamma$, which gives the isosteric heat of adsorption, Eq. (32), and $\chi \equiv \varphi$, which gives the equilibrium heat of adsorption, Eq. (52).

1.4   Relationships between Isothermal Heats of Adsorption

Expressions for four different heats of adsorption have been derived:

$q_{st} \equiv (h_G - \bar{h}_s)$, the isosteric heat of adsorption;
$q_d \equiv (e_G - \bar{e}_s)$, the differential heat of adsorption;
$(h_G - \mathbf{h}_s)$, the equilibrium heat of adsorption;
$(e_G - \mathbf{e}_s)$, the equilibrium energy of adsorption.

The relationship between them will now be developed.

The difference between the isosteric and differential heat of adsorption is

$$q_{st} - q_d = h_G - e_G - [(\partial H_s/\partial n_s)_{P, T, \alpha} - (\partial E_s/\partial n_s)_{V_s, \alpha, T}]. \tag{59}$$

Using $H_s \equiv E_s + PV_s$ and $V_s \equiv n_s v_s$, we get

$$q_{st} - q_d = RT - (\partial E_s/\partial n_s)_{P, T, \alpha} + (\partial E_s/\partial n_s)_{V_s, \alpha, T}$$
$$- Pv_s - Pn_s(\partial v_s/\partial n_s)_{P, T, \alpha}. \tag{60}$$

As previously mentioned, however, $(\partial E_s/\partial n_s)_{V_s, \alpha, T} \cong (\partial E_s/\partial n_s)_{P, T, \alpha}$. Also, the last two terms usually make a negligible contribution. As a good approximation, we write

$$q_{st} \cong q_d + RT. \tag{61}$$

The difference between the equilibrium heat of adsorption and the equilibrium energy of adsorption is

$$(h_G - e_G) - (\mathbf{h}_s - \mathbf{e}_s) = pv_G - pv_s \cong RT. \tag{62}$$

An expression for the difference between the equilibrium heat of adsorption and the isosteric heat of adsorption is obtained as follows:

We add and subtract $T\bar{s}_s$ to Eq. (51), giving

$$h_G - h_s = Ts_G - T\bar{s}_s + T\bar{s}_s - Ts_s = q_{st} + T(\bar{s}_s - s_s). \tag{63}$$

To evaluate $T(\bar{s}_s - s_s)$, we equate the expressions for $d\mu_s$ given in Eqs. (26) and (44) at constant P, n_s, α:

$$-\bar{s}_s \, dT = -s_s \, dT + (1/\Gamma) \, d\varphi, \tag{64}$$

or

$$(1/\Gamma)(\partial\varphi/\partial T)_{P, n_s, \alpha} = (1/\Gamma)(\partial\varphi/\partial T)_{P, \Gamma} = -(\bar{s}_s - s_s),$$

and usually at equilibrium $P = p$.

Substituting in Eq. (63) gives

$$h_G - h_s = q_{st} - (T/\Gamma)(\partial\varphi/\partial T)_{\Gamma, P} \tag{65}$$

for the relationship between the equilibrium heat of adsorption and the isosteric heat of adsorption.

1.5   Adiabatic Heat of Adsorption

Heats of adsorption are often conveniently measured adiabatically. Kington and Aston (5) have developed the relationship between q_a, the adiabatic heat, and the isothermal heats. Since a reversible change in an adiabatic process takes place at constant entropy, q_a is defined by

$$q_a \equiv (C_c + n_G c_{pG} + n_s c_{ps})(\partial T/\partial n_s)_{\alpha, S}, \tag{66}$$

where C_c is the heat capacity of the calorimeter and adsorbent at constant pressure, c_{pG} is the molar specific heat at constant pressure of the gas and c_{ps} is the molar specific heat at constant pressure and area of the adsorbed phase and $(\partial T/\partial n_s)_{\alpha, S}$ is at constant total entropy and area of the adsorbed phase. Following the derivation of Young and Crowell (6), we consider an adsorption process in an adiabatic calorimeter equipped with a movable piston. Initially, the system is in equilibrium and the total energy of the calorimeter and its contents is

$$E = E_c + E_G + E_s. \tag{67}$$

Then the piston is moved infinitely slowly so that V_G changes by dV_G and a new equilibrium state is reached with the transfer of dn_s moles to the surface phase. The calorimeter and adsorbent are assumed to be incompressible. The change for the adiabatic process is

$$dE + p \, dV_G = dE_c + dE_G + dE_s + p \, dV_G = 0. \tag{68}$$

Assuming that the adsorbent and calorimeter are inert and that the adsorbed phase has constant area α, we obtain for reversible processes

$$dE_c = C_c \, dT, \tag{69}$$

$$dE_G = T \, dS_G - p \, dV_G + \mu_G \, dn_G \tag{70}$$

$$dE_s = T \, dS_s - p \, dV_s + \mu_s \, dn_s \tag{71}$$

However, $\mu_G = \mu_s$, and $dn_G = -dn_s$ so that Eq. (68) becomes

$$C_c \, dT + T \, dS_G + T \, dS_s = 0. \tag{72}$$

It is necessary to obtain expressions for dS_G and dS_s.
   Taking $S_G = S_G(T, p, n_G)$ and $S_s = S_s(T, p, n_s, \alpha)$ gives

$$dS_G = (\partial S_G/\partial T)_{p, n_G} \, dT + (\partial S_G/\partial p)_{T, n_G} \, dp + (\partial S_G/\partial n_G)_{T, p} \, dn_G, \tag{73}$$

$$dS_s = (\partial S_s/\partial T)_{p, n_s, \alpha} \, dT + (\partial S_s/\partial p)_{T, n_s, \alpha} \, dp$$
$$+ (\partial S_s/\partial n_s)_{p, T, \alpha} \, dn_s \qquad \text{(at constant } \alpha\text{)}. \tag{74}$$

But,

$$\left(\frac{\partial S_G}{\partial T}\right)_{p, n_G} = (1/T)(\partial Q_G/\partial T)_{p, n_G} = n_G \, c_{pG}/T, \tag{75}$$

or

$$c_{pG} = (T/n_G)(\partial S_G/\partial T)_{p, n_G},$$

and

$$(\partial S_s/\partial T)_{p, n_s, \alpha} = (1/T)(\partial Q_s/\partial T)_{p, n_s, \alpha} = n_s \, c_{ps}/T, \tag{76}$$

or

$$c_{ps} = (T/n_s)(\partial S_s/\partial T)_{p, n_s, \alpha},$$

where Q_G and Q_s are reversible heats.
   We also need the following well-known relationships:

$$(\partial S_G/\partial p)_{T, n_G} = -(\partial V_G/\partial T)_{p, n_G} = -V_G/T, \tag{77}$$

$$(\partial S_s/\partial p)_{T, n_s, \alpha} = -(\partial V_s/\partial T)_{P, n_s, \alpha} \cong 0. \tag{78}$$

Substituting Eqs. (75)–(78) into Eqs. (73) and (74) gives

$$dS_G = n_G \, c_{pG} \, dT/T - (V_G/T) \, dp - s_G \, dn_s \tag{79}$$

and

$$dS_s = n_s \, c_{ps} \, dT/T + \bar{s}_s \, dn_s \qquad \text{(constant area)}. \tag{80}$$

Substituting Eqs. (79) and (80) into Eq. (72), we obtain

$$C_c \, dT + n_G \, c_{pG} \, dT + n_s \, c_{ps} \, dT - V_G \, dp + T\bar{s}_s \, dn_s - Ts_G \, dn_s = 0. \quad (81)$$

Using q_a from Eq. (66) and q_{st} from Eq. (31) gives

$$q_a = q_{st} + V_G(\partial p/\partial n_s)_{S, \alpha}, \quad (82)$$

where S is the *total* entropy of the system. Equation (82) is put into a more convenient form by noting that at equilibrium

$$n_s = n_s(p, T, \alpha) \quad \text{or} \quad p = p(T, \alpha, n_s),$$

so that, at constant adsorbent area,

$$dp = (\partial p/\partial T)_{n_s, \alpha} \, dT + (\partial p/\partial n_s)_{T, \alpha} \, dn_s. \quad (83)$$

Also, T may be expressed, $T = T(n_s, \alpha, S)$, so that $dT = (\partial T/\partial n_s)_{\alpha, S} \, dn_s$ at constant α and S. Therefore,

$$(\partial p/\partial n_s)_{\alpha, S} = (\partial p/\partial T)_{n_s, \alpha}(\partial T/\partial n_s)_{\alpha, S} + (\partial p/\partial n_s)_{T, \alpha}, \quad (84)$$

or

$$(\partial p/\partial n_s)_{\alpha, S} = (\partial p/\partial T)_\Gamma(\partial T/\partial n_s)_{\alpha, S} + (\partial p/\partial n_s)_{T, \alpha}.$$

Substituting Eq. (84) into Eq. (82) gives

$$q_a = q_{st} + V_G(\partial p/\partial n_s)_{T, \alpha} + V_G(\partial p/\partial T)_\Gamma(\partial T/\partial n_s)_{\alpha, S}. \quad (85)$$

Kington and Aston (5), studying the adsorption of nitrogen on rutile at 77°K, obtained both the adiabatic heat of adsorption q_a and, independently, the isosteric heat of adsorption q_{st}. Using relation, Eq. (85), they showed that the results of the two methods differed by less than 1.0%. Thus, they clarified for the first time, as Hill (2) pointed out, the relationship between these quantities.

REFERENCES

1. Hill, T. L., *J. Chem. Phys.* **13**, 520 (1949).
2. Hill, T. L., *Advan. Catalysis* **4**, 211 (1952).
3. Hill, T. L., *J. Chem. Phys.* **18**, 246 (1950).
4. Everett, D. H., *Trans. Faraday Soc.* **46**, 453 (1950).
5. Kington, G. L., and Aston, J. G., *J. Am. Chem. Soc.* **75**, 1929 (1951).
6. Young, D. M., and Crowell, A. D., "Physical Adsorption of Gases." Butterworth, London and Washington, D.C., 1962.
7. Halsey, G.D., Jr., *J. Colloid Interface Sci.* **21**, 358 (1966).
8. Van Ness, H. C., *Ind. Eng. Chem. Fundam.* **8**, 464 (1969).

II

LOCALIZED ADSORPTION—INDEPENDENT SYSTEMS

2.1 Description of Localized Adsorption

The phenomenon of adsorption may be classified in several different ways (*1*). It may be classified as physical or chemical, where the distinction is based on the magnitude of the heat of adsorption. Although widely used because of its convenience, this classification is not very precise. It may also be classified as monolayer or multilayer. Here again the distinction is not very exact because the building of multilayers often begins before the first layer is complete. Capillary condensation is considered to be a special case of multilayer adsorption, where multilayers growing from the sides of a small pore meet in the middle, thus filling the entire pore.

Adsorption may also be classified as localized or nonlocalized. In localized adsorption, molecules are considered to be adsorbed at discrete sites where they reside in positions of minimum potential energy. We shall illustrate localized adsorption by a monatomic gas on the surface of a regular lattice of a solid. The potential energy of interaction of a molecule of gas with the particles of the lattice is the sum of all the individual interactions

$$U = \sum_i u(r_i), \tag{1}$$

where $u(r_i)$ is the potential energy of interaction of the molecule of gas with the ith particle of the lattice at a distance r_i. U is a function of x, y, and z, where x and y are coordinates of the adsorbed molecule in the plane of the surface parallel to the rows and columns respectively of the particles of the solid, and z is the coordinate perpendicular to the surface. If x and y are held constant while z varies, $U(z)$ will have the shape of a potential energy curve with a minimum such as that given by the Lennard-Jones intermolecular pair potential (see Chapter VI). Other choices of x and y will give different values of $U(z)$. For example, $U(z)$ directly above a particle of the solid will be different from $U(z)$ between two particles. Specifically, U_m the potential minimum of $U(z)$ will vary periodically in both the x and y direction. A plot of U_m, the potential minimum of $U(z)$, against either x or y will give a curve with maxima and minima. The difference between the energy of the minima and that of molecules in the gas phase is the energy of adsorption ($\underline{U}_0$), and the difference between the energy of the minima and the maxima is the energy required for surface diffusion ($\underline{V}_0$). In the simplest case of localized adsorption, an

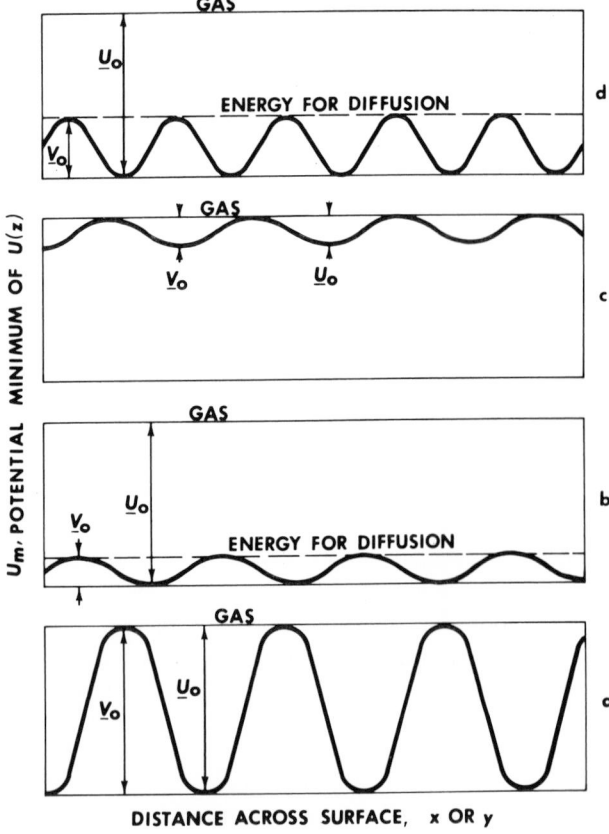

FIG. 2.1. Energy diagrams for localized adsorption on uniform surfaces.

adsorbed molecule is considered to be held at the bottom of a potential well whose depth is much greater than kT ($\underline{V}_0$, $\underline{U}_0 \gg kT$), the thermal energy of the molecule. All potential wells at the surface are assumed to be equal in depth and regularly spaced, as in Fig. 2.1a. Molecules in such positions have practically no chance of lateral movement across the surface. The three translational degrees of freedom that they possess in the gas phase have been transformed into three vibrational degrees of freedom, two in the plane of the surface (normal coordinates for the two-dimensional motion) and one perpendicular to the surface. In the other extreme of localized adsorption, potential wells again are equal, regularly spaced, and just as deep with respect to the gas phase as before ($\underline{U}_0 \gg kT$), but with a much lower energy barrier against lateral movement across the surface ($\underline{V}_0 \cong kT$). The situation is pictured in Fig. 2.1b. The lower barrier between potential wells which allows high diffusion rates, though the energy of adsorption is great, may be due to overlapping of potential force fields as suggested by Gilliland (*2*). High surface mobility may also be the result of small adsorption energies ($\underline{U}_0$, $\underline{V}_0 \cong kT$), as shown in Fig. 2.1c. There is an intermediate case of localized adsorption, where the adsorbed molecule is neither completely fixed nor completely mobile. Although the adsorbed molecule in this case spends a considerable part of its time at the bottom of a potential well, the well is not sufficiently deep to prevent some lateral movement of the molecule as a result of its thermal energy from one potential minimum to another. Figure 2.1d depicts this situation. Other cases of localized adsorption result when potential minima vary in depth over the surface. We shall call a surface *nonuniform* when its potential minima are not equal. An example of one possible kind of localized

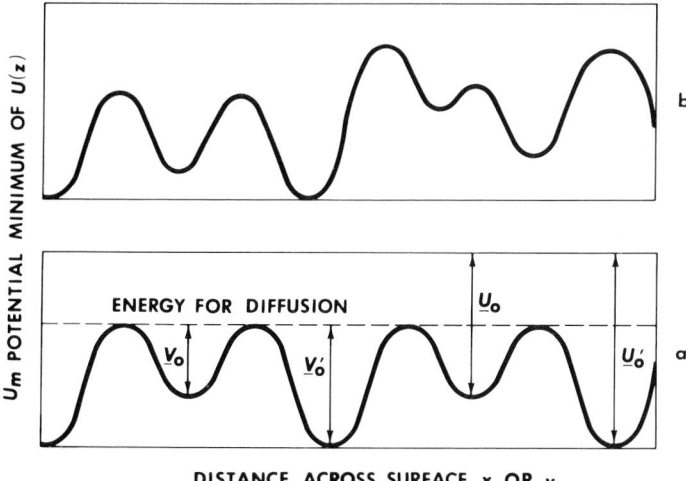

FIG. 2.2. Energy diagrams for localized adsorption on nonuniform surfaces.

adsorption on a nonuniform surface is shown in Fig. 2.2a, where the potential minima are regularly spaced. There are two different potential energy minima ($\underline{U}_0$, $\underline{U}_0'$) and two different energies necessary to raise molecules out of their respective minima to the level where diffusion can take place ($\underline{V}_0$, $\underline{V}_0'$). In this particular case, however, the energy level for surface diffusion is constant over the entire surface. The most complex type of localized adsorption occurs on a nonuniform surface where there are not only many different potential minima, but the minima are also spaced randomly over the surface. This is shown in Fig. 2.2b. In this case, molecules can exist in many different situations depending on their location on the surface.

When there are no favored positions of minimum potential energy on a surface, we have nonlocalized adsorption. The surface is not necessarily uniform, thus there may be regions of higher energies of adsorption than others. Localized adsorption with shallow potential minima as in Figs. 2.1b and c approximates nonlocalized adsorption. In fact, nonlocalized adsorption may be considered as the limit of localized adsorption as the distances between potential minima approach zero, becoming a continuum.

Adsorption may also be classified as mobile or immobile. In general, nonlocalized adsorption will be mobile unless, of course, the adsorbed molecules condense at high surface coverages. However, localized adsorption, as we have seen, may be mobile or immobile depending on the characteristics of the surface.

In this chapter, we shall discuss the theory of localized adsorption of independent systems, that is, systems in which a molecule adsorbed on a site does not interact with molecules on nearest-neighbor or more distant sites. As in Chapter I, we shall continue to assume that the adsorbent is inert, that it supplies a potential field for adsorption without being perturbed itself. Both the immobile case and the intermediate case (hopping molecules) will be treated here. The mobile case, however, will be treated in Chapter V since it approximates the nonlocalized case. The divisions between these cases are not sharp, and real systems are sometimes difficult to classify. A completely immobile system of adsorbed particles is paradoxical, for we are concerned with equilibrium systems, and equilibrium cannot be established without at least some small exchange with the gas phase. An idea of the lifetimes of adsorbed molecules as a function of the heat of adsorption is given in Table 2.1 for 25 and 400°C. If $\underline{V}_0$ is low enough, the adsorbed molecules may move around on the surface considerably during the lifetime of adsorption. The average adsorbed life is related to the heat of adsorption by the expression $\tau = \tau_0 \exp(Q/RT)$, where τ_0 is a property of the solid related to the frequency of vibration of the atoms of the solid (1).

Through the methods of statistical mechanics, partition functions will be derived for monatomic and polyatomic molecules, for adsorption of more

TABLE 2.1

AVERAGE LIFETIME OF ADSORPTION τ AT 25°C
AND 400°C[a]

Q, kcal/mole	τ, sec, 25°C	τ, sec, 400°C
0.1	6.0×10^{-14}	5.0×10^{-14}
1.0	2.7×10^{-13}	1.0×10^{-13}
10.0	1.6×10^{-6}	8.5×10^{-11}
15.0	9.0×10^{-3}	3.5×10^{-9}
20.0	50.0	1.4×10^{-7}
25.0	3.0×10^{5}	1.2×10^{-6}
30.0	2.0×10^{9}	2.5×10^{-4}
40.0		4.0×10^{-1}
50.0		1.4×10^{2}

[a] $\tau_0 = 5 \times 10^{-14}$ sec.

than one molecule per site, and for adsorption on nonuniform surfaces. From the partition functions, the standard thermodynamic functions for localized molecules on surfaces are derived. The increased fund of knowledge obtained over the more general results of the preceding chapter will be evident.

2.2   Immobile Adsorption of Monatomic Molecules on Uniform Surfaces

First, the partition function of a single adsorbed atom is derived with the use of the harmonic oscillator approximation; and it is then employed to obtain the partition function Q_s for a system of N_s vibrating atoms on B sites, where $N_s \leqslant B$. From Q_s, the Langmuir isotherm equation for immobile adsorption with all sites of the same adsorption energy, no interaction between adsorbed molecules, and no more than one molecule per occupied site will be derived by the methods of statistical thermodynamics. Fowler (3) was the first to make such a derivation, thereby clearing up some of the assumptions involved in the Langmuir equation and also giving an explicit definition of the empirical constant appearing in the original kinetic derivation of Langmuir. Finally, the equation of state and other thermodynamic functions for this particular case are derived from Q_s.

We assume that the three degrees of freedom of the adsorbed atom are all vibrational, one perpendicular to the surface with a frequency v_z and two in the plane of the surface with frequencies v_x and v_y. Furthermore, we assume that the three modes are separable (normal coordinates) and that the harmonic oscillator approximation holds. With these assumptions, the partition function for an adsorbed atom is

$$q_s(T) = q_x q_y q_z \exp(-U_0/kT), \tag{2}$$

where q_x, q_y, and q_z are one-dimensional harmonic oscillator partition functions with classical frequencies v_x, v_y, and v_z, respectively and U_0 is the potential energy minimum of the adsorbed atom. The energy levels of a one-dimensional harmonic oscillator are given by

$$\epsilon_n = (n + \tfrac{1}{2})hv, \qquad n = 0, 1, 2, \ldots, \tag{3}$$

where v is the classical frequency. The partition function for a single harmonic oscillator is

$$q = \sum_{n=0}^{\infty} \exp(-\epsilon_n/kT) = \exp(hv/2kT)/[\exp(hv/kT) - 1]. \tag{4}$$

At high temperatures, the classical case, when $T \gg hv/k$,

$$q \to kT/hv. \tag{5}$$

At low temperatures, when $T \ll hv/k$,

$$q \to \exp(-hv/2kT), \tag{6}$$

where $hv/2$ is called the zero-point energy, and the oscillator is in its ground state. For the Langmuir isotherm, the classical or high-temperature approximation is sometimes used and the classical frequencies are assumed equal (4),

$$v_x = v_y = v_z = v. \tag{7}$$

The partition function for an adsorbed molecule with these assumptions becomes

$$q_s(T) = (kT/hv)^3 \exp(-U_0/kT). \tag{8}$$

Ultimately, we are interested in the equilibrium between atoms of the gas phase and the adsorbed phase, therefore, the same zero of energy must be chosen for both phases. We select as the zero an atom at rest an infinite distance from the surface. With this selection, $U_0 = -U_0$, and is the potential minimum in the plot of U_m the potential minimum of $U(z)$ against x or y. In Fig. 2.3, a schematic plot of $U(z)$ against z at an x, y position of minimum potential is shown. In order to relate this energy to the heats of adsorption of Chapter I, the differential energy of adsorption q_d, is also indicated. The quantity U_0 is negative, but the vibrational energy is taken as positive and measured upward from the potential minimum. The heat of adsorption at absolute zero is, then,

$$U_0 + [h(v_x + v_y + v_z)/2] \cong U_0 + \tfrac{3}{2}hv.$$

In general, for a system of N vibrating particles on N sites—for example, the Einstein model of a crystal—the canonical ensemble partition function is $Q = q^N$. But in the present case, we have N_s atoms which are distributed

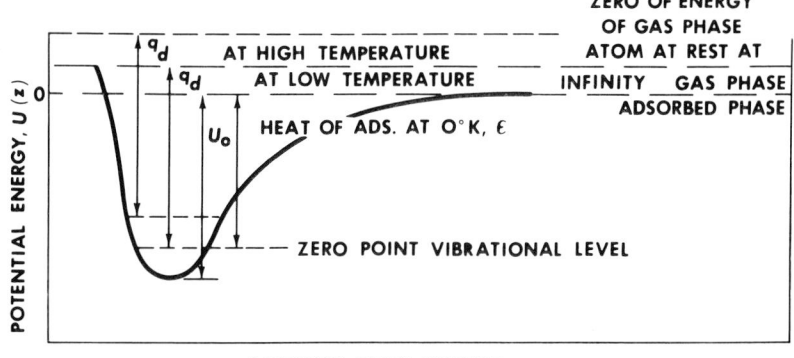

Fig. 2.3. Potential energy curve for an atom approaching an adsorption site.

among B sites $(B \gg N_s)$, one per occupied site. The number of possible arrangements of N_s identical objects on B equivalent but distinguishable sites is given by the well-known probability expression $B!/N_s!(B - N_s)!$. This is the number of degenerate eigenstates for each energy level of the system. Therefore, the canonical ensemble partition function for the adsorption case is

$$Q_s = B! q_s(T)^{N_s}/N_s!(B - N_s)!,$$

or using Stirling's approximation for factorials,

$$\ln Q_s = B \ln B - N_s \ln N_s - (B - N_s) \ln(B - N_s) + N_s \ln q_s. \tag{9}$$

Now the basic thermodynamic equation for adsorption from Eq. (12), Chapter I is

$$dE_s = T\, dS_s - \varphi\, d\alpha + \mu_s\, dn_s, \tag{10}$$

where we have omitted the insignificant term $P\, dV_s$. Since the surface area, α, may be assumed proportional to the number of sites B, $B = C\alpha$,[1] we may rewrite Eq. (10) in more appropriate form for localized adsorption,

$$dE_s = T\, dS_s - \pi\, dB + \mu_s\, dN_s, \tag{11}$$

where $\pi = \varphi/C$, and we have replaced n_s (number of moles) by N_s (number of atoms), so that μ_s is now the chemical potential per atom. Also,

$$dA_s = -S_s\, dT - \pi\, dB + \mu_s\, dN_s. \tag{12}$$

Using Eq. (12) and the well-known equation $A_s = -kT \ln Q_s$, we find that

$$-(\partial \ln Q_s/\partial N_s)_{B,\, T} = \mu_s/kT. \tag{13}$$

[1] Hill (*4*) discusses the case in which $B = C(T)\alpha$.

Taking the derivative of the logarithm of Q_s with respect to N_s in Eq. (9), we get

$$\mu_s/kT = \ln[\theta/(1 - \theta)q_s], \tag{14}$$

where $\theta = N_s/B$. In order to derive an expression for the isotherm, we set μ_s/kT equal to μ_G/kT at equilibrium

$$\mu_s/kT = \ln[\theta/(1 - \theta)q_s] = \mu_G/kT = [\mu^0(T)/kT] + \ln p, \tag{15}$$

in which

$$\mu^0(T) = -kT \ln[(2\pi mkT/h^2)^{3/2}kT] \tag{16}$$

is the chemical potential of an atom in a system at unit pressure (perfect gas) with three degrees of freedom, all translational, and h is Planck's constant and m the mass of the atom. As we shall see later (Section 2.4), μ^0 for polyatomic molecules will have additional terms for rotational and vibrational contributions.

Solving Eq. (15) for θ, we obtain

$$\theta(p, T) = \frac{q_s(T)\exp[\mu^0(T)/kT]p}{(1 + q_s(T)\exp[\mu^0(T)/kT]p)}, \tag{17}$$

which has the form of the equation derived kinetically with empirical constants, $\theta = a(T)p/(1 + a(T)p)$. When Eqs. (8) and (16) are substituted in Eq. (17), we obtain the following expression for $a(T)$, when $T \gg h\nu/k$,

$$a(T) = \frac{h^3(kT/h\nu)^3 \exp(-U_0/kT)}{(2\pi m)^{3/2}(kT)^{5/2}} = \frac{(kT)^{1/2}\exp(-U_0/kT)}{\nu^3(2\pi m)^{3/2}}. \tag{18}$$

If two different species of monatomic molecules are adsorbed on a surface, then the above equations must be modified. In this case, the number of degenerate adsorbed states on the surface may be calculated as follows: Consider that atoms of type 1 are adsorbed. There are $B!/N_{s_1}!(B - N_{s_1})!$ equivalent but distinguishable arrangements. For *each* of these arrangements, there are $(B - N_{s_1})!/N_{s_2}!(B - N_{s_1} - N_{s_2})!$ different arrangements of N_{s_2} atoms of type 2 on the remaining sites. The total number of ways of arranging N_{s_1} atoms of type 1 and N_{s_2} atoms of type 2 on B surface sites is

$$[B!/N_{s_1}!(B - N_{s_1})!][(B - N_{s_1})!/N_{s_2}!(B - N_{s_1} - N_{s_2})!]$$
$$= B!/N_{s_1}!N_{s_2}!(B - N_{s_1} - N_{s_2})!. \tag{19}$$

The canonical partition function is

$$Q_s = [B!/N_{s_1}!N_{s_2}!(B - N_{s_1} - N_{s_2})!]q_{s_1}^{N_{s_1}}q_{s_2}^{N_{s_2}}. \tag{20}$$

Proceeding in a manner similar to the previous case, we take the derivative of the logarithm of Q_s first with respect to N_{s_1} then with respect to N_{s_2}.

This gives two expressions, one for μ_{s_1}/kT and another for μ_{s_2}/kT. Setting each of these equal to the respective gas phase chemical potentials μ_{G_1}/kT and μ_{G_2}/kT, we get two expressions which can be solved simultaneously for θ_1 and θ_2, the fractions of sites covered by type 1 and type 2, respectively, at equilibrium. These expressions are:

$$\theta_1 = a_1 p_1/(1 + a_1 p_1 + a_2 p_2), \qquad a_1(T) = \exp(\mu_1{}^0/kT)q_{s_1}(T);$$
$$\theta_2 = a_2 p_2/(1 + a_1 p_1 + a_2 p_2), \qquad a_2(T) = \exp(\mu_2{}^0/kT)q_{s_2}(T). \tag{21}$$

From Eqs. (9) and (12) and $A_s = -kT \ln Q_s$, we find that

$$\pi/kT = (\partial \ln Q_s/\partial B)_{N_s, T} = -\ln(1 - \theta) = \theta + \tfrac{1}{2}\theta^2 + \tfrac{1}{3}\theta^3 + \cdots, \tag{22}$$

which is the equation of state for an adsorbed monatomic gas. Note that as $\theta \to 0$, $\pi \to \theta kT$. This is the two-dimensional analog of $p \to \rho kT$ in three dimensions, where $\rho = N/V$.

In general, the entropy of a system is expressed by

$$S = kT(\partial \ln Q/\partial T)_{V, N} + k \ln Q, \tag{23}$$

and specifically for localized adsorption,

$$S_s = kT(\partial \ln Q_s/\partial T)_{B, N} + k \ln Q_s. \tag{24}$$

Using Eq. (9) and performing the indicated operations of Eq. (24) gives

$$S_s = N_s k[\ln q_s + T(d \ln q_s/dT)] + k \ln[B!/N_s!(B - N_s)!]. \tag{25}$$

The first term on the right-hand side of Eq. (25) is the vibrational entropy of the adsorbed atom, S_v. The second term is called the configurational entropy, and it is independent of temperature, depending only on the number of configurations for given N_s with respect to the total number of sites. The configurational entropy, using Stirling's approximation, may be expressed as

$$S_c = k(B \ln[B/(B - N_s)] + N_s \ln[(B - N_s)/N_s]), \tag{26}$$

which holds within the limitations of the approximation, that is, $B!$, $N_s!$, and $(B - N_s)!$ must all be large numbers. In terms of $\theta = N_s/B$, we have

$$S_c = -N_s k(\ln \theta + [(1 - \theta)/\theta] \ln(1 - \theta)). \tag{27}$$

A plot of S_c against N_s, Eq. (26), gives a maximum at half-coverage, $N_s/B = \theta = \tfrac{1}{2}$.

The Helmholz free energy is given by

$$A_s = -kT \ln Q_s = +N_s kT[\ln \theta + [(1 - \theta)/\theta] \ln(1 - \theta) - \ln q_s], \tag{28}$$

and the energy E_s by

$$E_s = kT^2(\partial \ln Q_s/\partial T)_{B, N_s}. \tag{29}$$

Using Eqs. (2) and (9), the expression for E_s becomes

$$E_s = N_s kT^2[\partial \ln(q_x q_y q_z)/\partial T] + N_s U_0. \tag{30}$$

From Eq. (30), the differential energy of adsorption, $q_d = e_G - (\partial E/\partial n_s)_{V_s, \alpha, T}$, may be found:

$$q_d = e_G - N_0 U_0 - RT^2[\partial \ln(q_x q_y q_z)/\partial T], \tag{31}$$

where $N_0 = N_s/n_s$ is Avogadro's number. It will be observed that in this case, $(\partial E_s/\partial n_s)_{V_s, \alpha, T} = E_s/n_s$. For a monatomic gas, $e_G \equiv E_G/n_s = \frac{3}{2}RT$. If we assume that the adsorbed atom behaves as a classical three-dimensional harmonic oscillator, then $q_x q_y q_z = (kT/h\nu)^3$, and q_d in Eq. (31) becomes

$$q_d = -(N_0 U_0 + \tfrac{3}{2}RT), \tag{32}$$

and

$$q_{st} = q_d + RT = -(N_0 U_0 + \tfrac{1}{2}RT). \tag{33}$$

We now have a complete set of thermodynamic functions, A_s, S_s, E_s, π, μ_s along with the independent variables of $Q_s(N_s, B, T)$. From these, all others can be derived.

2.3   Adsorption of More Than One Molecule per Site

Brunauer et al. (5) have classified adsorption isotherms according to five types. These are shown in Fig. 2.4. Type I is often referred to as the Langmuir type because it corresponds to experimental isotherms of mono-layer adsorption. The rest of the curves are associated with multilayer ad-sorption. Since chemisorption never exceeds a monolayer, its isotherms are restricted to Type I. Types IV and V are obtained in multilayer adsorption on highly porous adsorbents and the flattening of the isotherms at near satura-tion pressure (p_0) is attributed to capillary condensation phenomena. Type II is the common S-shaped isotherm with an asymptotic approach to p_0. Types III and V are rare.

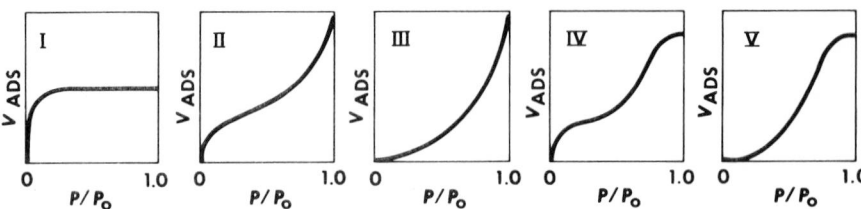

Fig. 2.4. Five types of isotherms according to Brunauer et al. (5). Reprinted by permission of the American Chemical Society.

The problem of multilayer adsorption is extremely difficult from a theoretical point of view. A multilayered adsorbed phase combines the formidable problems of a free liquid with the nonisotropic forces of adsorption. The simplest idealized case of multilayer adsorption occurs on a plane surface whose potential field has no periodicity in the x and y directions, but is the same at every point, varying only with z (uniform nonlocalized adsorption). Even this simplest case is far more complicated than the ordinary theory of liquids. On the surface of an idealized crystalline solid, the potential U not only varies with z, but also with x and y (uniform localized adsorption). On the surface of a real crystalline solid, the fluctuations of the potential in the x and y directions are extremely erratic (nonuniform localized adsorption). There is no hope at present of constructing a practical general theory of multilayer adsorption.

Wheeler (*6*) and Ono (*7–9*) have dealt with the simplest case mentioned above—uniform nonlocalized adsorption. They have delineated the rigorous approach and have developed *general* equations, but numerical calculations are not yet practical. More practical approximations have been developed by Frenkel (*10*), Halsey (*11*), Hill (*12*), McMillan and Teller (*13*), and Barrer and Robbins (*14*). All of these will be discussed in Chapter V.

The most important theory of multilayer adsorption is that of Brunauer *et al.* (*15*) and it is based on a model of localized adsorption. Their equation, the BET equation, was the first and still is the most useful covering the complete range of pressures up to p_0, the saturation pressure. We shall discuss the inconsistencies and shortcomings of the BET equation later in this section. Here we stress that in spite of the crudity of the assumptions involved in its derivation, the BET equation is an extremely useful qualitative guide for experimental work. Furthermore, it is still the basis for the most widely used method of determining surface areas. The original derivation given by Brunauer *et al.* may be considered as a generalization of Langmuir's kinetic derivation for a monolayer. Cassie (*16*) and Hill (*17*) have provided a statistical mechanical derivation by analogy with Fowler's derivation (*3*) of the Langmuir isotherm. Keii (*18*) has provided a concise derivation using the grand partition function. With the statistical derivation comes an explicit knowledge of all constants and parameters in terms of molecular properties and a clear picture of the model. The model from which the BET equation is obtained rests on the following assumptions:

(1) B equivalent sites are available for localized adsorption in the first layer.

(2) Each molecule adsorbed in the first layer is considered as a possible "site" for adsorption in a second layer; each molecule adsorbed in the second layer is considered as a possible "site" for adsorption in a third layer, etc.

(3) All molecules in the second and higher layers are assumed to have the same partition function as in the liquid state, which is different from the partition function of the first layer.

(4) Horizontal interactions between molecules are ignored for all layers.

Assumptions (1) and (4) provide Fowler's model (3) for the Langmuir isotherm. The model thus consists of a collection of piles of molecules built up on the molecules in the first layer with no interaction between piles but with a distribution in pile heights governed by the minimization of the free energy of the system. In this unnatural model of a liquid each molecule has only two nearest neighbors with which it interacts (one above and one below in the pile) instead of 10 or 12 for a real liquid. Also in this model the heat of adsorption per mole in the second and higher layers is equal to the heat of liquefaction e_L which, in general, is different and less than the heat of adsorption of the first layer, e_1.

In the derivation of the BET equation, we limit the discussion to monatomic molecules and essentially follow Keii (18). The method employs an extension of the original partition function Q_s for adsorption of a single atom on a site to adsorption of vertical piles of atoms according to the rules given above. For convenience, the derivation is switched to the use of the grand canonical partition function which involves Q_s. Certain physical inconsistencies are attributable to the thermodynamic expressions obtained because of the peculiar nature of the model. For example, the surface pressure φ erroneously goes to infinity as the gas pressure reaches its saturation value and bulk liquid starts to condense on the surface. Various modifications of the original BET equation have been proposed in order to eliminate inconsistencies, but little real progress has been made for a strictly localized model.

We may rewrite the canonical ensemble partition function of Eq. (9) in the following way, using the site partition functions,

$$Q_s(N_s, B, T) = \sum_{\mathbf{a}} [B! \, q_s(0)^{a_0} q_s(1)^{a_1}/a_0! \, a_1!], \qquad (34)$$

where $q_s(0) = 1$ and $q_s(1) = q_s$ are the site partition functions with zero and one molecule adsorbed, respectively, on the site, and $a_0 = B - N_s$ and $a_1 = N_s$ are the number of sites with zero and one molecule adsorbed on them. The summation is over all sets $\mathbf{a}$ which are consistent with

$$a_0 + a_1 = B, \qquad a_1 = N_s. \qquad (35)$$

In this, the simplest case, there is only one such set for fixed B and N_s.

If as many as two molecules may be bound to a site, the expression becomes

$$Q_s(N_s, B, T) = \sum_{\mathbf{a}} [B! \, q_s(0)^{a_0} q_s(1)^{a_1} q_s(2)^{a_2}/a_0! \, a_1! \, a_2!] \qquad (36)$$

and the summation is over all sets consistent with

$$a_0 + a_1 a_2 + = B, \qquad a_1 + 2a_2 = N_s. \tag{37}$$

In this case, it is evident that there is more than one set consistent with Eq. (37).

In general, for a maximum of m molecules on a site, we may write

$$Q_s(N_s, B, T) = \sum_a \frac{B! \, q_s(0)^{a_0} q_s(1)^{a_1} q_s(2)^{a_2} \cdots q_s(m)^{a_m}}{a_0! \, a_1! a_2! \cdots a_m!}, \tag{38}$$

where the summation is over all sets consistent with

$$\sum_{r=0}^{m} a_r = B, \qquad \sum_{r=0}^{m} r a_r = N_s, \tag{39}$$

and $q_s(r) = \sum_j \exp(-\epsilon_j(r)/kT)$ is the site partition function when r molecules are bound to the site. How the molecules are bound to the site is not stipulated yet in the equation.

The grand partition function is

$$\Xi \equiv \sum_{N_s=0}^{mB} Q_s(N_s, B, T) \lambda_s^{N_s}$$

$$= \sum_a \frac{B! \, q_s(0)^{a_0} [q_s(1)\lambda_s]^{a_1} [q_s(2)\lambda_s^2]^{a_2} \cdots [q_s(m)\lambda_s^m]^{a_m}}{a_0! \, a_1! a_2! \cdots a_m!}, \tag{40}$$

where $\lambda_s = \exp(\mu_s/kT)$ and $\lambda_s^{\sum_{r=0}^{m} r a_r} = \lambda_s^{N_s}$ from Eq. (39). The only restriction now on the sets **a** is the first of Eqs. (39), because in Eq. (40) we have summed over N_s for given B.

For the case, $r = 0, 1$, we find from Eqs. (34) and (40) using the binomial theorem,

$$\Xi = \sum_a \frac{B! \, q_s(0)^{a_0} [q_s(1)\lambda_s]^{a_1}}{a_0! \, a_1!} = \sum_{N_s=0}^{B} \frac{B! (q_s \lambda_s)^{N_s}}{(B - N_s)! \, N_s!}$$

$$= (1 + q_s \lambda_s)^B. \tag{41}$$

For the case $r = 0, 1, \ldots, m$, we find from Eq. (40) using the multinomial theorem, a generalization of the binomial theorem,

$$\Xi = [1 + q_s(1)\lambda_s + q_s(2)\lambda_s^2 + q_s(3)\lambda_s^3 + \cdots + q_s(m)\lambda_s^m]^B. \tag{42}$$

Now using the assumption for the BET model that the molecules are adsorbed in definite layers, we find for the site partition function,

$$q_s(r) = q_1 q_2 \ldots q_r, \tag{43}$$

where q_1 is the partition function of a molecule in the first layer; q_2, in the second layer, etc. Using another assumption of the BET model, namely, that all molecules in the second and higher layers have equal partition functions, $q_2 = q_3 = \cdots = q_m = q_L$, where q_L is the partition function of the liquid state, we may rewrite Eq. (42),

$$
\begin{aligned}
\Xi &= (1 + q_1\lambda_s + q_1 q_L \lambda_s^2 + q_1 q_L^2 \lambda_s^3 + \cdots + q_1 q_L^{m-1} \lambda_s^m)^B \\
&= [1 + q_1\lambda_s(1 + q_L \lambda_s + q_L^2 \lambda_s^2 + \cdots + q_L^{m-1}\lambda_s^{m-1})]^B \\
&= (1 + [q_1\lambda_s(1 - q_L^m \lambda_s^m)/(1 - q_L \lambda_s)])^B
\end{aligned} \tag{44}
$$

In the last step of Eq. (44), the geometric series has been summed.

The BET isotherm is now easily obtained by using the well-known statistical equation for the average number of molecules in a system $\bar{N}_s$,

$$
\bar{N}_s = kT(\partial \ln \Xi/\partial\mu_s)_{B,\,T} = \lambda_s(\partial \ln \Xi\, \partial\lambda_s)_{B,\,T}, \tag{45}
$$

where at equilibrium,

$$
\lambda_s \equiv \exp(\mu_s/kT) = \exp(\mu_G/kT) \equiv \lambda_{gas}. \tag{46}
$$

Differentiating the logarithm of Ξ in Eq. (44), we get,

$$
\bar{N}_s/B = \frac{q_1\lambda_s[1 - (m+1)q_L^m\lambda_s^m + mq_L^{m+1}\lambda_s^{m+1}]}{(1 - q_L\lambda_s + q_1\lambda_s - q_1 q_L^m\lambda_s^{m+1})(1 - q_L\lambda_s)} \tag{47}
$$

$$
\equiv \frac{cx[1 - (m+1)x^m + mx^{m+1}]}{(1 - x + cx - cx^{m+1})(1 - x)}, \tag{48}
$$

where

$$
c \equiv q_1/q_L, \qquad x \equiv q_L\lambda_s = q_L\lambda_{gas} = q_L \exp(\mu_0/kT)p. \tag{49}
$$

Equation (48) is the BET adsorption isotherm and reduces to the Langmuir isotherm when $m = 1$. It has proved useful in the evaluation of surface areas. It is customary to use the result obtained with m infinite

$$
\bar{N}_s/B = cx/(1 - x + cx)(1 - x). \tag{50}
$$

On thermodynamic grounds, $\bar{N}_s/B \to \infty$ as $p \to p_0$, the saturation pressure. That is, when saturation pressure is reached bulk liquid starts to condense on the surface. (We assume that the temperature is below the critical.) From Eq. (50), $x \to 1$ as $\bar{N}_s/B \to \infty$. Therefore,

$$
1/p_0 = q_L \exp(\mu_0/kT), \qquad x = p/p_0. \tag{51}
$$

When, in Eq. (50), $c \gg 1$, adsorption is favored in the first layer. This accounts for the "knee" in the isotherm as shown in Fig. 2.5 for $c = 200$. The curve has the shape of a Type II isotherm of which there are many experimental examples. Thus with large c, the Langmuir monolayer equation holds approximately up to $\bar{N}_s/B \cong 1.0$.

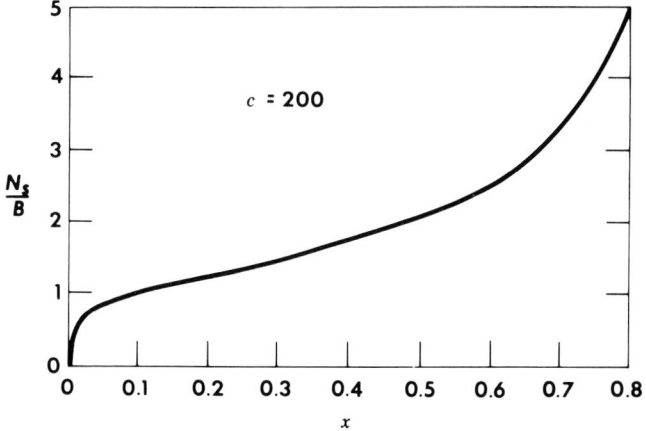

FIG. 2.5. BET adsorption isotherm for $c = 200$.

The surface pressure φ which is proportional to π [Eqs. (10), (11)] may be determined by the well-known expression for the grand partition function:

$$\Xi = \exp(\pi B/kT) = [1 + q_1 \lambda_s/(1 - q_L \lambda_s)]^B,$$

or

$$\pi/kT = \ln[1 + q_1\lambda_s/(1 - q_L \lambda_s)] = \ln[(1 - x + cx)/(1 - x)]. \qquad (52)$$

The final expression in Eq. (52) is obtained by taking the sum of the geometric series of Eq. (44) with $m = \infty$, and c and x are defined in Eq. (49).

It is not surprising that faults develop in the physical picture obtained with such a simplified model. The model, with its noninteracting piles of varying heights, is unnatural for either localized or nonlocalized adsorption.

A fault is evident from Eq. (52), which shows that $\pi \to \infty$ as $x \to 1$. But, thermodynamically, the surface pressure φ, which is proportional to π must be finite at $x = 1$. Since $\varphi = \gamma_0 - \gamma$ [see Chapter I, Eqs. (9), (10)], Eq. (52) implies that $\gamma = -\infty$. In general, the BET isotherm, when fitted to the points of a typical experimental isotherm of Type II in the region of a monolayer ($N_s/B = 1$), deviates above and below that region, as shown in Fig. 2.6. The discrepancy at lower pressure is often attributed to nonuniformity of the adsorbent surface (*13*).

Some progress has been made in explaining the discrepancy at higher pressures. The experimental isotherms approach $x = 1$ in such a way that, although $N_s/B \to \infty$ as $x \to 1$, the value of φ remains finite. Cassel (*19*) attributes the unrealistic behavior of the BET equation at higher pressures to the neglect of horizontal interactions. Hill (*20*) has taken these interactions into account using the Bragg–Williams approximation (see Chapter III) and

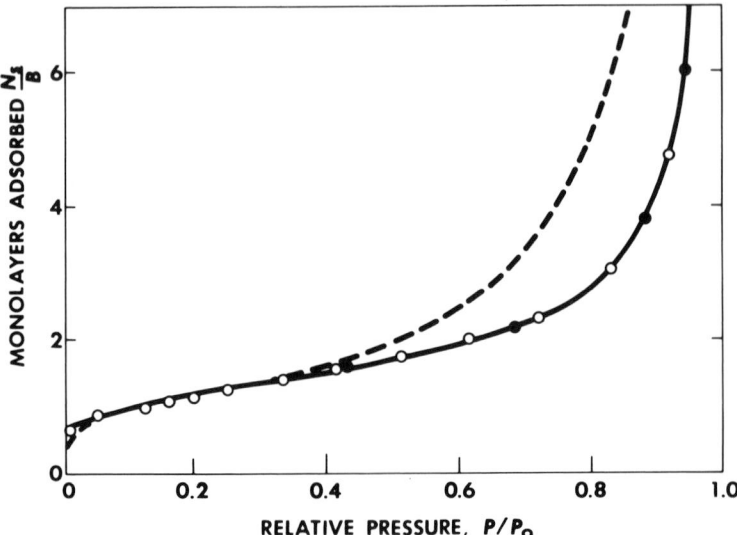

FIG. 2.6. Comparison of the BET isotherm (dashed curve) with the experimental (solid curve) in a typical case, *from* McMillan and Teller (*13*). Reprinted by permission of The American Institute of Physics.

shows that the effect is in the right direction, that is, to lower the isotherm. Halsey (*11*) in an equivalent procedure for taking interactions into account concluded that multilayer adsorption on a uniform surface would take place in stepwise fashion. The fact that experimentally stepwise isotherms are rarely seen, he attributed to the nonuniformity of most real surfaces, which would tend to smear the effects into a smooth isotherm.

Numerous modifications of the BET equation have been made involving variations in heats of the adsorption from one layer to another or as a result of nonuniformity of the adsorbent surface. For example, Pickett (*21*) attempted to take into account the decrease in probability of the escape of a molecule from the nth layer as more of the surface was covered with n layers. Hill (*22*) showed, however, that the treatment violated the principle of microscopic reversibility. Anderson (*23*) assumed that the heat of adsorption differed by a constant from the heat of liquefaction in the first few layers (excluding the first). Smith and Pierce (*24*), having found that isosteric heats of adsorption of ethyl chloride and ammonia on graphite decreased linearly with increasing relative pressure, obtained a modification of the BET equation having four constants when they took this fact into account. Steele (*25*) tested the case when $e_1 \neq e_2$ and $e_3 = e_4 \cdots = e_L$ where e_n is the heat of adsorption in the nth layer and e_L is the heat of liquefaction. McMillan (*26*) and Walker and

Zettlemoyer (*27*) introduced the concept of a nonuniform surface into the BET equation. Cook (*28*) introduced both interactions and variable heat of adsorption in an empirical manner. Since most of these modifications contain two or more arbitrary constants, it is not surprising that they show better agreement with experimental isotherms than the original BET equation which has only one arbitrary constant.

Hill (*29*) has shown that it is the configurational term of the BET equation which causes the spreading pressure φ to increase without limit as $x = p/p_0 \rightarrow$ 1. In this demonstration, he uses the canonical partition function (*17*), developing molecular partition functions immediately instead of site partition functions [Eq. (38)]. The partition function for a molecule in the first layer is

$$q_1 = q_{s_v} \exp(-U_{01}/kT), \tag{53}$$

where q_{s_v} represents the vibration of a monatomic molecule with respect to the surface and may be approximated by a harmonic oscillator model and U_{01} is the potential energy minimum of adsorption. The partition function for a molecule in higher layers is

$$q_L = q_{L_v} \exp(-U_{0L}/kT), \tag{54}$$

where, in accordance with the BET assumptions, q_{L_v} and U_{0L} correspond to values for the liquid state, although the molecules in these layers are arranged in vertical piles with only two interacting neighbors each. We suppose that there are B localized sites with X molecules adsorbed on them (the first layer) and $N_s - X$ molecules in all higher layers. The canonical partition function for molecules in the first layer is

$$Q_s = \frac{B! \, [q_{s_v} \exp(-U_{01}/kT)]^X}{(B - X)! \, X!}, \tag{55}$$

where, as before [Eq. (9)], the configurational factor $B!/(B - X)! \, X!$ is the number of arrangements of X identical molecules on B distinguishable sites. For higher layers,

$$Q_L = [(N_s - 1)!/(N_s - X)!(X - 1)!][q_{L_v} \exp(-U_{0L}/kT)]^{N_s - X}, \tag{56}$$

where the configurational factor[2] is the number of arrangements of $(N_s - X)$ identical molecules on X distinguishable sites with no restriction on the number per site up to the total number, $(N_s - X)$, available.

---

[2] The number of ways of placing n indistinguishable objects in g boxes is given by well-known permutation formula, $(n + g - 1)!/n!(g - 1)!$.

The complete partition function of the adsorbed film is

$$Q = \sum_{X=1}^{N_s \text{ or } B} Q_s Q_L$$

$$= \sum_{X=1}^{N_s \text{ or } B} \frac{N_s! \, B! \, [q_{s_v} \exp(-U_{01} \, kT)]^X [q_{L_v} \exp(-U_{0L} kT)]^{N_s - X}}{(N_s - X)!(B - X)!(X!)^2}. \quad (57)$$

where unity has been neglected in comparison with N_s and X, and the upper limit of the summation is N_s if $N_s < B$ and B if $N_s > B$.

We shall now use the canonical partition function of Eq. (57) to calculate the chemical potential of the entire adsorbed film, μ_{N_s}. We are interested in $\mu_{N_s} = (\partial F/\partial N_s)_{T, p}$ because it is related to the discrepancy at high pressures between the BET equation and experimental isotherms (see Fig. 2.6). The BET theory gives too low a vapor pressure (and, thus, too low a free energy to the adsorbed phase relative to the liquid state) for a given $\theta = N_s/B$.

We simplify Q in Eq. (57) by the usual approximation of finding the maximum term. The value of X in the maximum term is obtained from

$$\partial \ln Q_s Q_L/\partial X = 0, \quad (58)$$

which gives for the value of X corresponding to the maximum term using Stirling's approximation, the solution of the following equation:

$$(N_s - X)(B - X) = (q_{L_v}/q_{s_v}) \exp[-(U_{0L} - U_{01})/kT]X^2. \quad (59)$$

The chemical potential of the adsorbed molecules is

$$\mu_{N_s}/kT = -(\partial \ln Q_s Q_L/\partial N_s)_{B, T} = \ln[(N_s - X)/N_s] + U_{0L}/kT - \ln q_{L_v}, \quad (60)$$

with X given in Eq. (59). The chemical potential of the gas phase μ_G is given by Eq. (15). Equating μ_G and μ_{N_s}, we have

$$\mu_{N_s}/kT = U_{0L}/kT - \ln q_{L_v} + \ln[(N_s - X)/N_s] = \mu^0/kT + \ln p. \quad (61)$$

For the pure liquid (a collection of desorbed one-dimensional piles of molecules identical with the structure of the adsorbed film), we have

$$\mu_L/kT = U_{0L}/kT - \ln q_{L_v} = \mu^0/kT + \ln p_0. \quad (62)$$

Therefore,

$$\ln x = \ln(p/p_0) = \ln(N_s - X)/N_s = \mu_{Ns}/kT - \mu_L/kT. \quad (63)$$

After the first layer is filled, we have

$$x = (N_s - B)/N_s = (\theta - 1)/\theta, \quad (64)$$

or

$$\theta = 1/(1 - x),$$

which is the expression to which the BET equation [Eq. (50)] reduces where $x \to 1$ for large c. The function $1/(1 - x)$ is responsible for the infinite value of π in Eq. (52); its configurational origin is evident from the argument just presented.

The most successful modification of the BET equation calls for drastic measures. First, attractive forces of the adsorbent beyond the first layer are not ignored. The simple BET assumption that $e_1 > e_2 = e_3 = \cdots = e_L$ is thus replaced by $e_1 > e_2 > e_2 \cdots > e_L$. This replacement alone, while still maintaining the "pile" structure of the BET theory, is not sufficient. In fact, with only this change, the result predicts an adsorption exceeding that of the BET equation, which itself already exceeds experimental values. Second, it is assumed that for, say, $\theta > 2$, the effects of the detailed structure of the adsorbent have been smoothed out and the adsorbate in the third and higher layers is essentially like bulk liquid in density and with many interacting neighbors. The adsorbate is considered to be a slab of liquid of uniform thickness with the normal adsorbate–adsorbate interactions of the liquid and situated in the potential energy field of the adsorbent. Adsorption behavior is therefore predominantly determined by the potential energy field in which the molecule is adsorbed compared with the field in bulk liquid. This theory has been developed by Frenkel (10), Halsey (11), Hill (12), and McMillan and Teller (13). It is referred to as the slab theory. The result is far different in form from the simple BET equation,

$$\ln x = \ln(p/p_0) = -\kappa/\Gamma^3, \tag{65}$$

where κ is a constant and Γ is the number of molecules adsorbed per square centimeter of adsorbent surface. Since the assumptions in this theory bring it into the category of nonlocalized adsorption, the subject will be treated in detail in Chapter V.

2.4 Adsorption of Polyatomic Molecules

The adsorption of polyatomic molecules introduces two additional physical factors into monolayer adsorption: (1) the possibility of dissociative adsorption, and (2) the existence of internal degrees of freedom in the molecule. First we shall develop the isotherm equation for dissociative adsorption, using the diatomic molecule as an example. We shall then discuss the nature of the degrees of freedom of a polyatomic molecule in localized adsorption and develop their partition functions. The partition functions will be used to derive the adsorption isotherm and the equation of state. They will be used also in obtaining theoretical expressions for the various entropy functions which will be compared with experimental entropies. One of the objectives in comparing theoretical and experimental entropies is to determine whether adsorption is localized or nonlocalized. Studies over a broad enough range of surface coverages to make such comparisons meaningful are rare.

We assume that the diatomic molecule in the gas phase is dissociated upon adsorption, each atom occupying a single site and possessing no mobility. The derivation closely follows that for a monatomic molecule (Section 2.2) except that Eq. (14) for equilibrium now becomes

$$\mu_s/kT = \ln[\theta/(1-\theta)q_s] = \mu_G/2kT = [\mu^0(T)/2kT] + \tfrac{1}{2}\ln p,$$

$$\theta(p, T) = \frac{q_s(T)\exp[\mu^0(T)/2kT]p^{1/2}}{1 + q_s(T)\exp[\mu^0(T)/2kT]p^{1/2}}, \tag{66}$$

which may be compared to Eq. (17), the isotherm for a monatomic molecule. In Eq. (66), $q_s(T)$ is the partition function for a single adsorbed atom [see Eq. (2)],

$$q_s(T) = q_{s_v}\exp[-(U_0 + D/2)/kT] \equiv q_{s_v}\exp(-U_0'/2kT), \tag{67}$$

where q_{s_v} is the vibrational partition function of a single adsorbed atom measured from the bottom of the potential well; and U_0, a negative number, is the potential energy minimum of an adsorbed atom; and $(-D)$, a negative number, is the potential energy minimum for the diatomic molecule. Thus the zero of energy is the bottom of the potential well of the isolated diatomic molecule at rest, and $U_0' = 2U_0 + D$ is the heat of adsorption (including the zero-point vibrational energies for both the molecule and the adsorbed atom) when one molecule is adsorbed dissociatively at $0°K$. We note that the diatomic molecule in the gas phase now has a rotational (q_{G_r}) and an internal vibrational (q_{G_v}) partition function as well as a translational (q_{G_t}) one, so that μ_G in Eq. (66) becomes

$$\begin{aligned}\mu_G/kT &= -\ln(q_G/N_G) = -\ln(q_{G_t}q_{G_v}q_{G_r}/N_G)\\ &= -\ln[(2\pi mkT/h^2)^{3/2}kTq_{G_v}q_{G_r}] + \ln p\\ &= \mu^0(T)/kT + \ln p,\end{aligned} \tag{68}$$

which may be compared with Eqs. (15) and (16). Substituting Eq. (68) into Eq. (66), we obtain

$$\theta(p, T) = \frac{a(T)p^{1/2}}{1 + a(T)p^{1/2}},$$

$$a(T) = \frac{h^{3/2}q_{s_v}\exp(-U_0'/2kT)}{(2\pi m)^{3/4}(kT)^{5/4}q_{G_{int}}^{1/2}}, \tag{69}$$

where $q_{G_{int}} = q_{G_v}q_{G_r}$, the internal partition function of a gas phase diatomic molecule, and q_{G_v} is measured from the bottom of the potential well of the isolated diatomic molecule.

In connection with the dissociative adsorption of a diatomic molecule, Rossington and Borst (*30*) have recently reconsidered an interesting problem originally discussed by Roberts (*31*, *32*), Roberts and Miller (*33*), and Miller (*34*). When an immobile film is formed on a square array of adsorption sites by dissociative adsorption of a diatomic molecule, some individual sites will be surrounded by four filled sites. Such sites will not be available for adsorption, so that there will exist gaps in the film. Unoccupied sites of this nature have been regarded as important in certain catalytic processes. Roberts (*31*, *32*) using a model of 400 sites, and later, Roberts and Miller (*33*), carrying out a mathematical analysis, determined that a minimum of 7–8 % of the sites would not be available for adsorption into the first layer. Rossington and Borst repeated the work of Roberts, employing a model surface of 10,000 sites, with the aid of a computer. Making a total of 201 individual runs, they determined the mean value of the final percentage of vacant sites to be 9.15%, with a standard deviation of 0.25%.

We now turn to the consideration of the nature of the degrees of freedom of a polyatomic molecule in localized adsorption. A molecule composed of N atoms will have $3N$ degrees of freedom, three for each atom which would all be translational if the atoms were isolated. All molecules have three translational degrees of freedom. Linear molecules have two rotational degrees, and nonlinear molecules have three. Therefore, linear molecules have $3N - 5$ vibrational degrees of freedom, and nonlinear molecules $3N - 6$. When a polyatomic molecule is adsorbed, we must account not only for the transformation of translational degrees of freedom, as in the case of the monatomic molecule, but also for the rotational and vibrational degrees. In the extreme case of immobile localized adsorption, a polyatomic molecule will lose all translational and rotational degrees of freedom. The three translational degrees will be transformed into three vibrational degrees, one normal to the surface and two parallel to the surface in the x and y directions, respectively. The rotational degrees—for example, the three rotations associated with the ammonia molecule—will also be transformed into vibrations, two bending or rocking of the x and y axes, and one torsional about the vertical or z axis Thus, each molecule adsorbed in this manner will have a total of $3N$ vibrational degrees of freedom. In the other extreme, a molecule may, upon adsorption, lose only one translational degree of freedom and retain all of its rotational degrees. In that case, the translational degree of freedom will have been replaced by a vibration normal to the adsorbent surface. In general, the internal vibrational degrees of freedom of a molecule are relatively unaffected on adsorption, since surface forces are usually weak compared with the restoring forces of the internal vibrations of a molecule. The rotational motion, however, may be seriously affected by adsorption. It may be completely transformed into vibrational motion or it may change into some intermediate state of hindered rotation, where the potential barrier is not sufficient to completely prevent rotation, a situation which will be discussed in Chapter V in relation to nonlocalized adsorption. There is also the case of hindered translation, where the adsorbed molecule does not completely lose its translational motion, a case which will be discussed in Section 2.6.

Thus, in the Langmuir isotherm for a polyatomic molecule, the partition functions for both the gas phase and the adsorbed molecule will contain terms for internal degrees of freedom. If the partition function for the adsorbed molecule is separable into terms corresponding to the modes of motion, then it may be written

$$q_s(T) = q_{s_v} q_{s_{v'}} q_{s_r} \exp(-U_0/kT) \equiv q_{s_v} q_{s_{int}} \exp(-U_0/kT), \qquad (70)$$

where $q_{s_{v'}}$ refers to the internal modes of vibration of the molecule, and q_{s_v} refers to the vibration of the molecule with respect to the surface of the adsorbent. For immobile, localized adsorption, q_{s_v} will contain at least three vibrational degrees of freedom, corresponding to the three translational degrees of freedom of the gas phase; and may contain an additional three, if the rotational degrees are transformed to vibration. We may write the isotherm,

$$\theta(p, T) = \frac{a(T)p}{1 + a(T)p},$$

$$a(T) = \frac{h^3 q_{s_v} q_{s_{int}} \exp(-U_0/kT)}{(2\pi m)^{3/2}(kT)^{5/2} q_{G_{int}}}. \qquad (71)$$

All vibrational partition functions, q_{s_v}, $q_{s_{v'}}$, and q_{G_v}, are measured from the bottom of their respective potential wells, that is, include the zero-point vibrational term. It will be observed from Eq. (71) that if $q_{s_{int}}$ and $q_{G_{int}}$ have terms in common, they will cancel. Usually, the partition functions for internal vibrations, $q_{s_{v'}}$ and $q_{G_{v'}}$, will cancel because these vibrations are not changed much upon adsorption. However, rotations of the molecule are often strongly affected in localized adsorption, and, therefore, q_{G_r} and q_{s_r} usually do not cancel. For an adsorbed monatomic molecule only q_{s_v} remains of the ratio $q_{s_v} q_{s_{int}}/q_{G_{int}}$.

It is obvious that the state of a polyatomic molecule on a surface represents a complex set of phenomena. Some investigators (35–39) have attempted to make use of surface entropies to determine the state of the adsorbed phase. Comparisons of the entropies of adsorbed layers derived from experimental measurements with values calculated for mobile and immobile layers by means of statistical mechanics may be made using either integral or differential molar values. As Everett (35) points out, "it is largely a matter of personal choice whether we discuss the differential or integral quantities." It will be recalled from Chapter I that the relationship between these quantities is

$$S_s = n_s s_s, \qquad (\partial S_s/\partial n_s)_{T, p, \alpha} \equiv \bar{s}_s = n_s(\partial s_s/\partial n_s)_{T, p, \alpha} + s_s, \qquad (72)$$

so that s_s and $\bar{s}_s$ are equal only if $(\partial s_s/\partial n_s)_{T, p, \alpha} = 0$.

The differential experimental entropies, which are more readily determined from isotherms than integral entropies, may be obtained from (see Chapter I)

$$\overline{\Delta f} \equiv \bar{f}_s - f_G = (\bar{h}_s - h_G) - T(\bar{s}_s - s_G), \qquad (73)$$

where $(\bar{h}_s - h_G) = -q_{s_t}$, the isosteric heat of adsorption; and

$$\overline{\Delta f} = RT \ln p, \qquad (74)$$

provided we take 1 atm as the standard state of the gas and p as the equilibrium pressure. Hence,

$$\bar{s}_s = s_G - R \ln p - (q_{st}/T). \qquad (75)$$

To obtain the integral molar entropy, we use

$$s_s = (1/n_s) \int_0^{n_s} \bar{s}_s \, dn_s, \qquad (76)$$

which requires a study of coverages down to very low values.

The experimental entropies are compared with values calculated from statistical mechanics assuming that the partition functions of the various degrees of freedom are separable, and, in the case of rotations and translations, that the classical approximations hold.

Calculations of this kind exclude such possibilities as hindered rotation and translation. Also, no completely satisfactory *a priori* methods of evaluating the frequencies of vibrations of the adsorbed molecules relative to the adsorption sites are available, so that the corresponding entropies are usually obtained by difference using experimental values of total entropies, which introduces a serious handicap in the method.

In general, for gaseous and adsorbed phases, the entropy is calculated from [see Eq. (23)]

$$S = kT(\partial \ln Q/\partial T)_{L^n, N} + k \ln Q, \qquad (77)$$

where L^n is length, area, or volume as n, the number of translational degrees of freedom, is 1, 2, or 3, respectively. Also,

$$Q = q^N/N!, \qquad q = q_t \, q_r \, q_v' \, q_{s_v} \, q_{G_e} \, q_{s_p}, \qquad (78)$$

where the second equation of Eq. (78) shows a breakdown of the possible partition functions for gaseous or adsorbed phases. Those not marked by subscripts "G" or "s" refer to either phase. If there are no translational degrees of freedom, $N!$ is dropped.

The translational partition function for a molecule is

$$q_t = (2\pi m k T/h^2)^{n/2} L^n. \tag{79}$$

The partition function for a rigid rotator is

$$q_r = \frac{1}{\sigma \pi} \left[\frac{8\pi^3 \left(\prod_{i=1}^{n} I_i \right)^{1/n} kT}{h^2} \right]^{n/2}, \tag{80}$$

where I_i is the moment of inertia of the ith rotational degree of freedom, and, in general, the I_i are not necessarily all different, and σ is the symmetry number.

The vibrational partition function (including the zero-point vibrational term) for a harmonic oscillator with n degrees of vibrational freedom is

$$q_v' = \prod_{i=1}^{n} \frac{\exp(-h\nu_i/2kT)}{1 - \exp(-h\nu_i/kT)]}, \tag{81}$$

which represents internal vibrations of the molecule. The vibrations of an adsorbed molecule with respect to the surface q_{s_v}, which represent the translational and rotational degrees of freedom that have been transformed upon adsorption, are also given by the same harmonic oscillator approximation, but there are no completely satisfactory methods for evaluating the ν_i in the latter case.

The electronic partition function of the gas phase molecule is

$$q_{Ge} = \omega e^{D/kT}, \tag{82}$$

where D (greater than 0) is the depth of the minimum of the potential energy surface relative to the separated atoms at rest, and ω, which is usually 1, is the degeneracy of the ground state. This is consistent with choosing the bottom of the potential energy surface as the zero of vibrational energy. In adsorption studies, however, as we have already noted, the zero of energy is taken as the potential minimum D of the isolated molecule at rest, therefore, this term contributes nothing so long as $\omega = 1$. There is, of course, the potential energy term $q_{s_p} = \exp(-U_0/kT)$ for the adsorbed molecule, and in this case for want of better information the adsorbed ground state is assumed to be nondegenerate.

The translational entropies, integral molar and differential molar, are given as follows:

$$_n s_t = R \ln[(2\pi m k T/h^2)^{n/2}(L^n/N)e^{(n+2)/2}] \tag{83}$$

which becomes for $n = 3$,

$$_3 s_{G_t} = R \ln[(2\pi m k T/h^2)^{3/2}(V_G/N_G)e^{5/2}]$$
$$= R \ln[(2\pi m k T/h^2)^{3/2}(kT/p)e^{5/2}] \quad \text{(perfect gas)}, \tag{84}$$

or, for a standard state of 1 atm,

$$_3s_{G_t} = R \ln M^{3/2}T^{5/2} - 2.30 \text{ cal deg}^{-1} \text{ mole}^{-1}, \tag{85}$$

where M is molecular weight.

For $n = 2$, we have

$$
\begin{aligned}
_2s_{s_t} &= R \ln[(2\pi mkT/h^2)(\alpha/N_s)e^2] \\
&= R \ln MTA + 65.8 \text{ cal deg}^{-1} \text{ mole}^{-1},
\end{aligned} \tag{86}
$$

where $A = \alpha/N_s$ the area occupied per molecule. With α constant, so that N_s/α, the two-dimensional density, varies, we have

$$
\begin{aligned}
_2\bar{s}_{s_t} &= R \ln MTA + 63.8 \\
&= R \ln(MTb/\theta) + 63.8 \text{ cal deg}^{-1} \text{ mole}^{-1} \\
&= {}_2s_{s_t} - R,
\end{aligned} \tag{87}
$$

where b is the surface area per molecule in a filled monolayer, and $\theta = b/A$, the fractional coverage of the surface. Equation (87) can be put into the form

$$_2\bar{s}_{s_t} = {}_2\bar{s}_{0t} + R \ln(A/A_0), \tag{88}$$

where $_2\bar{s}_{0t}$ and A_0 refer to a standard state.

For a nonideal gas, the translational entropy usually must be modified. For example, if the equation of state is

$$\varphi(\alpha - N_s b) = N_s kT, \tag{89}$$

which is often called Volmer's equation (40) and corrects for the size of the adsorbed molecules, then the molecular partition function becomes

$$_2q_{s_t} = (2\pi mkT/h^2)(\alpha - N_s b).$$

Therefore, we obtain for the entropy,

$$
\begin{aligned}
_2s_{s_t} &= R \ln[(A - b)MT] + 65.8 \\
&= R \ln[(1 - \theta)bMT/\theta] + 65.8,
\end{aligned} \tag{90}
$$

and

$$_2\bar{s}_{s_t} = {}_2s_{s_t} - R[A/(A - b)] = {}_2s_{s_t} - [R/(1 - \theta)].$$

The equation of state may include a molecular interaction term, for example, the van der Waals' equation,

$$[\varphi + (a_v/A^2)](\alpha - N_s b) = N_s kT, \tag{91}$$

where $a_v \equiv 2\pi\epsilon_m r^{*3}/3$ and $-\epsilon_m$ is the minimum potential energy between an interacting pair of molecules and r^* is the distance of separation at this minimum. Including such an interaction term causes the differential heat of adsorption, q_{st}, to vary linearly with coverage, but the molar entropy is unaffected by molecular interactions. (See Chapter V for a detailed discussion of entropy and equations of state for mobile films.)

The internal vibrational entropy is

$$s_v' = \bar{s}_v' = R \sum_{i=1}^{n} \left\{ \frac{hv_i/kT}{\exp(hv_i/kT) - 1} - \ln[1 - \exp(-hv_i/kT)] \right\}. \tag{92}$$

In general, it is a small contribution to the total entropy of the gas. For example, it accounts for no more than 2% of the total entropy of ammonia. As mentioned previously, it changes relatively little upon adsorption of the molecule. The entropies of vibration with respect to the surface, s_{s_v}, due to the loss of one, two or three degrees of rotational and translational freedom are given by an identical expression and are normally small in chemisorption (less than 3 cal deg^{-1} mole^{-1}), but may be appreciably larger when adsorption is weak. So long as the adsorption sites are considered identical, $s_{s_v} = \bar{s}_{s_v}$ for the v_i are constant.

The rotational entropy is

$$s_r = \bar{s}_r = R \ln \left[(1/\pi\sigma) \left\{ 8\pi^3 \left(\prod_{i=1}^{n} I_i \right)^{1/n} e\, kT \right\}^{n/2} \right]. \tag{93}$$

The potential energy term of adsorption, $\exp(-U_0/kT)$, contributes nothing to the entropy unless it is degenerate, which is always assumed not to be the case.

The integral molar configurational entropy is [see Eq. (27)]

$$s_{s_c} = -R[\ln \theta + [(1 - \theta)/\theta] \ln(1 - \theta)], \tag{94}$$

where $\theta = N_s/B$. Upon differentiating the total entropy given in Eq. (26) with respect to N_s, we obtain for the differential molar configurational entropy

$$\bar{s}_{s_c} = -R \ln[\theta/(1 - \theta)], \tag{95}$$

which becomes zero at one-half coverage.

Everett (35) has compared experimental and calculated differential molar entropies of the adsorbed phase on charcoal using the data of Homfray (41) for argon, of Richardson (42) for ammonia, and of Smith (43) for carbon dioxide. In all three cases, the isosteric heats of adsorption are essentially constant over a broad range, so that the charcoal surface may be considered uniform. He calculated three differential molar entropies as a function of the amount adsorbed: translational for a two-dimensional ideal gas [Eq. (87)], translational for a two-dimensional nonideal gas [Eq. (90)], and the configurational for localized adsorption [Eq. (95)]. All other differential molar entropies, vibrational and rotational, do not change with coverage for uniform surfaces, and therefore add only a constant to the calculated quantities. Since the theoretical equations contain $\theta(=V/V_m)$, the *fraction* of a monolayer adsorbed, it is necessary to know the monolayer capacity (V_m cm³ gm⁻¹ STP) in order to make comparisons with experimental entropies, which are available only as functions of the *amount* adsorbed (V cm³ gm⁻¹ STP). Everett found, for argon, that a good fit (within a constant) with the experimental curve could be obtained for the nonideal gas model using a monolayer capacity V_m of 83 cm³ gm⁻¹, and for the localized model using a monolayer capacity V_m of 50 cm³ gm⁻¹. The ideal gas model deviates severely at the higher concentrations. The curves are shown in Fig. 2.7, where an arbitrary

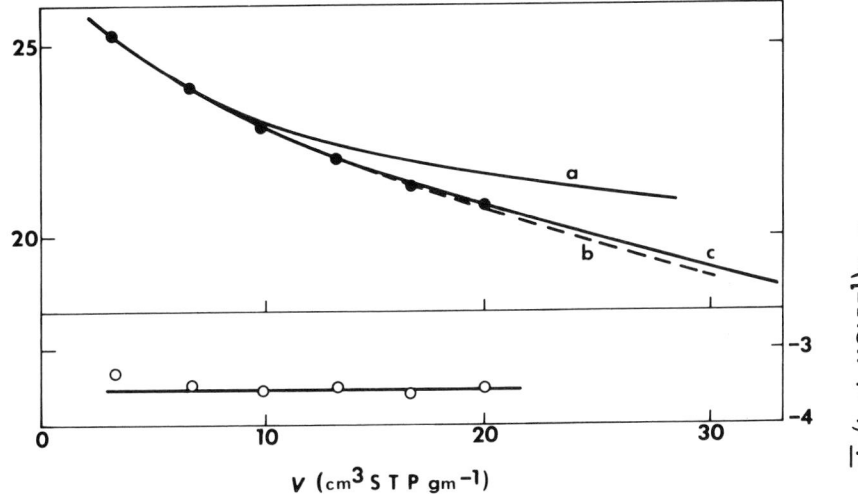

FIG. 2.7. Variation of differential entropy and heat of adsorption of argon adsorbed on charcoal with amount adsorbed, compared with theoretical curves. (a) Ideal two-dimensional gas; (b) nonideal two-dimensional gas with $V_m = 83$ cm³ gm⁻¹; (c) localized monolayer with $V_m = 50$ cm³ gm⁻¹. From Everett (*35*), reprinted by permission of The Chemical Society, London.

constant has been added to the theoretical curves to make them coincide with the experimental curve as closely as possible. Since argon is a monatomic gas, the arbitrary constant contains only entropy of vibration of the adsorbed atom with respect to the surface. Up to approximately 30 cm³ gm⁻¹, the curves for the nonideal gas model and the localized model coincide. Over this range, no clear distinction can be made between the models.

Comparisons may also be made at a standard surface state of $\theta = 0.5$. For the nonideal gas model, where $\theta = 0.5$ corresponds to a coverage of 41.5 cm³ gm⁻¹, it is calculated that the translational differential entropy is 12.5

cal deg^{-1} mole^{-1}, which is 2.9 less than the value for an ideal gas model. The area occupied by an argon atom was taken as 15 Å^2. The experimental value of the differential entropy at half-coverage (41.5 cm^3 gm^{-1}) may be read from the plot in Fig. 2.7 (after an extrapolation), and is 17.2 cal deg^{-1} mole^{-1}. The difference, 4.7 cal deg^{-1} mole^{-1}, may be attributed to the vibration of the atom normal to the surface and corresponds to a frequency of about 1.23 × 10^{12} sec^{-1} at 227°K. For the localized model, the differential configurational entropy is zero at $\theta = 0.5$ from Eq. (95). The experimental value at half-coverage (25 cm^3 gm^{-1}) is 20 cal deg^{-1} mole^{-1}, which may be attributed to three vibrations with respect to the surface. If we assume the motions parallel and normal to the surface to be equal then each vibration corresponds to 6.7 cal deg^{-1} mole^{-1} or a frequency of 0.47 × 10^{12} sec^{-1}. If we had selected 83 cm^3 gm^{-1} as the monolayer capacity for both models, the frequencies would have been 1.23 × 10^{12} sec^{-1} (as before) and 0.73 × 10^{12} sec^{-1}, respectively. Theoretical estimates indicate that all of the values of these frequencies are reasonable. Therefore, without an independent determination of the frequencies, we cannot distinguish between the two models.

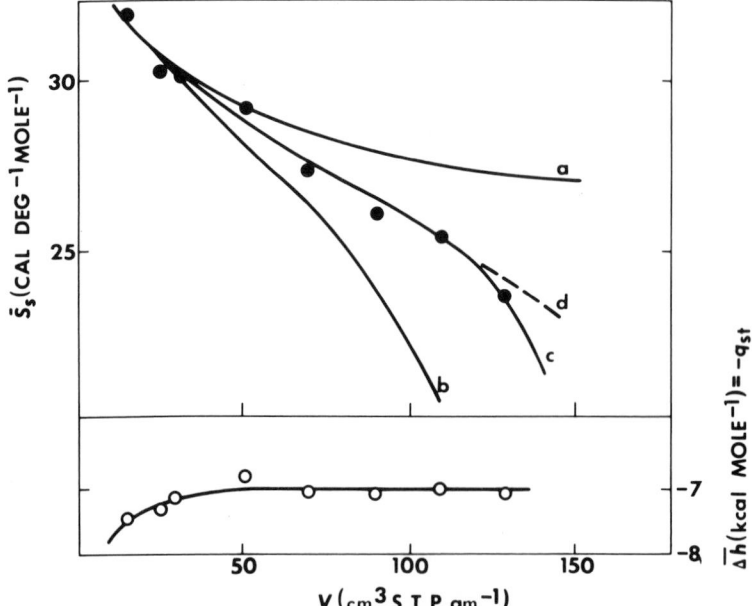

FIG. 2.8. Variation of differential entropy and heat of adsorption of ammonia adsorbed on charcoal with amount adsorbed, compared with theoretical curves. (a) Ideal two-dimensional gas; (b) nonideal two-dimensional gas with $V_m = 150$ cm^3 gm^{-1}; (c) localized monolayer with $V_m = 150$ cm^3 gm^{-1}; (d) as (b) but with $V_m = 250$ cm^3 gm^{-1}. From Everett (35), reprinted by permission of The Chemical Society, London.

For ammonia, the story is similar. Good fits with the experimental curve are obtained for the nonideal gas model with a monolayer capacity of 250 cm^3 gm^{-1} and for the localized model with a monolayer capacity of 150 cm^3 gm^{-1}, as may be seen in Fig. 2.8.

All three curves practically coincide up to 130 cm^3 gm^{-1}, the highest concentration measured. Over this range, no distinction can be made. Again the translational differential entropy for a nonideal gas is calculated for $\theta = 0.5$. We now have to consider rotational entropy. If we consider that the adsorbed molecules rotate freely, then 10.9 cal $deg^{-1}mole^{-1}$ must be added to the translational entropy. The difference between this sum and the experimental value (24.4 cal deg^{-1} $mole^{-1}$) at half-coverage (125 cm^3 gm^{-1}) is 2.7 cal deg^{-1} $mole^{-1}$, which is to be attributed to vibration normal to the surface and corresponds to a frequency of 4.5×10^{12} sec^{-1}. If we assume free rotation in the localized model, we must subtract 10.9 cal deg^{-1} $mole^{-1}$ of rotational entropy from the experimental entropy (27.2 cal deg^{-1} $mole^{-1}$) at half-coverage (75 cm^3 gm^{-1}) giving 16.3 cal deg^{-1} $mole^{-1}$ or 4.15 cal deg^{-1} $mole^{-1}$ for each vibrational mode, equivalent to a frequency of 1.02×10^{12} sec^{-1}. Again, it is difficult to distinguish between the two models. These frequencies represent upper limits because rotations are, in all likelihood, restricted in comparison with the gas phase. The theoretical study of the hindered rotation of adsorbed polyatomic molecules presents many problems, and the amount of progress which has been made is small. There do not appear to be any theoretical studies of the rotation of polyatomic molecules in a state of localized adsorption. The studies which have been made in relation to nonlocalized adsorption will be discussed in Chapter V.

For carbon dioxide, the results are clearer, and mostly because experimental measurements were available over the broadest possible range at 181.6°K. From Fig. 2.9, it would appear that the nonideal gas model can be rejected in this case because of the variation of entropy near saturation vapor pressure. From the precipitous drop of the experimental curve at a coverage of 81 per 100 gm of adsorbent (413 cm^3 $gm^{-1} = V_m$), it is obvious that the gas is at saturation vapor pressure p_0, and that the monolayer capacity cannot be greater than 413 cm^3 gm^{-1}. The localized model with this value of V_m fits the experimental curve extremely well, whereas the nonideal gas model does not. Other evidence indicates that at 181.6°K the carbon dioxide molecules retain little or no rotational motion. Therefore, at half-coverage (207 cm^3 gm^{-1}) for localized adsorption a differential entropy of 18.7 cal deg^{-1} $mole^{-1}$ is attributed to vibrational degrees of freedom, equivalent to an average lower limit of vibrational frequency for each vibrational mode of 0.5×10^{12} sec^{-1}.

The general conclusion is that it is extremely difficult to distinguish between localized and nonlocalized adsorption by consideration of the form of

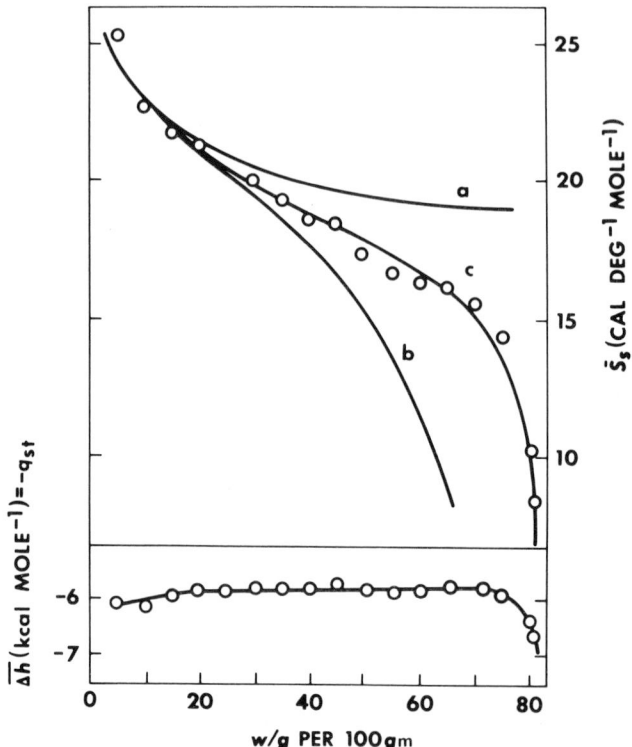

FIG. 2.9. Variation of differential entropy and heat of adsorption of carbon dioxide adsorbed on charcoal with amount adsorbed, compared with theoretical curves. (a) Ideal two-dimensional gas; (b) nonideal two-dimensional gas with $w_m = 81$ gm; (c) localized monolayer with $w_m = 81$ gm. From Everett (*35*), reprinted by permission of The Chemical Society, London.

the entropy against coverage or the values in some standard state. The chances of differentiation are improved when the range of surface coverage studied experimentally is broad. Such studies are rare.

2.5   Immobile Adsorption on Nonuniform Surfaces

Most real surfaces are nonuniform, and therefore more difficult to treat theoretically. We are not concerned at this point with the causes of nonuniformity or heterogeneity as it is often called. We are concerned here with the effects of heterogeneity on the adsorption isotherm and the thermodynamic properties of the adsorbed monolayer. We are also concerned with the experimental determination of differential entropies as a function of surface coverage, the determination of distribution functions, and how their forms affect the nature of the adsorption isotherm. A serious difficulty in these studies is the inherent insensitivity of the theoretical isotherm to the form of the distribution function within the accuracy of the experimental data. First we shall derive the generalized Langmuir isotherm for localized adsorption with a distribution of adsorption energies. We follow Hill's derivation (44).

Of the total B sites, let B_i have an adsorption potential energy minimum U_{0i}, where B_i for all i is a large number. It is assumed that a molecule on any of these B_i sites has the same partition function $q_s(T)$ for all vibrational degrees of freedom with respect to the surface and internal degrees of freedom. Of the total N_s molecules on B sites, let $N_{si} \leqslant B_i$ be the number on sites of potential energy minimum U_{0i}. It is assumed that the N_{si} are also large numbers. For each distribution N_{si} among the B_i sites (for all i) there exists a partition function $Q_{N_{si}}$

$$Q_{N_{si}} = \prod_i [B_i!/N_{si}!(B_i - N_{si})!]q_{si}^{N_{si}} \exp(-N_{si}U_{0i}/kT), \qquad (96)$$

which may be compared with Eq. (9) for a uniform surface. The total partition function Q_s is the sum of partition functions $Q_{N_{st}}$, one for each distribution N_{si},

$$Q_s = \sum_{N_{si}} Q_{N_{si}}. \qquad (97)$$

The usual approximation of replacing the sum by the maximum term in Q_s is made. This is done by using the standard method of Lagrange undetermined multipliers, which maximizes N_{si} for all i consistent with $N_s = \sum_i N_{si}$. Using Stirling's approximation and taking the derivative of $Q_{N_{si}} - \beta \sum N_{si}$ [Eq. (96)], with respect to N_{si}, where β is the undetermined multiplier, and setting the result equal to zero, we find as the maximum term, the distribution

$$N_{si} = B_i/[(1/q_{si}) \exp[\beta + (U_{0i}/kT)] + 1], \qquad i = 1, 2, 3, \ldots. \qquad (98)$$

Substituting Eq. (98) into Eq. (96), we find

$$\ln Q_s = \sum_i B_i \left\{ \frac{\beta}{(1/q_{si}) \exp[\beta + (U_{0i}/kT)] + 1} \right.$$
$$\left. - \ln\left[\frac{(1/q_{si}) \exp[\beta + (U_{0i}/kT)]}{(1/q_{si}) \exp[\beta + (U_{0i}/kT)] + 1} \right] \right\}. \qquad (99)$$

It is now necessary to find the relationship between β and thermodynamic quantities. To do this, we note first that summing Eq. (98), $N_s = \sum_i N_{si}$, gives $\beta = \beta(N_s, B, T)$. We then take the derivative $(\partial \ln Q_s/\partial N_s)_{B, T}$ in Eq. (99)

using the facts that $N_s = \sum_i N_{si}$ gives $\sum_i (dN_{si}/dN_s) = 1$, and $N_{si}(\partial\beta/\partial N_{si})_{B,\,T} = B_i/(N_{si} - B_i)$. The result is

$$-\frac{\mu_s}{kT} = \left(\frac{\partial \ln Q_s}{\partial N_s}\right)_{B,\,T} = \left[\frac{\partial(-A_s/kT)}{\partial N_s}\right]_{B,\,T} = \beta, \qquad (100)$$

so that β is identified with the chemical potential, μ_s. At equilibrium, $\mu_G/kT = \mu_s/kT = -\beta$. Making this substitution in Eq. (98) and summing, we find

$$N_s = \sum_i N_{si} = \sum_i \frac{B_i}{(1/q_{si})\exp[-(\mu_G - U_{0i})/kT] + 1}, \qquad (101)$$

which is the adsorption isotherm.

If the distribution B_i is such that the summation over the B_i can be replaced by integration (45, 46), then let $Bf(U_0)\,dU_0$ be the number of sites with values of U_{0i} between U_0 and $U_0 + dU_0$, where $\int f(U_0)\,dU_0 = (1/B)$ $\int [dB(U_0)/dU_0]\,dU_0 = 1$, and $0 < B(U_0) < B$. The isotherm, Eq. (101), now becomes, after substituting the expression for μ_G from Eq. (68),

$$\frac{N_s}{B} = \theta(p, T) = \int \frac{a(T, U_0)p}{1 + a(T, U_0)p}\, f(U_0)\,dU_0,$$

$$a(T, U_0) = \frac{h^3 q_{s_v}(T, U_0)q_{s_{int}}(T, U_0)\exp(-U_0/kT)}{(2\pi m)^{3/2}(kT)^{5/2}q_{G_{int}}}, \qquad (102)$$

which may be compared with Eq. (71) for a uniform surface. The dependence of $q_{s_{int}} = q_{s_v}q_{s_r}$ on U_0 resides almost entirely in q_{s_r}, since $q_{s_v} \cong q_{G_{v'}}$, and is relatively unaffected by variations of U_0. Equation (102) may, of course, be written in terms of the variable $\epsilon = -[U_0 + \sum_i (h\nu_i/2)]$, the energy of adsorption (greater than 0) at absolute zero, where $\sum_i (h\nu_i/2)$ is the sum of the zero-point energies of vibration with respect to the surface. In that case,

$$(1/B)\int [dB(U_0)/dU_0]\,dU_0 = \int f(U_0)\,dU_0$$

$$= \int f[U_0(\epsilon)][dU_0(\epsilon)/d\epsilon]\,d\epsilon$$

$$= \int g(\epsilon)\,d\epsilon = 1, \qquad (103)$$

and zero-point vibrational terms are excluded from *all* vibrational partition functions appearing in the isotherm equation. This is equivalent to making the zero of energy the ground state (zero-point vibrational level) of the isolated molecule instead of the bottom of the potential well; and the potential drop of the adsorbed molecule is measured down to the zero point level instead of to the bottom of the potential well.

The various thermodynamic functions may be calculated for localized adsorption on nonuniform surfaces in a manner similar to that employed in deriving the adsorption isotherm [Eq. (102)]. The Helmholz free energy A_s and the energy E_s are given by

$$A_s = -kT \ln Q_s$$

$$= B \int \left\{ \frac{\mu_G}{(1/q_s) \exp[-(\mu_G - U_0)/kT] + 1} \right.$$

$$\left. + kT \ln \left[\frac{(1/q_s) \exp[-(\mu_G - U_0)/kT]}{(1/q_s) \exp[-(\mu_G - U_0)/kT] + 1} \right] \right\} f(U_0) \, dU_0 ,$$

$$\tag{104}$$

$$E_s = kT^2 \left(\frac{\partial \ln Q_s}{\partial T} \right)_{N_s, B}$$

$$= B \int \left[kT^2 \frac{\partial \ln q_s}{\partial T} + U_0 \right] \frac{f(U_0) \, dU_0}{(1/q_s) \exp[-(\mu_G - U_0)/kT] + 1} ,$$

where μ_G is given by Eq. (68).

The entropy may be calculated from $S_s = (E_s - A_s)/T$, but it is interesting to separate the entropy into a configurational entropy S_{s_c} and a nonconfigurational entropy S_s^*,

$$S_{s_c} = k \ln \prod_i \frac{B_i !}{N_{si} ! (B_i - N_{si})!}$$

$$= kB \int \left\{ \ln \frac{(1/q_s) \exp[-(\mu_G - U_0)/kT] + 1}{(1/q_s) \exp[-(\mu_G - U_0)/kT]} \right.$$

$$\left. + \frac{\ln(1/q_s) \exp[-(\mu_G - U_0)/kT]}{(1/q_s) \exp[-(\mu_G - U_0)/kT] + 1} \right\} f(U_0) \, dU_0 \tag{105}$$

and

$$S_s^* = kB \int \frac{T(\partial \ln q_s/\partial T) + \ln q_s}{(1/q_s) \exp[-(\mu_G - U_0)/kT] + 1} f(U_0) \, dU_0 . \tag{106}$$

In comparison with the molar configurational entropy S_{s_c}/n_s for a uniform surface, that for a nonuniform surface is reduced. The reason is that on a nonuniform surface those configurations in which molecules are adsorbed on the strongest sites are more probable than those involving weaker sites. The differential molar configurational entropy $(\partial S_{s_c}/\partial n_s)_{T, p, \alpha} \equiv \bar{s}_{s_c}$ falls below that for a uniform surface at less than half coverage and is greater at higher coverages.

The limits of integration in the expressions for θ, A_s, E_s, and S_s given above should be the lowest and the highest adsorption energies of the surface. However, these quantities are rarely determinable, therefore, the limits are usually taken from 0 to $\pm\infty$.[3] No serious error is introduced with these limits, for, under normal conditions of pressure and temperature of adsorption experiments, the very weak sites contribute little to the total adsorption; and the number of sites with energies above the highest measured values of U_0 is usually very small. Sips (45) and Halsey and Taylor (46) have used $-\infty$ to $+\infty$ as limits of integration in their isotherm equations. Hill (44), however, feels that these limits are unrealistic, for ordinarily $f(U_0) = 0$ for $U_0 > 0$, and in most cases $f(U_0)$ is essentially zero except in a range $U_{01} < U_0 < U_{02}$, where $U_{01} > -\infty$ and $U_{02} < 0$.

Everett (47) has derived the isotherm in terms of enthalpy and entropy functions. Taking the standard state of the gas as 1 atm (perfect gas), we find from, Eqs. (73) and (74),

$$RT\ln p = \overline{\Delta h} - T\overline{\Delta s}, \tag{107}$$

where $\overline{\Delta h} = -q_{st}$ is the differential change in enthalpy and $\overline{\Delta s} = (\bar{s}_s - s_G)$ is the differential change in entropy when 1 mole of gas at 1 atm is transferred to the surface at a constant coverage corresponding to equilibrium pressure p. The term $\bar{s}_s$, the differential molar surface entropy may be broken down into the sum of a configurational term $\bar{s}_{s_c}$ and a term including contributions from vibrational and rotational degrees of freedom $\bar{s}_s{}^*$,

$$\bar{s}_s = \bar{s}_{s_c} + \bar{s}_s{}^* = -R\ln[\theta/(1 - \theta)] + \bar{s}_s{}^*, \tag{108}$$

where we have used Eq. (95) for a uniform surface to express the configurational part. We now write

$$\Delta s = (\bar{s}_{s_c} + \bar{s}_s{}^* - s_G) \equiv \bar{s}_{s_c} + \overline{\Delta s}^* = -R\ln[\theta/(1 - \theta)] + \overline{\Delta s}^*, \tag{109}$$

where $\overline{\Delta s}^*$ is the differential change of entropy on adsorption excluding configurational entropy. Substituting Eq. (109) in Eq. (107), we obtain

$$\theta = \frac{\exp[-(\overline{\Delta h} - T\overline{\Delta s}^*)/RT]p}{1 + \exp[-(\overline{\Delta h} - T\overline{\Delta s}^*)/RT]p}, \tag{110}$$

which is Langmuir's isotherm when $\overline{\Delta h}$ and $\overline{\Delta s}^*$ are independent of θ, as they will be for localized adsorption on a uniform surface with no interaction

[3] The limit is $+\infty$ when energy >0, $-\infty$ when energy <0, as U_0.

between adsorbed molecules. For nonuniform surfaces, we have

$$N_s = \sum_i N_{si} = \sum_i \frac{B_i \exp[-(\overline{\Delta h_i} - T\overline{\Delta s_i}^*)/RT]p}{1 + \exp[-(\overline{\Delta h_i} - T\overline{\Delta s_i}^*)/RT]p}, \tag{111}$$

which may be put into integral form,[4]

$$\theta = \frac{N_s}{B} = \int \frac{\exp(\overline{\Delta s}^*/R) \exp(q_{st}/RT)p}{1 + \exp(\overline{\Delta s}^*/R) \exp(q_{st}/RT)p} h(q_{st}) \, dq_{st}, \tag{112}$$

where the substitution $q_{st} = -\overline{\Delta h}$ has been made, and

$$\int h(q_{st}) \, dq_{st} = (1/B) \int [dB(q_{st})/dq_{st}] \, dq_{st} = 1, \qquad 0 \leqslant B(q_{st}) \leqslant B.$$

In order to use the statistical mechanical isotherm equation, Eq. (102), effectively, it is necessary to know the distribution function $f(U_0)$ or its equivalent $g(\epsilon)$, the partition function of vibration with respect to the surface $q_{s_v}(U_0)$, and the partition function of rotations and internal vibrations of the adsorbed molecule $q_{s_{int}}(U_0) = q_{s_{v'}} q_{s_r}$. There does not appear to have been any success in deriving an expression for q_{s_r} as a function of U_0. The term $q_{s_{v'}}$, as explained previously, is assumed to be unaffected by adsorption, and therefore cancels with $q_{G_{v'}}$. The only attempts to obtain expressions for the variations of surface partition functions with U_0 have been limited to mona-tomic molecules, so that $q_{s_{int}}$ drops out and $q_{s_v}(U_0)$ is composed of one verti-cal and two horizontal vibrations corresponding to the three translational degrees of freedom of the gas-phase molecule.

Frequently, $\prod_i \exp(-hv_i/2kT)$, the zero-point term of the vibrational partition function q_{s_v}, Eq. (81), is combined with $\exp(-U_0/kT)$, to give $\exp(\epsilon/kT)$, and then it is assumed that the remainder $\prod_i [1 - \exp(-hv_i/kT)]^{-1}$ is unity. This is equivalent to assuming that the v_i are sufficiently large or T sufficiently low that $\exp(-hv_i/kT) \ll 1$ for all i, so that the adsorbed atom is in the zero-point level. Under certain conditions, appreciable errors may be introduced with this assumption. Hill (*44*) gives reasonable justification on theoretical grounds that the frequencies of the two horizontal and one vertical vibrations of an adsorbed monatomic molecule with respect to the

---

[4] Any appropriate adsorption energy term X may be used in place of q_{st} so long as

$$(1/B)\int[dB(q_{st})/dq_{st}] \, dq_{st} = \int h(q_{st}) \, dq_{st} = \int h[q_{st}(X)] \, (dq_{st}/dX) \, dX$$
$$= \int J(X) \, dX = (1/B)\int[dB(X)/dX] \, dX$$

is determinable.

surface are proportional to $\epsilon^{1/2}$, the square root of the energy of adsorption at absolute zero, $\epsilon = -[U_0 + (h/2)(v_x + v_y + v_z)]$; and for rough calculations for argon on KCl assumes $v_x = v_y = v_z$. At sufficiently high temperatures, $q_{s_v} = (kT/hv_x)(kT/hv_y)(kT/hv_z)$, or q_{s_v} is proportional to $\epsilon^{-3/2}$.

When using the classical thermodynamic isotherm equation (112), it has been customary to set the term $\exp(\overline{\Delta s^*}/R)$ equal to a constant. Since this term is related to the vibrational partition function q_{s_v}, we would expect that errors of the same nature would be introduced by assuming that its value did not change with U_0. Work of Clark and Holm (48) for ammonia on alumina, and of Drain and Morrison (49) for argon on rutile indicates variation of nonconfigurational entropy with coverage, that is, with U_0 or ϵ. Halsey (50) has assumed a linear variation of the differential entropy of adsorption with the differential heat of adsorption, $\overline{\Delta s^*} = c_1 + c_2 q_{st}$, where c_1 and c_2 are constants.

The distribution function $f(U_0)$ or $g(\epsilon)$ has been determined in a variety of ways: by calorimetrical measurements and from isotherms, by introducing trial distribution functions into the integral form of the Langmuir isotherm until good agreement with the experimental isotherm is obtained, and by the inverse process of solving the integral Langmuir equation for that distribution function which brings about an equality between the Langmuir equation and an empirical equation developed from the experimental data.

We shall now review briefly a few specific examples in which the above methods and assumptions have been used to obtain explicit theoretical isotherms and the various thermodynamic functions. It is emphasized that in all these examples localized adsorption is assumed.

The distribution $g(\epsilon)$ may be determined experimentally by measuring the heat of adsorption as a function of coverage at low temperatures, then extrapolating to absolute zero, making use of heat capacity measurements of the adsorbed phase. At absolute zero, molecules will tend to fill sites in the order of decreasing ϵ, so that $g(\epsilon) = (1/B)[dB(\epsilon)/d\epsilon] = (1/B)(dN_s/d\epsilon)$. Drain and Morrison (49) have determined $g(\epsilon)$ in this manner for argon adsorbed on rutile. They found that at the low temperatures which they were using that the dependence of q_{s_v} on coverage was not marked, since satisfactory agreement between the experimental differential entropy and the sum of the calculated values of $\bar{s}_{s_c}$ and $\bar{s}_{s_v}$ was obtained with q_{s_v} constant. However, they state that their results were also consistent with Hill's theory that the three frequencies v_x, v_y, and v_z (all assumed equal) were proportional to $\epsilon^{1/2}$ within experimental error. Including this variation did not appreciably affect the configurational entropy, but introduced a slight dependency of the nonconfigurational entropy (vibrational) on surface coverage. The results are shown in Fig. 2.10. The experimental differential entropy is represented by the circles, and was

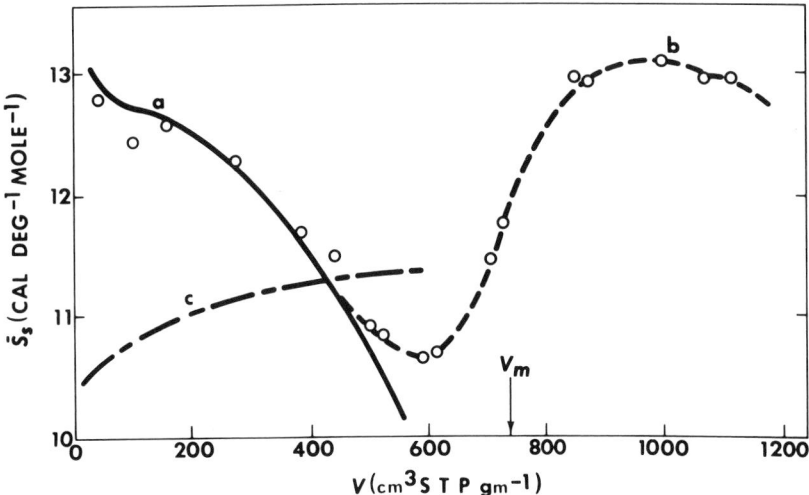

FIG. 2.10. Differential entropy of argon adsorbed on rutile as function of coverage at 85°K. Curve (a), calculated differential entropy or argon adsorbed on rutile; circles and curve (b), experimental data; curve (c), calculated thermal entropy. Monolayer capacity $V_m = 755$ cm³. [After Drain and Morrison, *Trans. Faraday Soc.* **49**, 654 (1953).] Reprinted by permission of the Faraday Society.

calculated from Eq. (107) and from $\overline{\Delta s} = \bar{s}_s - s_G$. In order to obtain numerical agreement at low coverages, an arbitrary constant of 1.6 cal deg⁻¹ mole⁻¹ was added to theoretical entropy values, which reflects an uncertainty in the absolute values of the vibrational frequencies. The theoretical curve and the experimental points coincide closely up to a coverage estimated to be about 0.6 of a monolayer. The deviation beyond this point may be explained by the onset of a second layer of adsorption. A second layer will not form so long as the energy of adsorption on surface sites is very much greater in relation to kT than the energy of adsorption on top of a group of adsorbed molecules. When the formation of a second layer does become energetically feasible, then there are more possible arrangements, and consequently the configurational entropy increases.

Experimental differential entropies $\bar{s}_s$ obtained in the work of Clark and Holm (*48*), Fig. 2.11, are seen to rise sharply with increasing coverage, whereas in the work of Drain and Morrison (*49*) for argon on rutile, Fig. 2.10, the opposite behavior may be observed. In the latter case, the range of heat of adsorption values as the coverage increases from approximately zero up to 0.6 of a monolayer is small (approximately 1100 cal mole⁻¹), and the average heat of adsorption is relatively low (approximately 2500 cal mole⁻¹). Under these conditions, one would expect a rather small increase in the nonconfigurational (vibrational) differential entropy with coverage since $v \sim \epsilon^{1/2}$. On the other

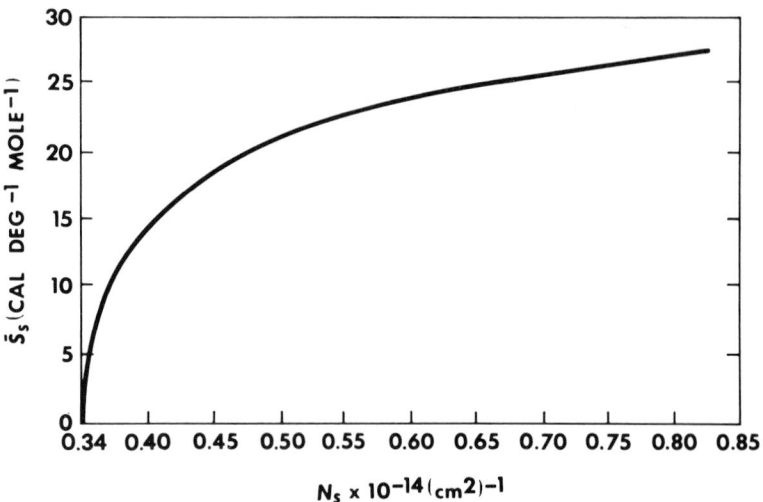

Fɪɢ. 2.11. Variation of differential entropy $\bar{s}_s$ with coverage for NH_3 on Al_2O_3 at 300°C (48).

hand one would expect the differential configurational entropy to decrease with coverage and to approach the values for a uniform surface as the range of heat of adsorption values decreases. The net result is a total differential entropy which decreases with coverage. In the work of Clark and Holm, the range of heat of adsorption values is from about 40 down to 12 kcal mole^{-1} as coverage increases from about 0.36×10^{14} to 1.6×10^{14} molecules cm^{-2}. With such a precipitous drop in heat of adsorption with coverage, the differential configurational entropy, though it would be expected to decrease, should have very small values. However the nonconfigurational entropy would be expected to rise rapidly with the sharp drop in ϵ. The net result is a total differential entropy which *increases* with coverage.

Halsey and Taylor (46) correlated θ (p, T) in Eq. (110) with $h(q_{st})$, the distribution function, by the method of trial functions. They assumed $\exp(\overline{\Delta s^*}/R)$ to be constant, and also, as previously mentioned, used the limits of integration, $-\infty$ to $+\infty$. With this combination of conditions, they showed that an exponential distribution, $h(q_{st}) = a_0 \exp(-a_1 q_{st})$, where a_0 and a_1 are constants, leads to the Freundlich adsorption isotherm, $\theta = A(T)p^{c(T)}$ which fits many experimental data. Note that integration of $h(q_{st})$ over the limits zero to infinity leads to $a_0 = a_1$. Furthermore, Halsey (50) later showed that if $\exp(\overline{\Delta s^*}/R)$ was not assumed constant, but that Δs^* varied linearly with q_{st}, $\overline{\Delta s^*} = b_0 + b_1 q_{st}$, where b_0 and b_1 are constants, then again an exponential distribution leads to a Freundlich isotherm. With regard to the limits of integration used, Halsey states that using the limits zero to infinity gives the same result within experimental accuracy. However, the equation no longer strictly retains the form of the Freundlich isotherm. Clark and Holm (48) have shown that an exponential distribution in the case of ammonia on alumina leads experimentally to a Freundlich isotherm.

Various isotherms can be obtained by assuming different distribution functions. Consider, for example, the linear function,

$$q_{st} = q_0[1 - aB(q_{st})],$$ (113)

where $B(q_{st})$ is the number of adsorption sites with heats of adsorption greater than q_{st}, a is a constant, and q_0 is the maximum heat of adsorption. From Eq. (113), the distribution function is

$$h(q_{st}) = (1/B)[dB(q_{st})/dq_{st}] = -1/Baq_0.$$ (114)

Substituting this into the isotherm, Eq. (112), and integrating with $\exp(\Delta s^*/R)$ constant gives

$$\theta = \frac{RT}{Baq_0} \ln \frac{(p/b) \exp(q_0/RT) + 1}{(p/b) + 1},$$ (115)

where $b = \exp(-\overline{\Delta s^*}/R)$. If we assume that p is small enough that $p/b \ll 1$ and yet large enough that $(p/b) \exp(q_0/RT) \gg 1$, then we obtain for the middle range of θ,

$$\theta = (RT/Baq_0) \ln[(p/b) \exp(q_0/RT)],$$ (116)

which is called the Temkin isotherm (51). This isotherm frequently fits experimental data. Note that with the linear variation of q_{st} with $B(q_{st})$, a finite upper limit q_0 for q_{st} must be selected.

The inverse procedure for determining the distribution function under the integral sign was first introduced by Temkin and Levich (52) in 1946. It consists in rewriting the integral form of the Langmuir isotherm [Eq. (112)], as a Fourier integral and then inverting it by standard mathematical procedures in order to solve for the unknown distribution $h(q_{st})$. Independently, Sips (45, 53) showed that the integral isotherm equation could be reformulated as a Stieltjes transform of the distribution function which could be solved by standard methods (54). He used the isotherm equation (112) in the general form

$$\theta(p, T) = \int_0^\infty \frac{f(q) \, dq}{1 + [b(T)/p] \exp(-q/RT)},$$ (117)

where q is an appropriate energy of adsorption. The term $b(T)$ is assumed to be independent of q. The general procedure is to set Eq. (117) equal to an empirical equation that faithfully represents the experimental data, and then solve by means of a Stieltjes transform for the distribution function $f(q)$. In his earlier work (45), Sips used the integration limits $-\infty$ to $+\infty$ and the Freundlich isotherm as the empirical equation, and obtained an exponential distribution function. However, integrating an exponential form of $f(q)$ between $-\infty$ and $+\infty$ gives an infinite value instead of unity. Consistent with this, but also physically unrealistic, the Freundlich isotherm, $\theta = Ap^c$, increases to infinity with increasing p instead of leveling out asymptotically to unity.

In a second paper, Sips (53) made more physically realistic assumptions, starting with Eq. (117). We shall review this work briefly in order to illustrate the elegance of the method.

In Eq. (117), setting $(b/p) + 1 = y$ and $\exp(q/RT) - 1 = x$, we obtain

$$\frac{\theta[b/(y-1)]}{RT} = \int_0^\infty \frac{f[RT \ln(x+1)]\, dx}{(x+y)}. \tag{118}$$

The expression on the right is known as the Stieltjes transform of the function $f[RT \ln(x+1)]$. From the theory of this transform (54), it is known that if

$$g(y) = \int_0^\infty \frac{\varphi(x)\, dx}{x+y}, \tag{119}$$

then, provided the definite integral Eq. (119) exists,

$$\varphi(x) = [g(xe^{-\pi i}) - g(xe^{\pi i})]/2\pi i. \tag{120}$$

In order to apply the transform, it is necessary to set Eq. (118) equal to an experimentally determined equation of the isotherm. One of the functions selected by Sips is

$$\theta(p) = [p/(p+b)]^c, \tag{121}$$

with $0 < c < 1$. Eq. (121) satisfies $\theta(0) = 0$, $\theta(\infty) = 1$, and at low pressure reduces to the Freundlich isotherm. Thus, in this case,

$$g(y) = \theta/RT = y^{-c}/RT = \int_0^\infty f[RT \ln(x+1)]\, dx/(x+y), \tag{122}$$

and

$$\varphi(x) \equiv f[RT \ln(x+1)] = \{[(xe^{-\pi i})^{-c} - (xe^{\pi i})^{-c}]/2\pi i\}(1/RT), \tag{123}$$

or

$$f(q) = (1/RT)(\sin \pi c/\pi)[\exp(q/RT) - 1]^{-c}. \tag{124}$$

At $q = 0$, Eq. (124) shows that $f(q) = \infty$. However, for small values of q, $\exp(q/RT) - 1$ may be replaced approximately by q/RT, so that

$$\int_0^q (dq/q^c) = q^{1-c}/(1-c), \tag{125}$$

which tends to zero as $q \to 0$, provided $0 < c < 1$. As $q \to \infty$, $f(q) \to 0$. Thus, no difficulty is encountered with the requirement, $\int_0^\infty f(q)\,dq = 1$, which may be seen directly by letting $p \to \infty$ in

$$\lim_{p \to \infty} [p/(p+b)]^c = \lim_{p \to \infty} \int_0^\infty f(q)\,dq/[1 + (b/p)\exp(-q/RT)]$$

$$= \int_0^\infty f(q)\,dq = 1. \tag{126}$$

Using the isotherm $\theta = [p/(p+A)]^c$, where the constant $A < b$, merely shifts the distribution without actual change. The expression for the distribution function, $f(q)$, in this case contains the constants A, c, and b. Constants A and c are known experimentally, and the constant b is determined from the relation $\int_0^\infty f(q)\,dq = 1$.

It should be emphasized again that agreement between the theoretical isotherms with distribution functions determined by the various procedures outlined above and experimental isotherms does not guarantee that the true physical picture has been discovered. For the initial assumptions were made that the adsorption was localized and restricted to a monolayer, that no perturbation of the adsorbent occurred, that a fixed number of sites was present, and no lateral interactions between adsorbed molecules existed. Changing any of these assumptions could very well lead to new forms of equations in agreement with the experimental adsorption isotherms. Another difficulty is the inherent insensitivity of the theoretical isotherm to the form of the distribution function within the accuracy of experimental data. Honig (55) showed, for example, that for nitrogen adsorption on rutile at 77°K, several radically different trial-distribution functions fitted the isotherm data equally well over a limited range of surface coverage ($\theta > 0.4$). His calculations emphasize the necessity for obtaining extremely accurate experimental data over the broadest possible range of coverage. Even though the inversion method determines a unique distribution function, a small change in the empirical adsorption equation might produce an entirely different function.

2.6 Hopping Molecules

When $\underline{V}_0$ (see Figs. 2.1 and 2.2) is not high enough to prevent appreciable thermal motion of adsorbed molecules, the Langmuir equation no longer strictly applies. In order to treat this case, which is intermediate between the immobile case that has concerned us up to now and the mobile case (see Chapter V), an expression for $U_m(x, y)$, the potential minimum of $U(z)$ as a function of x and y, is needed in terms of the parameters U_0 and $\underline{V}_0$. No theoretical derivation of such an expression has been made yet. A reasonable expression was used by Hill (*56*) to illustrate the procedure for this intermediate case. A simple square lattice is assumed with a nearest-neighbor distance *a* between adsorption sites. The surface area is $\alpha = L^2 = Ba^2$, where *B* is the number of sites. The *x, y* motion is assumed independent of the *z* motion, and the latter is represented by a one-dimensional harmonic oscillator partition function q_z. The expression representing $U_m(x, y)$ is taken to be

$$U_m(x, y) = U_0 + \tfrac{1}{2}\underline{V}_0[1 - \cos(2\pi x/a)] + \tfrac{1}{2}\underline{V}_0[1 - \cos(2\pi y/a)], \quad (127)$$

where U_0, as before, is negative and $\underline{V}_0$ is positive. Potential minima U_0 occur at locations directly above each site, $(x, y) = (na, ma)$; potential maxima, $U_0 + 2\underline{V}_0$, occur at locations equidistant from four sites, $[(n + \tfrac{1}{2})a, (m + \tfrac{1}{2})a]$; and potential saddle points, $U_0 + \underline{V}_0$, occur at locations equidistant from only two points, $[(n + \tfrac{1}{2})a, ma]$ and $[na, (m + \tfrac{1}{2})a]$; where in all cases $n, m = 0, 1, 2, 3, \ldots$.

At low temperatures, the adsorbed molecules vibrate about U_0, the minima of $U_m(x, y)$. The frequency of the vibration is obtained from the second derivative of $U_m(x, y)$ with respect to either *x* or *y*,

$$\partial^2 U_m(x, y)/\partial x^2 = \partial^2 U_m(x, y)/\partial y^2 = 2\pi^2 \underline{V}_0/a^2 \equiv f,$$
$$v_x = v_y = (1/2\pi)(f/m)^{1/2} = (\underline{V}_0/2ma^2)^{1/2}. \tag{128}$$

At high temperatures, the adsorbed molecule surmounts the potential barrier and moves freely in the *x, y* plane, and the partition function for two degrees of translational freedom, Eq. (79), may be used.

At intermediate temperatures, Eq. (127) must be used as it stands. The problem is to find the quantum mechanical energy levels and partition function for the potential $U_m(x, y)$. This is an extremely complicated problem, which has been solved by Pitzer and Gwinn (*57*) for hindered rotation. These authors also introduced an excellent approximation which they tested against exact results. We present the approximation here applied to hindered translation. In the approximation, the *x, y* partition function is written

$$q_{xy} = q_{\text{class}} \times (q_{\text{har osc–quant}})/(q_{\text{har osc–class}}), \tag{129}$$

where q_{class} is the classical q_{xy} with the potential $U_m(x, y) - U_0$, $q_{\text{har osc–quant}}$ is the quantum mechanical harmonic oscillator partition function for motion about the potential minimum U_0, and $q_{\text{har osc–class}}$ is the classical limit of $q_{\text{har osc–quant}}$. Equation (129) obviously has the correct properties at the extremes of temperature,

$$T \to \infty, \qquad q_{\text{har osc–quant}} \to q_{\text{har osc–class}}, \qquad q_{xy} \to q_{\text{class}},$$
$$T \to 0, \qquad q_{\text{class}} \to q_{\text{har osc–class}}, \qquad q_{xy} \to q_{\text{har osc–quant}}.$$

The complete partition function for the system is written

$$Q_s = (1/N_s!)q_s^{N_s}, \tag{130}$$

where

$$q_s = q_{xy} q_z \exp(-U_0/kT).$$

This is the partition function for a dilute adsorbed phase, for it neglects interactions between the partially mobile adsorbed molecules.

The partition function q_{class} is obtained from the usual classical expression,

$$q_{class} = (1/h^2) \int_{-\infty}^{\infty} \int \int_0^L \int e^{-H/kT} \, dx \, dy \, dp_x \, dp_y, \tag{131}$$

where the Hamiltonian H is

$$H = (p_x^2/2m) + (p_y^2/2m) + \tfrac{1}{2}V_0[2 - \cos(2\pi x/a) - \cos(2\pi y/a)]. \tag{132}$$

We then have

$$q_{class} = (2\pi mkT/h^2) \exp(-\underline{V}_0/kT)\left\{\int_0^L \exp[(\underline{V}_0/2kT) \cos(2\pi x/a)] \, dx\right\}^2. \tag{133}$$

Substitution of $u = \underline{V}_0/2kT$ and $\theta = 2\pi x/a$, into Eq. (133) yields

$$q_{class} = (2\pi mkT/h^2)e^{-2u}\left[(a/2\pi) \int_0^{2\pi\sqrt{B}} e^{u \cos\theta} \, d\theta\right]^2$$

$$= (2\pi mkT/h^2)e^{-2u}\left[(L/2\pi) \int_0^{2\pi} e^{u \cos\theta} \, d\theta\right]^2. \tag{134}$$

The solution of the integral is $2\pi I_0(u)$, where $I_0(u)$ is a modified Bessel function of the first kind. Thus,

$$q_{class} = (2\pi mkT/h^2)\alpha e^{-2u} I_0^2(u) \tag{135}$$

The classical harmonic-oscillator partition function $q_{har\ osc-class}$ may be obtained from q_{class}, Eq. (135), using the limit $T \to 0$, $(u \to \infty)$

$$q_{har\ osc-class} = q_{class}(u \to \infty) = B(kT/hv_x)^2. \tag{136}$$

The factor B is expected because there are B sites upon which the oscillator may be found in the area α over which the phase integral, Eq. (131), extends. In deriving Eq. (136) from Eq. (135), we have used, $\lim_{u \to \infty} I_0(u) = e^u/(2\pi u)^{1/2}$.

The quantum mechanical harmonic oscillator partition function must be [see Eq. (81)]

$$q_{har\ osc-quant} = B\left[\frac{\exp(-hv_x/2kT)}{1 - \exp(-hv_x/kT)}\right]^2. \tag{137}$$

Equation (136) may also be obtained by taking the high temperature limit of Eq. (137).

Substituting Eqs. (135)–(137) into Eq. (129), we obtain

$$q_{xy} = 2\pi B u e^{-2u} e^{-2Ku} I_0^2(u)/(1 - e^{-Ku})^2, \tag{138}$$

where $K = (2h^2/ma^2 \underline{V}_0)^{1/2}$.

From the complete partition function Q_s [Eq. (130)], we can find the chemical potential,

$$\mu_s = -kT(\partial \ln Q_s/\partial N_s)_{B, T} \, ;$$

and as before, setting $\mu_s = \mu_G$ at equilibrium, we obtain the isotherm

$$N_s = q_{xy} q_z \exp(-U_0/kT) \exp(\mu^0/kT) p, \tag{139}$$

which is linear since the adsorbed phase is dilute. Other thermodynamic quantities are easily obtained.

REFERENCES

1. Dacy, J. R., *Ind. Eng. Chem.* **57**, No. 6, 26 (1965).
2. Gilliland, R. E., *Am. Inst. Chem. Eng. J.* **4**, 90 (1958).
3. Fowler, R. H., *Proc. Cambridge Phil. Soc.* **31**, 260 (1935).
4. Hill, T. L., *J. Chem. Phys.* **17**, 520 (1949).
5. Brunauer, S., Deming, L. S., Deming, W. E., and Teller, E. J., *J. Am. Chem. Soc.* **62**, 1723 (1940).
6. Wheeler, A., Paper given at meeting of American Chemical Society, Atlantic City, New Jersey, September (1949).
7. Ono, S., *J. Chem. Phys.* **18**, 397 (1950).
8. Ono, S., *J. Phys. Soc. Japan* **5**, 232 (1950).
9. Ono, S., *J. Phys. Soc. Japan* **6**, 10 (1951).
10. Frenkel, J., " Kinetic Theory of Liquids." Oxford Univ. Press, London and New York, 1946.
11. Halsey, G., *J. Chem. Phys.* **16**, 931 (1948).
12. Hill, T. L., *J. Chem. Phys.* **17**, 590, 668 (1949).
13. McMillan, W. G., and Teller, E., *J. Chem. Phys.* **19**, 25 (1951); *J. Phys. Colloid. Chem.* **55**, 17 (1951).
14. Barrer, R. M., and Robbins, A. B., *Trans. Faraday Soc.* **47**, 773 (1951).
15. Brunauer, S., Emmett, P. H., and Teller, E., *J. Am. Chem. Soc.* **66**, 309 (1938).
16. Cassie, A. B. D., *Trans. Faraday Soc.* **41**, 450 (1945).
17. Hill, T. L., *J. Chem. Phys.* **14**, 263 (1946); **17**, 772 (1949).
18. Keii, T., *J. Chem. Phys.* **22**, 1612 (1954).
19. Cassel, H. M., *J. Chem. Phys.* **12**, 115 (1944); *J. Phys. Chem.* **48**, 195 (1944).
20. Hill, T. L., *J. Chem. Phys.* **15**, 767 (1947).
21. Pickett, G., *J. Am. Chem. Soc.* **67**, 1958 (1945).
22. Hill, T. L., *J. Am. Chem. Soc.* **68**, 535 (1946).
23. Anderson, R. B., *J. Am. Chem. Soc.* **68**, 686 (1946).
24. Smith, R. N., and Pierce, C., *J. Phys. Chem.* **52**, 1115 (1948).
25. Steele, W. A., *J. Chem. Phys.* **25**, 819 (1956).

26. McMillan, W. G., *J. Chem. Phys.* **15**, 390 (1947).
27. Walker, W. C., and Zettlemoyer, A. C., *J. Phys. Chem.* **52**, 47 (1948).
28. Cook, M. A., *J. Am. Chem. Soc.* **30**, 2925 (1948).
29. Hill, T. L., *Advan. Catalysis* **4**, 232 (1952).
30. Rossington, D. R., and Borst, E., *Surface Sci.* **3**, 202 (1965).
31. Roberts, J. K., *Proc. Roy. Soc. (London), Ser. A* **152**, 455 (1935).
32. Roberts, J. K., *Proc. Cambridge Phil. Soc.* **34**, 399, 577 (1938).
33. Roberts, J. K., and Miller, A. R., *Proc. Cambridge Phil. Soc.* **35**, 293 (1934).
34. Miller, A. R., "The Adsorption of Gases on Solids." Cambridge University Press, London and New York, 1949.
35. Everett, D. H., *Proc. Chem. Soc.* **37**, (1957).
36. Kemball, C., *Proc. Roy. Soc. (London), Ser. A* **187**, 73 (1946).
37. Rideal, E. K., and Sweett, F., *Proc. Roy. Soc. (London), Ser. A* **257**, 291 (1960).
38. Kisliuk, P., *J. Chem. Phys.* **30**, 174 (1959).
39. de Boer, J. H., and Kruyer, S., *Proc. Koninkl. Ned. Akad. Wetenschap.* **55B**, 451 (1952).
40. Volmer, M., *Z. Phys. Chem.* **115A**, 253 (1925).
41. Homfray, R., *Z. Phys. Chem.* **74**, 129 (1910).
42. Richardson, L. B., *J. Am. Chem. Soc.* **39**, 1828 (1917).
43. Smith, F. W., Ph.D. Thesis, University, St. Andrews, Scotland (1952).
44. Hill, T. L., *J. Chem. Phys.* **17**, 762 (1949).
45. Sips, R., *J. Chem. Phys.* **16**, 490 (1948).
46. Halsey, G., and Taylor, H. S., *J. Chem. Phys.* **15**, 624 (1947).
47. Everett, D. H., *Trans. Faraday Soc.* **46**, 942 (1950).
48. Clark, A., and Holm, V. C. F., *J. Catalysis* **2**, 21 (1963).
49. Drain, L. E., and Morrison, J. A., *Trans. Faraday Soc.* **49**, 654 (1953).
50. Halsey, G. D., *Advan. Catalysis* **4**, 259 (1952).
51. Frumkin, A., and Slygin, A., *Acta Physicochim. U.R.S.S.* **3**, 791 (1935).
52. Temkin, M., and Levich, V., *Zh. Fiz. Khim.* **20**, 1441 (1946).
53. Sips, R., *J. Chem. Phys.* **18**, 1024 (1950).
54. Titchmarsh, E. C., "Introduction to the Theory of Fourier Integrals." Oxford Univ. Press (Clarendon) London and New York, 1959.
55. Honig, J. M., *Ann. N. Y. Acad. Sci.* **58**, 741 (1954).
56. Hill, T. L., "Introduction to Statistical Thermodynamics," p. 172. Addison–Wesley, Reading, Massachusetts, 1960.
57. Pitzer, K. S., and Gwinn, W. D., *J. Chem. Phys.* **10**, 428 (1942).

III

LOCALIZED ADSORPTION—DEPENDENT SYSTEMS

3.1 Phase Transitions

The discussion of localized adsorption is continued with the added stipulation that nearest-neighbor adsorbed molecules exert intermolecular forces on each other. The nature of the forces of adsorption will be treated in Chapters VI–VIII; here we are concerned only with the effects of these forces on the properties of adsorbed films. With the introduction of these forces, the geometric arrangement of adsorption sites—the distance of separation and the number of nearest neighbors—becomes important. We assume a regular lattice of sites in one, two, or three dimensions, in which each site is either occupied by an adsorbed molecule (state " a ") or is unoccupied (state " b "). Other systems are related closely to the same statistical model. For example, in the theory of a ferromagnetic system, each lattice site is occupied either by an up-spin (a) or a down-spin (b). In the theory of order-disorder transition in binary alloys, each lattice site is occupied either by an atom of " a " or an atom of " b." The potential energy of interaction between a pair of nearest-neighbors is w_{aa}, w_{ab}, or w_{bb} depending on how the pair is occupied. For the adsorption system, $w_{aa} = w$ is the potential energy of interaction between a nearest-neighbor pair of adsorbed molecules, and w_{ab} and w_{bb} are taken as zero. Second and higher-neighbor interactions may be important in some systems, but for simplicity we consider only nearest-neighbor interactions. Such regular lattice systems with nearest-neighbor interactions are frequently called "Ising models," and it will be shown that these models reflect many of the characteristics of real systems.

The introduction of intermolecular forces brings about the possibility of phase transitions. Among the common types of phase transitions, including localized and nonlocalized systems, are the condensation of gases, melting of solids, transitions of ferromagnetic substances through the Curie point, and order–disorder transitions in binary alloys. Many phase transitions other than those of binary alloy systems are order–disorder transitions. The transition between the highly ordered state of a crystal and the relatively disordered state of its melt is one example. Another is that of a ferromagnetic metal. Below the Curie point there is a strong preference for spins with a particular orientation; above the Curie point long-range order disappears. However, the condensation of a gas represents a transition between two phases neither of which possesses much long-range order. Many phase transitions represent changes of state, such as melting and condensation, and thus differ in this respect from ferromagnetic and binary alloy transitions. Phase changes have also been classified into first-order and second-order transitions. In first-order transitions, the first derivatives $(\partial \mu / \partial P)_T$ and $(\partial \mu / \partial T)_P$, using the language of a three-dimensional gas, have different values for the two phases, and thus change discontinuously at the phase transition. The discontinuity occurs at the point where both phases (A and B) are in equilibrium with each other. At this point, $\mu_A = \mu_B$, and since any change in temperature and pressure will bring about a new equilibrium between the phases, below the critical temperature, $d\mu_A = d\mu_B$. A change in pressure at constant temperature or in temperature at constant pressure at the equilibrium point will cause one of the phases to disappear. A plot of μ versus P at constant temperature in Fig. 3.1 illustrates

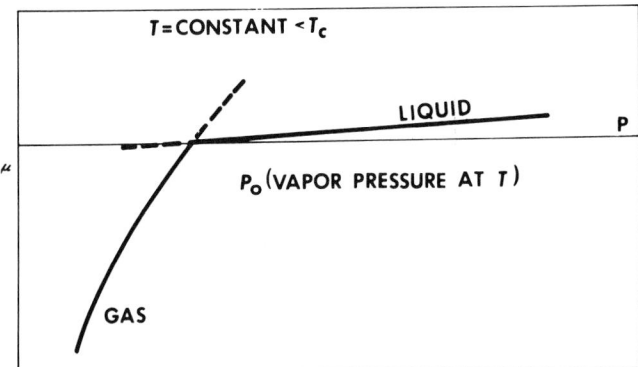

FIG. 3.1. Chemical potential μ versus pressure P at constant temperature showing a phase transition (liquid–gas). [After Hill, "An Introduction to Statistical Thermodynamics," 1960, Addison-Wesley, Reading, Massachusetts.]

the discontinuity in the slope $(\partial\mu/\partial P)_T$ at the equilibrium point for a first-order phase transition. Since,

$$(\partial\mu/\partial P)_T = V/N, \tag{1}$$

the volume per molecule, there is a discontinuous change in the volume below the critical temperature in a first-order phase transition.

Similarly,

$$E/N = T(S/N) - P(V/N) + \mu = \mu - P(\partial\mu/\partial P)_T - T(\partial\mu/\partial T)_P, \tag{2}$$

which shows that the energy is also discontinuous below the critical temperature in a first-order phase transition, and that such a system will have a latent heat.

Second-order phase transitions are defined as involving discontinuities in the second derivatives, $\partial^2\mu/(\partial T\,\partial P)$, $(\partial^2\mu/\partial T^2)_P$ and $(\partial^2\mu/\partial P^2)_T$. They exhibit no discontinuity in the volume or energy, and do not have a latent heat. They do, however, have discontinuities (or infinities) in the heat capacities $(\partial E/\partial T)$.

Transitions involving a change of state such as melting or condensation are always first order and always possess a latent heat. Magnetic and binary alloy transitions are generally second order. Despite the apparent difference between phase transitions involving a change of state and those which do not, it is often difficult to distinguish them experimentally. In the first place, it is often difficult to distinguish between a discontinuity and a rapid variation of the thermodynamic variable in a small interval. Secondly, at the critical point the latent heat of a first-order transition drops to zero and cannot be distinguished there in this respect from a second-order transition such as the Curie point of a ferromagnet which has no latent heat in the absence of an external magnetic field. Thus, the magnetic case might be looked upon as the limiting case of a transition involving a change of state, in which the latent heat has dropped to zero.

The one feature which all phase transitions possess, regardless of order, whether they represent changes of state or not, is the sharpness of the appearance of a new phase. A chemical reaction varies smoothly from one equilibrium point to another with change of pressure or temperature. For a simple reaction such as $A \rightleftharpoons B$, the probability that a particular molecule A reacts to form B (or vice versa) is independent of the behavior of the other molecules. But a phase change is a cooperative phenomenon in which many molecules tend to switch in unison from state A to state B, thus stabilizing each other through intermolecular attraction. It is this cataclysmic property of phase changes with which we shall be concerned in this chapter.

A two-dimensional, immobile adsorbed phase with intermolecular forces acting between nearest-neighbors corresponds to a two-phase liquid–gas system below the critical temperature. When $w_{aa} < 0$ (attractive forces between molecules), the system will split into two phases, one with high density and the other low. It will exhibit the characteristics of a first-order phase transition with a latent heat, and a discontinuity in $(\partial\mu_s/\partial\varphi)_T$ and in $(\partial\mu_s/\partial T)_\varphi$. Analogous to Eq. (1), we have

$$(\partial\mu_s/\partial\varphi)_T = \alpha/N_s,\qquad(3)$$

which is easily derived from Eq. (15), Chapter I,

$$d\mu_s = (\partial\mu_s/\partial T)_\varphi\, dT + (\partial\mu_s/\partial\varphi)_T d\varphi\,,\quad\text{and}\quad d(F_s + \varphi\alpha) = dF_s + \varphi\, d\alpha + \alpha\, d\varphi.$$

In Fig. 3.2, the equation of state is shown schematically above and below the critical temperature, where $1/\theta = B/N_s \sim \alpha/N_s$ and $\pi \sim \varphi$ as seen in Chapter II. It will be observed that the critical density occurs at $\theta = \frac{1}{2}$ which we shall show theoretically below. If the system is restricted to this critical density, and the temperature is raised through the critical point, the latent heat will be

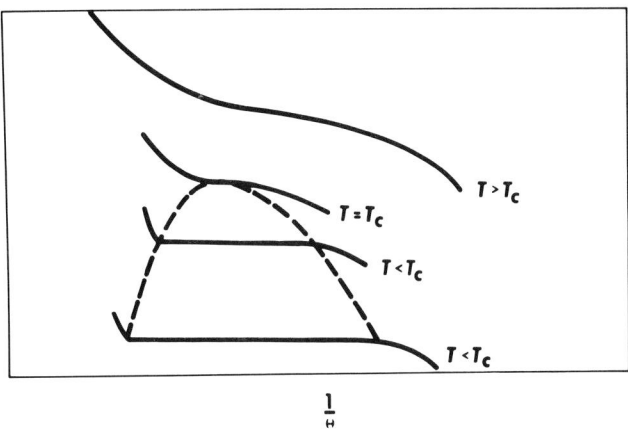

$$\frac{1}{\theta}$$

FIG. 3.2. Equation of state for lattice gas with $w < 0$ showing two-phase region ($T < T_c$).

zero and there will be no discontinuity in the first derivative of μ. There will, however, be a singularity in the heat capacity. In other words, the transition behaves as though it were second order, which is generally the behavior of first-order phase transitions at the critical point.

If $w_{aa} > 0$ (repulsive forces between molecules), molecules will tend to occupy alternate sites at low temperatures, without splitting into two phases.

When near-neighbor interactions are present, an extra term in the partition function for interaction energy is required. With this extra term, only partition functions for the whole system can be written. Expressions for the canonical and grand canonical partition functions will be developed. By use of the symmetry properties of the grand canonical partition function, it will be shown that if a phase transition exists, the critical density occurs when the surface is half-covered. After the partition functions are developed, they will be applied first to the case of the one-dimensional lattice gas. This is the only case for which a general analytic solution for the partition functions, and thus for the thermodynamic functions, can be obtained at present. All other cases are expressed in terms of series solutions, except for the special case of half-coverage, which will be developed in the next chapter for the two-dimensional lattice. For the one-dimensional lattice, there is no evidence of phase transitions. Closed approximate solutions for the two-dimensional case can be obtained, and the two most important of these, the Bragg–Williams and the quasi-chemical, will be derived. Both show phase transitions for two-dimensional systems; and, in addition, the Bragg–Williams approximation shows erroneously a phase transition for the one-dimensional case. A typical exact combinatorial solution of the two-dimensional case in series form will be derived, which reduces the solution of the partition function to finding the number of closed polygons that can be drawn on the lattice. Such series solutions do not allow the determination of numerical values for the critical constants. However, a derivation of the duality theorem of Kramers and Wannier will be presented, which determines the critical temperature for a square lattice without necessity of summing the partition function series. The value of the critical temperature obtained in this way is identical with the one obtained by the closed solution for half-coverage which will be derived in the next chapter. The chapter will be concluded with an expansion of the surface pressure for an imperfect adsorbed gas, developed from the grand canonical partition function.

3.2   Partition Functions

The basic thermodynamic equation for the canonical partition function is

$$dA_s = -S_s \, dT - \pi \, dB + \mu_s \, dN_s. \tag{4}$$

The partition function is

$$Q_s(N_s, B, T) = q_s^{N_s} \sum_{i=1}^{\Omega} \exp(-E_i/kT), \tag{5}$$

where q_s is the single-molecule partition function without interactions, encountered in Chapter II; and E_i is the total nearest-neighbor energy of interaction for the ith configuration of N_s molecules. The total number of configurations Ω is the number of ways of distributing N_s molecules on B sites, no more than one molecule per site,

$$\Omega = B!/[N_s!(B - N_s)!]. \tag{6}$$

We can rewrite Eq. (5) in the form

$$Q_s(N_s, B, T) = q_s^{N_s} \sum_{N_{aa}} g(N_s, B, N_{aa}) \exp(-wN_{aa}/kT), \tag{7}$$

where N_{aa} is the number of nearest-neighbor pairs both occupied, $g(N_s, B, N_{aa})$

is the number of configurations having N_{aa} occupied pairs of sites when there are N_s molecules and B sites, and w is the potential energy of interaction between two molecules adsorbed on two nearest-neighbor sites. The zero of potential energy is infinite separation, and w is a negative number for attractive forces and positive for repulsive forces. For subsequent use, we shall find it more convenient in the expression for Q_s to replace N_{aa} by N_{ab} the number of nearest-neighbor pairs of sites, one occupied and the other unoccupied. The relationships between N_{aa}, N_{ab}, and N_{bb} with $B \to \infty$ and for a particular configuration are

$$2 N_{aa} + N_{ab} = cN_s,$$
$$2 N_{bb} + N_{ab} = c(B - N_s),$$

(8)

where c is the number of nearest-neighbors for any given site ($c = 4$ for a square lattice). Equation (7) now becomes

$$Q_s(N_s, B, T) = [q_s \exp(-cw/2kT)]^{N_s} \sum_{N_{ab}} g(N_s, B, N_{ab}) \exp(wN_{ab}/2kT).$$

(9)

When $w = 0$, Eq. (9) reduces to the Langmuir expression [Eq. (9), Chapter II], with

$$\sum_{N_{ab}} g(N_s, B, N_{ab}) = B!/[N_s!(B - N_s)!].$$

In the following section, we shall use Q_s expressed as in Eq. (9) to derive the thermodynamic functions for the one-dimensional case. Only in the one-dimensional case has an analytic expression been derived for $g(N_s, B, N_{ab})$.

For higher-dimensional cases, the grand canonical partition function is more useful. The basic thermodynamic function is

$$d(\pi B) = S_s \, dT + \pi \, dB + N_s \, d\mu.$$

(10)

The partition function is

$$\Xi(\mu_s, B, T) = \sum_{N_s=0}^{B} \exp(N_s \mu/kT)Q_s(N_s, B, T),$$

(11)

which becomes on substituting from Eq. (9),

$$\Xi(\mu_s, B, T) = \sum_{N_s=0}^{B} \left[\sum_{N_{ab}} g(N_s, B, N_{ab}) \exp(wN_{ab}/2kT) \right]$$
$$\times \exp(N_s \mu/kT)q_s^{N_s} \exp(-cwN_s/2kT).$$

(12)

This expression has been solved in closed form for the two-dimensional case only when

$$\exp(N_s \mu_s/kT)q_s^{N_s} \exp(-cwN_s/2kT) = \exp[(\mu_s + kT \ln q_s - cw/2)N_s/kT] = 1,$$

or

(13)

$$\mu_s + kT \ln q_s - cw/2 = 0.$$

The condition expressed in Eq. (13) is equivalent to restricting surface coverage to $\theta = \frac{1}{2}$. This is shown as follows:

Let

$$z = q_s \exp(\mu_s/kT) \qquad \text{and} \qquad \exp(cw/2kT) = \sigma. \tag{14}$$

Then,

$$\Xi = \sum_{N_s=0}^{B} \left[\sum_{N_{ab}} g(N_s, B, N_{ab}) \exp(wN_{ab}/2kT) \right] \exp[N_s \ln(z/\sigma)]. \tag{15}$$

The quantity z is an activity. In fact, since at equilibrium, $\mu_s = \mu_G = \mu_0(T) + kT \ln p$ for a perfect gas, z is equal to the pressure in this case to within a proportionality constant which is a function only of T. To show that $\theta = \bar{N}_s/B = \frac{1}{2}$ when $\ln z/\sigma = 0$, we differentiate $\ln \Xi$ with respect to μ to obtain

$$\frac{kT}{B} \frac{\partial \ln \Xi}{\partial \mu} = \frac{kT}{B} \frac{\partial \ln \Xi}{\partial z} \frac{\partial z}{\partial \mu} = \theta\left(\ln \frac{z}{\sigma}\right)$$

$$= \frac{\sum_{N_s=0}^{B} \left[\sum_{N_{ab}} g(N_s, B, N_{ab}) \exp\left(\dfrac{wN_{ab}}{2kT}\right) \right] \dfrac{N_s}{B} \exp\left(N_s \ln \dfrac{z}{\sigma}\right)}{\Xi}, \tag{16}$$

which is the average value $\bar{N}_s/B$.

Then

$$\theta\left(-\ln \frac{z}{\sigma}\right)$$

$$= \frac{\sum_{N_s=0}^{B} \left[\sum_{N_{ab}} g(N_s, B, N_{ab}) \exp\left(\dfrac{wN_{ab}}{2kT}\right) \right] \dfrac{N_s}{B} \exp\left(-N_s \ln \dfrac{z}{\sigma}\right)}{\sum_{N_s=0}^{B} \left[\sum_{N_{ab}} g(N_s, B, N_{ab}) \exp\left(\dfrac{wN_{ab}}{2kT}\right) \right] \exp\left(-N_s \ln \dfrac{z}{\sigma}\right)}$$

$$= \frac{\sum_{(B-N_s)=0}^{B} \left[\sum_{N_{ab}} g(N_s, B, N_{ab}) \exp\left(\dfrac{wN_{ab}}{2kt}\right) \right]\left(1 - \dfrac{B-N_s}{B}\right) \exp\left[(B-N_s)\ln\dfrac{z}{\sigma}\right]\exp\left(-B\ln\dfrac{z}{\sigma}\right)}{\sum_{(B-N_s)=0}^{B} \left[\sum_{N_{ab}} g(B-N_s, B, N_{ab}) \exp\left(\dfrac{wN_{ab}}{2kT}\right) \right] \exp\left[(B-N_s)\ln\dfrac{z}{\sigma}\right]\exp\left(-B\ln\dfrac{z}{\sigma}\right)}$$

$$= \frac{\sum_{N_s=0}^{B} \left[\sum_{N_{ab}} g(N_s, B, N_{ab}) \exp\left(\dfrac{wN_{ab}}{2kT}\right) \right]\left(1 - \dfrac{N_s}{B}\right) \exp\left(N_s \ln \dfrac{z}{\sigma}\right)}{\sum_{N_s=0}^{B} \left[\sum_{N_{ab}} g(N_s, B, N_{ab}) \exp\left(\dfrac{wN_{ab}}{2kT}\right) \right] \exp\left(N_s \ln \dfrac{z}{\sigma}\right)}$$

$$= 1 - \theta\left(\ln \frac{z}{\sigma}\right), \tag{17}$$

where we have used the facts that $g(N_s, B, N_{ab}) = g(B - N_s, B, N_{ab})$, and that summing over the variable $(B - N_s)$ from 0 to B is equivalent to summing over N_s from 0 to B. Thus,

$$\theta[\ln(z/\sigma)] + \theta[-\ln(z/\sigma)] = 1, \tag{18}$$

and when $z = \sigma$,

$$2\theta(0) = 1, \qquad \theta(0) = \tfrac{1}{2},$$

which is what we set out to show. A plot of θ versus $\ln(z/\sigma)$ is essentially an adsorption isotherm, since z stands for activity. From Eq. (18), we see that such a plot is symmetrical about $\theta = \tfrac{1}{2}$ for all temperatures. Therefore, if a phase transition exists, it must occur at $\theta = \tfrac{1}{2}$.

In Chapter IV, it will be shown that closed expressions for the square lattice can be derived for Ξ when $\ln(z/\sigma) = 0$. Although this limitation restricts the usefulness of the derivation, information for the coverage $\theta = \tfrac{1}{2}$, the critical density, is nevertheless extremely important.

3.3   One-Dimensional Lattice Gas

A study of the one-dimensional case provides information of value for the higher-dimensional cases. This is the only case for which closed expressions of the thermodynamic functions can be obtained for all values of θ. The simplest approach is to use the canonical partition function, Eq. (9), which becomes for the one-dimensional case $(c = 2)$,

$$Q_s(N_s, B, T) = [q_s \exp(-w/kT)]^{N_s} \sum_{N_{ab}} g(N_s, B, N_{ab}) \exp(wN_{ab}/2kT). \tag{19}$$

The crux of the problem is the derivation of an analytic expression for $g(N_s, B, N_{ab})$. Ultimately, we shall find the maximum term in the summation of Eq. (19), and therefore we regard N_s, B, and N_{ab} as large numbers. To be specific, suppose that N_{ab} is odd and that the site on the left of the linear array of sites is an "a" type (occupied). Then the site on the right of the linear array will be a "b" type (unoccupied). Consider the following example:

$$\text{aaa} \mid \text{bb} \mid \text{a} \mid \text{b} \mid \text{a} \mid \text{bb} \mid \text{aa} \mid \text{b}$$

$$N_s = 7, \qquad N_{ab} = 7, \qquad B = 13, \qquad B - N_s = 6.$$

There are $(N_{ab} + 1)/2$ groups of a's and $(N_{ab} + 1)/2$ groups of b's with the right-hand group a "b" group and the left-hand group an "a" group. These statements follow because an "ab" pair occurs at each boundary between an "a" group and a "b" group. We now want to determine the number of ways of arranging N_s a's in $(N_{ab} + 1)/2$ groups. Each "a" group must have at least one "a" site in it, and, therefore, the desired number of arrangements is the number of ways of assigning the remaining $[N_s - (N_{ab} + 1)/2] (=X)$ a's among the $(N_{ab} + 1)/2 (=Y)$ groups with no restriction on the number of a's per group. This is the well-known statistical problem of determining the number of ways of arranging X indistinguishable objects in Y boxes,

$$\frac{(X + Y - 1)!}{(Y - 1)!X!} = \frac{N_s!}{(N_{ab}/2)![N_s - (N_{ab}/2)]!}, \tag{20}$$

where unity has been dropped in comparison with the large numbers in the second expression. By replacing N_s by $B - N_s$, we obtain the corresponding expression for the b's. Now $g(N_s, B, N_{ab})$ is twice the product of the expression for the a's and that for the b's, because we could have just as well selected the left group as a "b" group. Since we shall use only $\ln g(N_s, B, N_{ab})$ and not $g(N_s, B, N_{ab})$, the factor of two is negligible. Therefore,

$$g(N_s, B, N_{ab}) = \frac{N_s!(B - N_s)!}{[N_s - (N_{ab}/2)]![B - N_s - (N_{ab}/2)]![(N_{ab}/2)!]^2}. \tag{21}$$

When $\theta = \frac{1}{2}$ (see Section 3.2), that is, $N_s = B/2$, we find, using Stirling's approximation for $\ln g$,

$$g = B!/[(B - N_{ab})!N_{ab}!], \tag{22}$$

and

$$Q_s = [q_s \exp(-w/kT)]^{B/2} \sum_{N_{ab}} [B! \exp(wN_{ab}/2kT)/(B - N_{ab})!N_{ab}!]$$

$$= [q_s \exp(-w/kT)]^{B/2}[1 + \exp(w/2kT)]^B$$

$$= q_s^{B/2}[2 \cosh(w/4kT)]^B \exp(-Bw/4kT), \tag{23}$$

where we have used the binomial theorem and

$$\cosh(w/4kT) = [\exp(w/4kT) + \exp(-w/4kT)]/2.$$

The energy E_s at $\theta = \frac{1}{2}$ is given by

$$E_s = kT^2(\partial \ln Q_s/\partial T)_{B, N_s = B/2}$$

$$= (kT^2B/2)(\partial \ln q_s/\partial T)_{B, N_s = B/2} - (Bw/4) \tanh(w/4kT) + (Bw/4)$$

$$= E_s^* + E_{sc}, \tag{24}$$

where E_s^* is the nonconfigurational energy and E_{sc} is the configurational energy. The configurational heat capacity at $\theta = \frac{1}{2}$ is

$$C_{sc} \equiv (\partial E_{sc}/\partial T)_{B, N_s = B/2} = Bk[(w/4kT) \operatorname{sech}(w/4kT)]^2. \tag{25}$$

It was mentioned, in Section 3.1, that if the system with a first-order phase change is restricted to the critical density ($\theta = \frac{1}{2}$), and the temperature is raised through the critical point, then the latent heat will be zero, and there will be no discontinuity in the first derivatives of μ, but there will be a singularity in the heat capacity. In Fig. 3.3, we plot the configurational heat capacity against temperature from Eq. (25), and note that no discontinuity occurs. Thus, for the one-dimensional case, there is no indication of a phase transition.

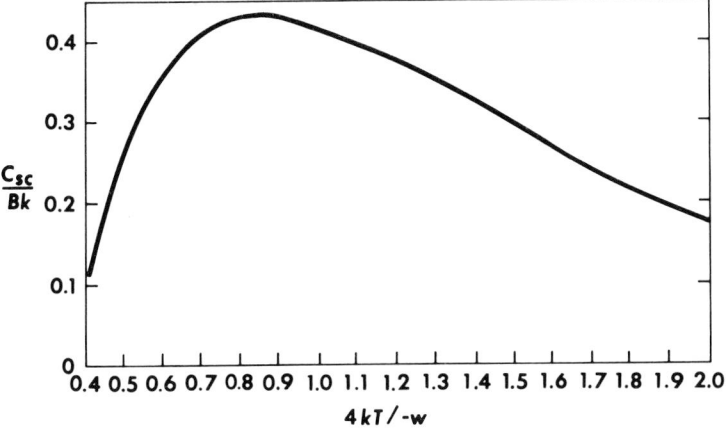

FIG. 3.3. Exact configurational heat capacity for one-dimensional lattice gas. [After Hill, "An Introduction to Statistical Thermodynamics," 1960, Addison-Wesley, Reading, Massachusetts.]

We return now to the general one-dimensional case. Into Eq. (19) for the one-dimensional partition function, we substitute Eq. (21), the expression for $g(N_s, B, N_{ab})$. Next, we determine the maximum term in the summation,

$$\sum_{N_{ab}} g(N_s, B, N_{ab}) \exp(wN_{ab}/2kT).$$

Let $t(N_{ab}, N_s, B, T) = g \exp(wN_{ab}/2kT)$. Then the condition

$$(\partial \ln t/\partial N_{ab}) = (\partial \ln g/\partial N_{ab}) + (w/2kT) = 0, \tag{26}$$

and Eq. (21) determines that the value of N_{ab} in the maximum term is

$$N_{ab}^*/2B = [2\theta(1 - \theta)]/(\beta + 1), \tag{27}$$

where $\beta = \{1 - 4\theta(1 - \theta)[1 - \exp(-w/kT)]\}^{1/2}$ and $\theta = N_s/B$. The partition function becomes

$$\ln Q_s = N_s \ln q_s \exp(-w/kT) + \ln t(N_{ab}^*, N_s, B, T), \tag{28}$$

from which we obtain the chemical potential,

$$-\frac{\mu_s}{kT} = \left(\frac{\partial \ln Q_s}{\partial N_s}\right)_{B,\,T}$$

$$= \ln q_s \exp\frac{-w}{kT} + \left(\frac{\partial \ln t}{\partial N_s}\right)_{N^*_{ab},\,B,\,T} + \left(\frac{\partial \ln t}{\partial N^*_{ab}}\right)_{N_s,\,B,\,T}\left(\frac{\partial N^*_{ab}}{\partial N_s}\right)_{B,\,T}$$

(29)

But the last term drops out because $(\partial \ln t/\partial N^*_{ab}) = 0$ from Eq. (26), and also

$$(\partial \ln t/\partial N_s)_{N^*_{ab},\,B,\,T} = (\partial \ln g/\partial N_s)_{N^*_{ab},\,B,\,T}.$$

From

$$-\mu_s/kT = \ln q_s \exp(-w/kT) + (\partial \ln g/\partial N_s)_{N^*_{ab},\,B,\,T},\qquad(30)$$

and Eq. (21) for g, we get

$$\exp\frac{\mu_s}{kT}\,q_s\exp\frac{-w}{kT} = \frac{1-\theta}{\theta}\left[\frac{\theta-(N^*_{ab}/2B)}{1-\theta-(N^*_{ab}/2B)}\right].\qquad(31)$$

Using Eqs. (14) and (27) to eliminate $N^*_{ab}/2B$ from Eq. (31), we obtain, finally,

$$z/\sigma \equiv \exp(\mu_s/kT)q_s\exp(-w/kT) = (\beta-1+2\theta)/(\beta+1-2\theta).\qquad(32)$$

We showed in Section 3.2 that z stands for activity (pressure for a perfect gas) and that σ is a function only of temperature for given w, so that Eq. (32) is really the adsorption isotherm. It reduces to the Langmuir adsorption isotherm when $w/kT \to 0$.

We now find the equation of state. From Eqs. (4), (28), and $A_s = -kT \ln Q_s$, we could find $(\partial \ln Q_s/\partial B)_{T,\,N_s} = \pi/kT$. However, since μ_s has already been found in Eq. (32), it will be more convenient to start with $F_s = A_s + \pi B = \mu_s N_s$, or

$$\pi/kT = (\mu_s N_s/BkT) - (A_s/BkT) = \mu_s N_s/BkT + (1/B)\ln Q_s.\qquad(33)$$

Using Eqs. (21), (27), (28), and (30) we find,

$$\pi/kT = \theta\ln[(\beta-1+2\theta)/(\beta+1-2\theta)] + (1/B)\ln g(N_s,\,B,\,N^*_{ab})$$

$$+ [2\theta(1-\theta)]/(\beta+1)(w/kT).\qquad(34)$$

Finally, with Eqs. (21) and (27), we eliminate N^*_{ab} from $\ln g$. After considerable algebra, we obtain

$$\pi/kT = \ln[(\beta+1)/(\beta+1-2\theta)].\qquad(35)$$

This reduces to Eq. (22) of Chapter II, the Langmuir equation of state, when $w/kT \to 0$.

A plot of π/kT against $1/\theta$ is the two-dimensional lattice gas analog of a P–V plot of a three-dimensional gas. It is obvious, from inspection of Eq. (35), that it does not contain a discontinuity over the entire range of values of θ. Neither does it contain a van der Waals loop which is sometimes associated with equations of state derived from partition functions (see Fig. 3.4). If it did contain a van der Waals loop, then $\partial\pi/\partial(1/\theta)$ would be positive or $\partial\pi/\partial\theta$ would be negative at the critical density, $\theta = \frac{1}{2}$. Equation (35) shows, however, that

$$\left[\frac{\partial\pi/kT}{\partial\theta}\right]_{\theta=1/2} = 2\exp\frac{w}{2kT}, \tag{36}$$

which can never be negative. Thus, we see once more that a one-dimensional lattice gas does not show a first-order phase transition. If there were a phase transition,

$$\left(\frac{\partial\pi/kT}{\partial\theta}\right)_{\theta=1/2}$$

would be zero at the critical temperature. But $2\exp(w/2kT) \to 0$ as $T \to 0$ for negative w. It might be said then that a "critical point" exists at $\theta_c = \frac{1}{2}$, $T_c = 0$. Two-dimensional lattices, we shall see, show first-order phase transitions at $T_c > 0$, $\theta_c = \frac{1}{2}$.

3.4 Approximate Methods

Before taking up the exact methods for the two-dimensional lattice gas, two approximations, the Bragg–Williams and the quasi-chemical, will be discussed. Approximate methods are important because closed, exact solutions for the two-dimensional case are not available except for $\theta = \frac{1}{2}$. The Bragg–Williams approximation is perhaps the simplest approximation that can be made while still retaining the salient features of a phase transition. In this approximation, we start with the canonical partition function [Eq. (7)]

$$Q_s = q_s^{N_s} \sum_{N_{aa}} g(N_s, B, N_{aa}) \exp(-N_{aa}w/kT)$$

and replace $N_{aa}w$ by the average interaction energy $\bar{N}_{aa}w$

$$Q_s = q_s^{N_s} \exp\frac{-\bar{N}_{aa}w}{kT}\sum g = \frac{B!q_s^{N_s}\exp(-\bar{N}_{aa}w/kT)}{(B-N_s)!N_s!}. \tag{37}$$

This is equivalent to giving all configurations of N_s molecules on B sites the same weight as they would have if $w = 0$. In other words, configurational degeneracy and nearest-neighbor interaction energy are treated as though the molecules were distributed randomly among the sites. An expression for $\bar{N}_{aa}$ is easily calculated. For random distribution, a molecule on a site has $c\theta = cN_s/B$ occupied nearest-neighbor sites, where c is the total number of nearest-neighbors ($c = 4$ for square lattice). Then

$$\bar{N}_{aa} = (cN_s/B)(N_s/2) = (cN_s^2/2B),$$

where the factor two prevents the counting of each aa pair twice. Thus in the Bragg–Williams approximation,

$$Q_s = \frac{B!\, q_s^{N_s} \exp(-cwN_s^2/2kTB)}{(B - N_s)!N_s!}. \tag{38}$$

Thermodynamic quantities are now easily obtained:

$$A_s/kT = -\ln Q_s = N_s \ln N_s + (B - N_s)\ln(B - N_s) - B \ln B$$
$$- N_s \ln q_s + (cwN_s^2/2kTB), \tag{39}$$

$$E_s/kT^2 = (\partial \ln Q_s/\partial T)_{B, N_s} = N_s(\partial \ln q_s/\partial T)_{B, N_s} - (cwN_s^2/2kT^2B), \tag{40}$$

$$\pi/kT = -[(\partial A_s/kT)/\partial B]_{T, N_s} = (\partial \ln Q_s/\partial B)_{T, N_s} = (cw\theta^2/2kT) - \ln(1 - \theta), \tag{41}$$

$$S_s/k = \ln Q_s + T(\partial \ln Q_s/\partial T)_{N_s, B}$$
$$= \ln[B!/N_s!(B - N_s)!] + N_s[\ln q_s + T(d \ln q_s/dT)]. \tag{42}$$

The last equation for the entropy is the same as that for the entropy of an ideal lattice gas [see Eq. (25), Chapter II]. From the chemical potential,

$$\frac{\mu_s}{kT} = -\left(\frac{\partial \ln Q_s}{\partial N_s}\right)_{B, T} = \ln \frac{\theta \exp(wc\theta/kT)}{(1 - \theta)q_s}, \tag{43}$$

we obtain the adsorption isotherm,

$$\frac{z}{\sigma} \equiv \exp\frac{\mu_s}{kT}\, q_s \exp\frac{-cw}{2kT} = \frac{\theta \exp[cw(2\theta - 1)/2kT]}{1 - \theta},$$

or

$$z = \exp\frac{\mu_s}{kT}\, q_s = \frac{\theta \exp(cw\theta/kT)}{(1 - \theta)}, \tag{44}$$

which may be compared with Eq. (32), the exact expression for the one-dimensional lattice gas.

In Fig. 3.4, we plot the equation of state [Eq. (41)] of a Bragg–Williams lattice gas for various values of the parameter cw/kT. The curve for $cw/kT = -5$ is below the critical condition and that for $cw/kT = -3$ is above, while the critical curve is $cw/kT = -4$. The van der Waals loop is evident in the

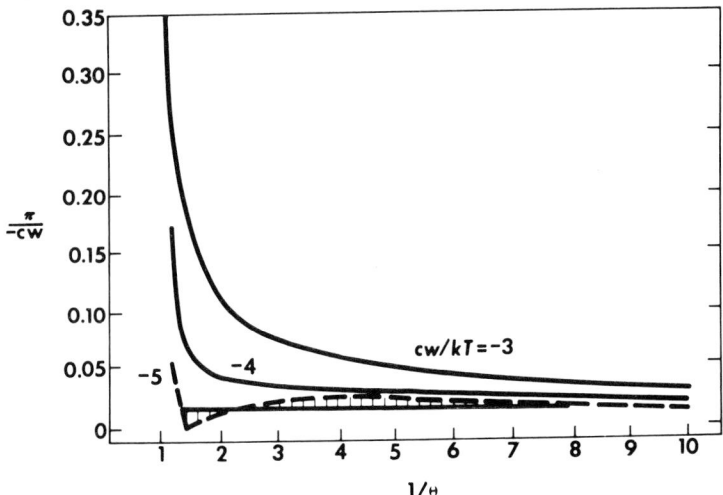

FIG. 3.4. Equation of state for a Bragg–Williams lattice-gas. [After Hill, "An Introduction to Statistical Thermodynamics," 1960, Addison-Wesley, Reading, Massachusetts]

curve which is below the critical condition. The critical curve shows a horizontal inflection at $\theta = \frac{1}{2}$.

Critical conditions for the Bragg–Williams lattice gas can be determined by using the mathematical conditions for a horizontal inflection point on Eq. (41),

$$[(\partial \pi/kT)/\partial \theta] = (cw\theta/kT) + [1/(1 - \theta)] = 0,$$
$$[(\partial^2 \pi/kT)/\partial \theta^2] = (cw/kT) + [1/(1 - \theta)^2] = 0. \tag{45}$$

Solving these equations simultaneously, we find $\theta = \frac{1}{2}$ and $cw/kT_c = -4$ for the critical conditions, which become for a square lattice $\theta_c = \frac{1}{2}$, $w/kT_c = -1$. A phase transition is, therefore, predicted for negative w. The exact values for a square lattice which will be derived in Chapter IV are $\theta_c = \frac{1}{2}$, $w/kT_c = -1.76$. It should be noted that the Bragg–Williams approximation incorrectly predicts a phase transition for the one-dimensional lattice gas ($c = 2$).

In Fig. 3.5, we plot the adsorption isotherm of a Bragg–Williams lattice gas in the form $\ln z = \mu_s/kT + \ln q_s$ versus θ from Eq. (44). Again we observe the characteristic van der Waals loop at $cw/kT = -6$, which is below the critical condition.

It is interesting to consider whether the grand canonical partition function shows the van der Waals loop (1). We start with the partition function in the form

$$\Xi = \sum_{N_s = 0}^{B} Q_s(B, N_s, T) \exp(N_s \mu_s/kT) \tag{46}$$

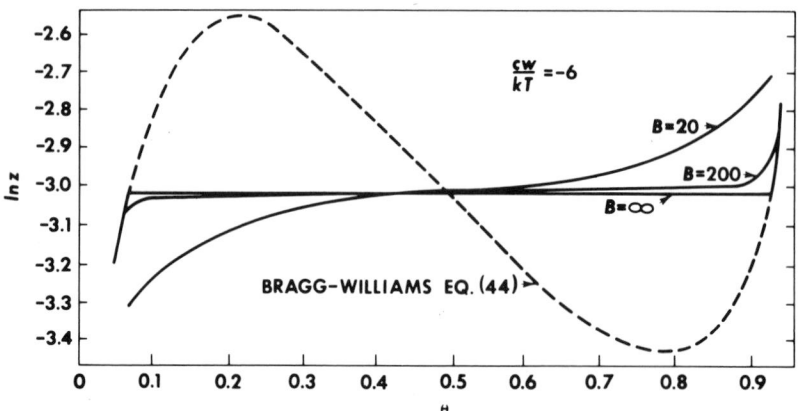

FIG. 3.5. Adsorption isotherms in the form $\ln z$ versus θ (1). [After Hill, *J. Phys. Chem.* **57**, 324, (1953). By permission of The American Chemical Society.]

Taking the derivative of $\ln \Xi$ with respect to μ_s, we find the average value of $N_s (= \overline{N}_s)$,

$$\left(kT \frac{\partial \ln \Xi}{\partial \mu} \right)_{B,T} = \frac{\sum_{N_s=0}^{B} N_s Q_s \exp(N_s \mu_s / kT)}{\sum_{N_s=0}^{B} Q_s \exp(N_s \mu_s / kT)} = \overline{N}_s, \tag{47}$$

or

$$\theta \Xi = (1/B) \sum_{N_s=0}^{B} N_s Q_s \exp(N_s \mu_s / kT).$$

Now if the second expression in Eq. (47) is differentiated again with respect to μ_s, we get

$$\partial \theta / \partial \mu_s = (1/kT) \times \overline{N_s^2} - (\overline{N}_s)^2 = \overline{(N_s - \overline{N}_s)^2}. \tag{48}$$

This equation states that the slope of the curve of θ versus $\mu_s (= \ln z$ to within a constant dependent only on temperature) is always positive. Therefore, a van der Waals loop does not exist. It further states that this is true whether the expression used for Q_s is exact or approximate. In short, it is impossible to get a loop using the grand canonical partition function. This was corroborated by Hill (1) for $B = 20$, 200 and ∞, using the Bragg–Williams Q_s in Eq. (47). The case for $B = 20$ was calculated directly from

$$\theta = \frac{\overline{N}_s}{B} = \frac{(1/B) \sum_{N_s=0}^{B} N_s Q_s \exp(N_s \mu_s / kT)}{\Xi},$$

where

$$\Xi = \sum_{N_s=0}^{B} Q_s \exp(N_s \mu_s / kT)$$

$$= \frac{B! \exp[-(cw/2kT)(1/B)]}{1!(B-1)!} z + \frac{B! \exp[-(cw/2kT)(4/B)]}{2!(B-2)!} z^2 + \cdots,$$

and

$$\sum_{N_s=0}^{B} N_s Q_s \exp(N_s \mu_s/kT)$$

$$= \frac{B! \exp[-(cw/2kT)(1/B)]}{1!(B-1)!} z + \frac{2B! \exp[-(cw/2kT)(4/B)]}{2!(B-2)!} z^2 + \cdots.$$

For $B = 200$, intervals of $\Delta N_s = 5$ and numerical integration were used. The results have been plotted in Fig. 3.5, $\ln z$ versus θ for $cw/kT = -6$, which is below the critical condition $cw/kT = -4$. It is evident that there is no loop, and that the curves become flatter at the phase transition as $B \to \infty$.

We have seen that an approximate Q_s, in particular the Bragg–Williams approximation, can give a loop. The question arises whether an exact Q_s gives a loop or not. Hill (2) has analyzed this problem in considerable detail. Here we shall only summarize his results briefly. He concludes that a loop will always be encountered in a theory which uses the canonical partition function Q_s and introduces implicitly or explicitly the restraint of uniform density over the entire system. Under these restraining conditions, it is not possible for two phases to coexist as they do experimentally and in the use of all grand canonical partition functions, exact or approximate. Thus, theories that force systems using a Q_s to be homogeneous under all conditions show a loop, whereas theories which allow phase inhomogeneities show a three-branched curve with a horizontal branch as in Fig. 3.5. The Bragg–Williams theory explicitly introduces the restraint of homogeneity. A three-branched curve (no loop) will result if Q_s is evaluated exactly[1], since all configurations will be present, homogeneous and heterogeneous. Even if Q_s is evaluated approximately, but with all possible densities allowed in any small region of the system, a three-branched curve will be obtained rather than a loop. In a strictly mathematical sense, sharp corners occur in a three-branched curve only when $B \to \infty$.

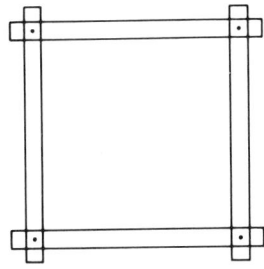

[1] Katsura (3) gives evidence for a loop in a finite system using an exact canonical partition function (Ising model). Hill (2) shows that this loop is not of the usual van der Waals type, but vanishes as $B \to \infty$.

We now turn to the quasi-chemical approximation, which is significantly better than the Bragg–Williams approximation. The important assumption in this method is that pairs of nearest-neighbor sites are treated as if they were independent of each other. This assumption, of course, is not true, because pairs overlap as illustrated on p. 77.

We start with Eq. (9) and derive an expression for $g(N_s, B, N_{ab})$ using the above assumption of independent nearest-neighbor pairs. Each pair of sites may be occupied in four different ways: aa, ab, ba, bb. The total number of pairs is $cB/2$, and the number of each type is (Eqs. 8):

Number of aa pairs $= N_{aa} = (cN_s/2) - (N_{ab}/2)$.
Number of ab pairs $= N_{ab}/2$.
Number of ba pairs $= N_{ab}/2$.
Number of bb pairs $= N_{bb} = [c(B - N_s)/2] - (N_{ab}/2)$.

The number of ways of assigning a total of $cB/2$ independent pairs to the four categories given above, with any number 0 through $cB/2$ per category consistent with the total, is

$$\omega(N_s, B, N_{ab}) = \frac{(cB/2)!}{[(cN_s/2) - (N_{ab}/2)]!\{[c(B - N_s)/2] - (N_{ab}/2)\}![(N_{ab}/2)!]^2}.$$
(49)

This cannot be set equal to $g(N_s, B, N_{ab})$, because treating the pairs as independent gives some impossible situations such as:

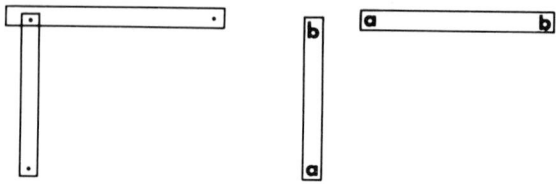

where the upper left-hand site cannot be both "a" and "b." Furthermore, we have seen that

$$\sum_{N_{ab}} g(N_s, B, N_{ab}) = B!/[N_s!(B - N_s)!].$$

But $\sum \omega(N_s, B, N_{ab})$ can be found by differentiation of Eq. (49), setting the result equal to zero and solving for N_{ab}^*, the value of N_{ab} in the maximum term of ω, to be

$$\sum_{N_{ab}} \omega(N_s, B, N_{ab}) = \omega(N_s, B, N_{ab}^*) = [B!/N_s!(B - N_s)!]^c, \qquad (50)$$

which is much larger than $g(N_s, B, N_{ab})$. If we set

$$g(N_s, B, N_{ab}) = C(N_s, B)\omega(N_s, B, N_{ab}), \qquad (51)$$

we find

$$\sum_{N_{ab}} g(N_s, B, N_{ab}) = B!/N_s!(B - N_s)! = C(N_s, B)\omega(N_s, B, N_{ab}^*)$$

$$= C(N_s, B)[B!/N_s!(B - N_s)!]^c, \qquad (52)$$

which gives

$$C(N_s, B) = [B!/N_s!(B - N_s)!]^{(1-c)}. \qquad (53)$$

For the one-dimensional case ($c = 2$), it can be shown that the expression in Eq. (51) is exact. This becomes evident immediately when Eqs. (49) and (53) with $c = 2$ are substituted into Eq. (51) and $g(N_s, B, N_{ab})$ set equal to the one-dimensional expression, Eq. (21). In this case, the overcounting in ω is done to the same factor for each N_{ab}. In multidimensional cases, this is not so, for then the factor C is not independent of N_{ab}. We now substitute

$$C(N_s, B)\omega(N_s, B, N_{ab}) \qquad \text{for} \qquad g(N_s, B, N_{ab})$$

in Eq. (9) to obtain

$$Q_s(N_s, B, T) = \left(q_s \exp\frac{-cw}{2kT}\right)^{N_s} \sum_{N_{ab}} \left(\frac{B!}{N_s!(B - N_s)!}\right)^{1-c}$$

$$\cdot \frac{(cB/2)! \exp(wN_{ab}/2kT)}{[(cN_s/2) - (N_{ab}/2)]!\{[c(B - N_s)/2] - (N_{ab}/2)\}![(N_{ab}/2)!]^2}, \qquad (54)$$

or

$$\ln Q_s = N_s \ln q_s \exp(-cw/2kT) + \ln t(N_{ab}^*, N_s, B, T),$$

where $t(N_{ab}, N_s, B, T) = C\omega \exp(wN_{ab}/2kT)$ and $t(N_{ab}^*, N_s, B, T)$ is the maximum term of the summation in Eq. (54). The procedure is exactly the same as in the one-dimensional case, except that N_{ab}^*/cB is used instead of $N_{ab}^*/2B$, Eqs. (26)–(29). As in Eqs. (30)–(32), we obtain

$$\frac{z}{\sigma} \equiv \exp\frac{\mu_s}{kT} q_s \exp\frac{-cw}{2kT} = \left(\frac{1 - \theta}{\theta}\right)^{c-1} \left(\frac{\theta - (N_{ab}^*/cB)}{1 - \theta - (N_{ab}^*/cB)}\right)^{c/2}, \qquad (55)$$

and

$$\frac{z}{\sigma} = \left[\frac{(\beta - 1 + 2\theta)(1 - \theta)}{(\beta + 1 - 2\theta)\theta}\right]^{c/2} \frac{\theta}{1 - \theta}$$

for the adsorption isotherm. The equation of state is

$$\frac{\pi}{kT} = \ln\left\{\left[\frac{(\beta+1)(1-\theta)}{\beta+1-2\theta}\right]^{c/2}\frac{1}{1-\theta}\right\},\tag{56}$$

which may be compared with Eq. (35). It is easily determined that the critical density is $\theta_c = \frac{1}{2}$, and the critical temperature may be determined as in Eq. (45) from $[(\partial\pi/kT)/\partial\theta]_{\theta_c=1/2} = 0$. In this case, the critical condition turns out to be $cw/kT_c = 2c\ln[(c-2)/c]$, or for a square lattice ($c = 4$), $w/kT_c = -1.38$, which is intermediate between the Bragg–Williams value (-1) and the exact value (-1.76).

Hill (*19*), using the quasi-chemical approximation with a distribution of site energies and sites of different energy scattered randomly over the surface, found that under certain conditions condensation occurs in two steps rather than in the usual one step. These two steps were confirmed by Gordon (*20*) in a computer simulation of adsorption on a nonuniform, hexagonal close-packed planar lattice with nearest-neighbor interactions. He also obtained the unanticipated result that the probability of a site being occupied is not a single-valued function of the adsorption energy, as indicated by Hill (*19*). On the contrary, it depends also on the adsorption energies of the neighboring sites.

3.5   A Combinatorial Solution

We now take up the discussion of exact multidimensional models with particular emphasis on the two-dimensional adsorption model. In this chapter, the discussion will be limited to open or series solutions, and in the following chapter closed or analytic solutions will be treated. In this section, we consider an important combinatorial solution for $\theta = \frac{1}{2}$. It is important not only because it is the basis of some of the series solutions, but also because the same approach leads to closed solutions.

Starting with Eq. (15), we find with $\theta = \frac{1}{2}$ ($z/\sigma = 1$),

$$\Xi \equiv \sum_{N_s=0}^{B}\left[\sum_{N_{ab}}g(N_s, B, N_{ab})\exp(wN_{ab}/2kT)\right]$$

$$= \sum_{N_s=0}^{B}\left[\sum_{i=1}^{\Omega}\exp(wN_{ab}^i/2kT)\right],$$

where we have replaced the sum over all configurations grouped according to the number of ab bonds by its equivalent, the sum over all configurations, $\Omega = B!/N_s!(B-N_s)!$, taken individually. We then multiply by $\exp(cBw/8kT)$, where c, as before, is the total number of nearest neighbors of a site,

$$\Xi = \exp(cBw/8kT) \sum_{N_s=0}^{B} \sum_{i} \exp[(wN_{ab}^i - cBw/4)/2kT]$$

$$= \exp(cBw/8kT) \sum_{N_s=0}^{B} \sum_{i} \exp(-E_i'/kT), \qquad (57)$$

with

$$E_i' = (cBw/8) - (wN_{ab}^i/2) = (w/4)(cB/2 - 2N_{ab}^i)$$

$$= (w/4)\{[(cB/2) - N_{ab}^i] - N_{ab}^i\}.$$

In the last expression for E_i', $cB/2$ is the total number of nearest-neighbor pairs on a lattice of B sites and c nearest-neighbor sites per site. Thus $(cB/2) - N_{ab}^i$ is the sum of aa and bb nearest-neighbor pairs in the ith configuration. A pair of auxiliary configuration variables $s_j s_k$ are now introduced; they describe the situation on a pair of nearest-neighbor sites j and k. The value of the configuration variable is $+1$ if the site is occupied (state a) and -1 if the site is unoccupied (state b). Thus $s_j s_k$ is $+1$ for an aa or bb pair and -1 for an ab pair. We may rewrite E_i',

$$E_i' = (w/4)\left(\sum_{\substack{nn \\ s_j = s_k}} s_j s_k + \sum_{\substack{nn \\ s_j = -s_k}} s_j s_k \right) = (w/4) \sum_{nn} s_j s_k. \qquad (58)$$

In the middle expression, the first summation is over all nearest-neighbor pairs both occupied and both unoccupied (aa and bb), and the second summation is over all nearest-neighbor pairs one occupied, the other unoccupied (ab), to give a total of $cB/2$ nearest-neighbor pairs for the ith configuration. Calling the double sum in Eq. (57) Z, we have

$$Z = \sum_{N_s=0}^{B} \sum_{i} \exp(-E_i'/kT) = \sum_{s_1 \cdots s_B = \pm 1} \exp\left[(-w/4kT) \sum_{nn} s_j s_k \right], \qquad (59)$$

where in the last expression, the double sum over all configurations for each value of N_s from 0 to B is replaced by its equivalent, the summation over the 2^B possible configurations generated by the set $s_1 \cdots s_B$ which also comprises all configurations for each value of N_s from 0 to B. Now

$$\exp[(-w/4kT)s_j s_k] = \cosh(-w/4kT)s_j s_k + \sinh(-w/4kT)s_j s_k$$

$$= \cosh(-w/4kT) + s_j s_k \sinh(-w/4kT)$$

$$= [1 + s_j s_k \tanh(-w/4kT)] \cosh(-w/4kT)$$

$$= [1 + s_j s_k x] \cosh(w/4kT), \qquad x = \tanh(-w/4kT), (60)$$

where we have used the facts that $\cosh(-u) = \cosh u$ and $\sinh(-u) = -\sinh u$. Then,

$$Z = \sum_{s_1 \cdots s_B = \pm 1} \prod_{nn} \exp\left(\frac{-w}{4kT} s_j s_k\right) = \sum_{s_1 \cdots s_B = \pm 1} \prod_{nn} \left[\cosh \frac{w}{4kT}(1 + s_j s_k x)\right]$$

$$= \left(\cosh \frac{w}{4kT}\right)^{cB/2} \sum_{s_1 \cdots s_B = \pm 1} \prod_{nn} (1 + s_j s_k x)$$

$$= \left(\cosh \frac{w}{4kT}\right)^{cB/2} \sum_{s_1 \cdots s_B = \pm 1} [(1 + s_1 s_2 x)(1 + s_2 s_3 x) \cdots]. \tag{61}$$

The last expression may be expanded to give

$$Z = \cosh(w/4kT)^{cB/2} \sum_{s_1 \cdots s_B} \left[1 + x \sum_{nn} s_j s_k + x^2 \sum (s_j s_k)(s_l s_m) + \cdots\right], \tag{62}$$

where each individual term under a summation contains a specific near-neighbor pair only once. Many of the terms under each of the summations vanish. For instance, any near-neighbor pair under the summation $\sum_{nn} s_j s_k$, say $s_1 s_2$, will vanish when summed over all possible values of s_1 and s_2 ($s_1 = \pm 1, s_2 = \pm 1$). In the next sum, $\sum_{nn} (s_j s_k)(s_l s_m)$, terms such as $(s_1 s_2)$ $(s_2 s_3)$ vanish because every term with, for example, $s_1 = 1$, there will be another which is identical except that for $s_1 = -1$. In short, it is clear that those terms will disappear in which any of the configuration variables s_j occur an odd number of times. That is,

$$\sum_{s_i = \pm 1} s_i^n = 0, \qquad n \text{ odd}; \qquad \sum_{s_i = \pm 1} s_i^n = 2, \qquad n \text{ even}. \tag{63}$$

In Fig. 3.6, some possible terms from the various summations in Eq. (62) are graphed. The simplest term that does not vanish when summed over $s_1 \cdots s_B$ is, for a square lattice, $x^4(s_1 s_2)(s_2 s_8)(s_8 s_7)(s_7 s_1)$, which is represented in Fig. 3.6A. A product such as

$$x^6(s_3 s_4)(s_4 s_{10})(s_{10} s_9)(s_9 s_3)(s_{11} s_5)(s_5 s_6)$$

when summed over $s_1 \cdots s_B$ vanishes because s_6 and s_{11} occur only once each (see Fig. 3.6B). A product such as

$$x^8(s_{13} s_{14})(s_{14} s_{20})(s_{20} s_{19})(s_{19} s_{13})(s_{20} s_{21})(s_{21} s_{27})(s_{27} s_{26})(s_{26} s_{20})$$

does not vanish when summed over the configurational variables (see Fig. 3.6C). However, one such as

$$x^7(s_{22} s_{23})(s_{23} s_{29})(s_{29} s_{28})(s_{28} s_{22})(s_{23} s_{24})(s_{24} s_{30})(s_{30} s_{29})$$

vanishes (see Fig. 3.6D).

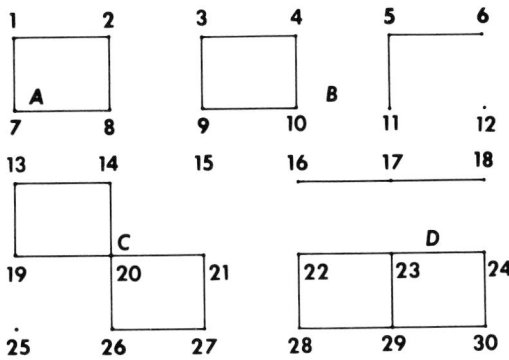

FIG. 3.6. Graphs representing products of configurational variables.

For a square lattice, the maximum number of bonds through any lattice point is four as in s_{20}, Fig. 3.6C. Only those graphs in which each point is connected by an even number of bonds contribute to the summation, which is just another way of saying that every configurational variable in a term must occur an even number of times in order that the term not vanish. Such graphs as in Fig. 3.6B where s_6 and s_{11} are each connected by only one bond, and as in Fig. 3.6D where points s_{23} and s_{29} are each connected by three bonds, contribute nothing.

Those graphs all of whose points are connected by an even number of bonds form closed polygons, whereas those which have at least one point connected by an odd number of bonds are "open." In fact, there is a correspondence between closed polygons and products of configurational variables that do not vanish when summed over $s_1 \cdots s_B$. However, the correspondence is not one-to-one. For example, the term (Fig. 3.6A)

$$x^4(s_1 s_2)(s_2 s_8)(s_8 s_7)(s_7 s_1)$$

forms a closed polygon (a square) and its value is always unity for values of all $s_j = \pm 1$. When it is summed over $s_1 \cdots s_B$, we get

$$\sum_{s_1 \cdots s_B = \pm 1} (s_1 s_2)(s_2 s_8)(s_8 s_7)(s_7 s_1) = 2^B,$$

a factor of 2 for each $s_j = \pm 1$. Similarly,

$$\sum_{s_1 \cdots s_B = \pm 1} 1 = 2^B,$$

and

$$\sum (s_{13} s_{14})(s_{14} s_{20})(s_{20} s_{19})(s_{19} s_{13})(s_{20} s_{21})(s_{21} s_{27})(s_{27} s_{26})(s_{26} s_{20}) = 2^B,$$

(Fig. 3.6C). Thus, we conclude that for each closed graph, there is a factor 2^B in the product of configurational variables when summed over $s_1 \cdots s_B = \pm 1$. If we let $\mathcal{G}(l, B)$ be the number of different closed graphs with l bonds (pairs of nearest-neighbor configurational variables) that can be drawn on B lattice sites, we can rewrite Eq. (62) as

$$Z = 2^B[\cosh(w/4kT)]^{cB/2} \sum_{l=0}^{cB/2} \mathscr{G}(l, B)x^l,$$

or from Eq. (57)   (64)

$$\Xi = \exp(cBw/8kT)Z.$$

The problem of finding the partition function has been reduced to finding the number of closed polygons which can be drawn on the lattice.

The result may be generalized somewhat by assuming different energies for vertical and horizontal bonds to give

$$Z = 2^B[\cosh(w/4kT) \cos(w'/4kT)]^{cB/4} \sum_l \sum_m \mathscr{G}(l, m, B)x^l y^m, \quad (65)$$

where $y = \tanh(-w'/4kT)$. Equations (64) and (65) will be used for the square lattice in developing series solutions later in this chapter and a closed solution in Chapter IV.

3.6   The Duality Theorem and Critical Temperature

Kramers and Wannier (4) were the first to find the critical temperature for a two-dimensional square lattice. They did not, however, obtain an analytic expression for the partition function by their method of locating the critical condition. Later, Onsager (5) developed an extremely sophisticated matrix method which yielded a closed expression for the partition function at $\theta = \frac{1}{2}$, and checked the critical condition of Kramers and Wannier as well as generalizing it for other two-dimensional lattices. Kaufman (6) then simplified Onsager's matrix method using spinor analysis, and Kac and Ward (7) starting with the combinatorial Eq. (64) showed that the same results could be obtained by reasonable topological arguments. An elegant algebraic approach also starting with Eq. (64) and limited to $\theta = \frac{1}{2}$ was carried through by Green and Hurst (8) and will be discussed in detail in the next chapter. We restrict our discussion here to the two-dimensional square lattice following Kramers and Wannier (4). Newell and Montroll (9), in an excellent survey article, consider other two-dimensional lattices as well. Runnells and Combs (21) extended the matrix method to lattice gases of "hard molecules." Specifically, they calculated the cases for square and triangular lattice gases consisting of molecules large enough to prevent simultaneous occupancy of adjacent, or nearest-neighbor sites. Interactions between molecules were restricted to infinite repulsion between overlapping molecules, the simplest model of a dense fluid retaining the excluded-volume effect. Computer calculations were made on lattices of infinite length and width up to 24 sites. Both the square and triangular lattice systems exhibit phase transitions, most likely second order, at a density of 74.2% for the square lattice and 83.7% for the triangular. Orban and Bellemans (22) carried out similar calculations corresponding to exclusion ranges up to second-, third-, and fourth-neighboring sites. In all three cases, they found good evidence for first order phase changes in contrast to the second-order change for the case of excluding first-neighboring sites only.

We use Eq. (64) with $c = 4$,

$$Z = 2^B[\cosh(w/4kT)]^{2B} \sum_{l=0}^{2B} \mathscr{G}(l, B) \tanh^l(-w/4kT), \qquad (66)$$

and Eq. (57) modified slightly as follows:

$$Z = \exp(-Bw/2kT)\Xi = \exp(-Bw/2kT) \sum_{N_s=0}^{B}$$

$$\left[\sum_{N_{ab}} g(N_s, B, N_{ab}) \exp(wN_{ab}/2kT) \right]$$

$$= \exp(-Bw/2kT) \sum_{N_{ab}}$$

$$\left(\sum_{N_s=0}^{B} g(N_s, B, N_{ab}) \right) \exp(wN_{ab}/2kT)$$

$$= \exp(-Bw/2kT) \sum_{N_{ab}} G(N_{ab}, B)$$

$$\times \exp(wN_{ab}/2kT), \qquad (67)$$

where $G(N_{ab}, B) = \sum_{N_s=0}^{B} g(N_s, B, N_{ab})$ is the number of ways of arranging N_{ab} "ab" bonds on B lattice sites using all values of N_s from 0 to B.

We now want to show that $G(N_{ab}, B) = \mathscr{G}(l, B)$ for a square lattice, and to do this construct a dual lattice. The original lattice as shown in Fig. 3.7 de-

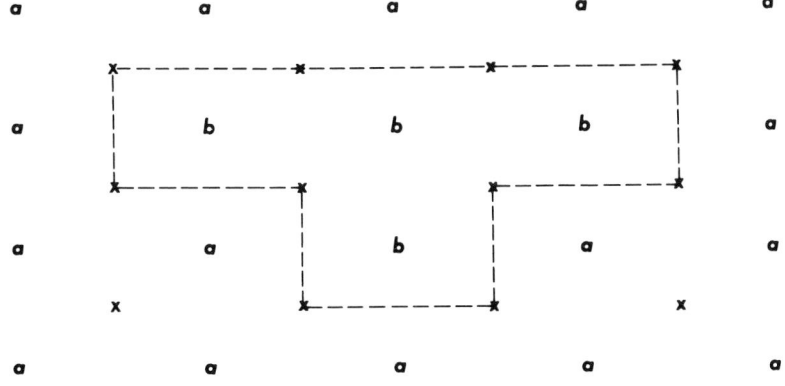

FIG. 3.7. Closed polygon on the dual lattice of the original "ab" lattice.

picts a configuration of occupied (a) and unoccupied (b) sites; the dual lattice is formed by bisecting all nearest-neighbor bonds of the original lattice and its lattice points are represented by X. Both lattices are square. If each ab bond of the original lattice is bisected by a dotted line (Fig. 3.7), there is obtained a closed polygon on the dual lattice. We observe that for each bond of the closed polygon on the dual lattice there is an ab nearest-neighbor pair on the original lattice. Also, there is a one-to-one correspondence between closed polygons on the dual lattice and possible configuration of ab pairs on the original lattice. In other words, $\mathcal{G}(l, B)$, the number of ways of drawing closed polygons with l bonds is identical for a square lattice with $G(N_{ab}, B)$, the number of configurations of N_{ab} ab pairs on B sites. This relationship does not hold, in general, for other lattices.

Since in the square lattice $l = N_{ab}$, we may write [Eqs. (66), (67)],

$$Z(K) = 2^B(\cosh K)^{2B} \sum_{l=0}^{2B} \mathcal{G}(l, B) \tanh^l(K) \tag{68}$$

$$= \exp 2KB \sum_{l=0}^{2B} \mathcal{G}(l, B) \exp(-2Kl), \tag{69}$$

where we have set $K = -w/4kT$.

We now define

$$\tanh K^* = e^{-2K}, \tag{70}$$

so that Eq. (69) becomes

$$Z(K)e^{-2KB} = \sum_{l=0}^{2B} \mathcal{G}(l, B) \tanh^l K^* = Z(K^*)/[2^B(\cosh K^*)^{2B}]. \tag{71}$$

Equation (71) may be interpreted to relate Z at any temperature $T_1 = -w/4kK$ to Z at another temperature $T_2 = -w/4kK^*$, where $K^*(K)$ is given by Eq. (70). If there is a singularity at T_1, there must also be a singularity at T_2. If, as intuitively expected, only one singularity exists, then $T_1 = T_2$ or $K = K^*$, and

$$\tanh K = e^{-2K}. \tag{72}$$

Equation (72) determines the critical temperature (if one exists),

$$\sinh 2K_c = 1$$

or

$$K_c = \tfrac{1}{2} \sinh^{-1} 1 = -w/4kT_c = 0.4407,$$
$$w/kT_c = -1.76,$$

which may be compared to the figure given for the Bragg–Williams approximation in Section 3.4, $w/kT_c = -1$ and for the quasi-chemical approximation, $w/kT_c = -1.38$ on the square lattice.

3.7 Expansion of Pressure (π)

Expansions of this type (10) were first introduced by Mayer (11) for the imperfect three-dimensional gas and later applied to the lattice problem by Fuchs (12). In order to obtain an expansion of the pressure (π), we start with Eq. (15) for the grand canonical partition function,

$$\Xi = \sum_{N_s=0}^{B} \left(\sum_{N_{ab}} g v^{N_{ab}} \right) u^{N_s} = \exp(\pi B/kT), \tag{73}$$

$$v = \exp(w/2kT), \qquad u = z/\sigma, \qquad \sigma = \exp(cw/2kT).$$

When $N_s = 0$, we have $N_{ab} = 0$ and $g(N_s, N_{ab}, B) = 1$. When $N_s = 1$, the only possible value of N_{ab} is c (which equals 4 for a square lattice), and $g(1, c, B) = B$. When $N_s = 2$, $N_{ab} = 2c$ (two molecules not on adjacent sites) or $N_{ab} = 2c - 2$ (two molecules on adjacent sites). For $N_{ab} = 2c$, $g(2, 2c, B) = B(B - c - 1)/2$ because the first molecule can be on any one of the B sites and the second can be on any sites except the $c + 1$ associated with the first molecule and its c nearest neighbors. For $N_{ab} = 2c - 2$, $g(2, 2c - 2, B) = Bc/2$ because the second molecule must be on a site adjacent to the first. Corresponding to $N_s = 0, 1, 2, \ldots$, we have $N_s = B, B - 1, B - 2, \ldots$ associated, respectively, with identical values of N_{ab} and g. For example, $g(B - 2, 2c, B) = g(2, 2c, B) = B(B - c - 1)/2$.

We may now write for the square lattice, $c = 4$,

$$\Xi = (1 + u^B) + Bv^4(u + u^{B-1}) + \{[B(B-5)/2] \cdot v^8 + 2Bv^6\}(u^2 + u^{B-2}) + \cdots. \tag{74}$$

For $z < \sigma$, u^B, u^{B-1}, u^{B-2}, etc. are negligible and

$$\Xi = 1 + Bv^4 u + \{[B(B - 5)/2]v^8 + 2Bv^6\}u^2 + \cdots. \tag{75}$$

Taking the logarithm of both sides of Eq. (75), expanding and dividing by B, we obtain

$$\pi/kT = v^4 u + (2v^6 - \tfrac{5}{2}v^8)u^2 + \cdots$$
$$= v^4 u + (2v^{-2} - \tfrac{5}{2})(uv^4)^2 + \cdots,$$

or since $u = z/\sigma = z/v^4$,

$$\pi/kT = z + (2v^{-2} + \tfrac{5}{2})z^2 + [6v^{-4} - 16v^{-2} + (31/3)]z^3$$
$$+ [v^{-8} + 18v^{-6} - 85v^{-4} + 118v^{-2} - (209/4)]z^4 + \cdots, \qquad z < \sigma, \tag{76}$$

where the additional terms have been given by Lee and Yang (13).

At $z = \sigma$ ($\theta = \tfrac{1}{2}$), a first-order phase transition occurs.

When $z > \sigma$, we may neglect u, u^2, u^3, etc. in Eq. (74),

$$\Xi = u^B + Bv^4u^{B-1} + \{[B(B-5)/2]v^8 + 2Bv^6\}u^{B-2} + \cdots$$
$$= u^B\{1 + Bv^4u^{-1} + \{[B(B-5)/2]v^8 + 2Bv^6\}u^{-2} + \cdots\}. \qquad (77)$$

As before, taking the logarithm of both sides, expanding in powers of u^{-1} and dividing by B, we obtain

$$\pi/kT = \ln u + v^4u^{-1} + [2v^6 - \tfrac{5}{2}v^8]u^{-2} + \cdots$$
$$= \ln u + v^8(v^4u)^{-1} + (2v^{14} - \tfrac{5}{2}v^{16})(v^4u)^{-2} + \cdots$$
$$= \ln(z/\sigma) + v^8z^{-1} + (2v^{14} - \tfrac{5}{2}v^{16})z^{-2} + [6v^{20} - 16v^{22} + (31/3)v^{24}]z^{-3}$$
$$+ [v^{24} + 18v^{26} - 85v^{28} + 118v^{30} - (209/4)v^{32}]z^{-4} + \cdots, \qquad z > \sigma.$$
$$(78)$$

Equation (78) can be used in the liquid region. Both Eqs. (76) and (78) converge if $v < 1$ (attractive forces between molecules adsorbed on adjacent sites).

With the statistical mechanical relationship $\theta = N_s/B = z[(\partial\pi/kT)/\partial z]$, it is possible to obtain an expansion of θ in terms of z, which may be inverted and substituted into Eq. (76) to obtain an expansion of π/kT in terms of θ for the gas region ($z < \sigma$). The latter expression is known as the virial expansion and is an equation of state. Virial expansions will be discussed in more detail in Chapter V in connection with mobile films.

Special low and high temperature series have been developed from the grand canonical partition function (*14–16*) in terms of v and $K = -w/4kT$, respectively for attractive and repulsive forces between adsorbed molecules. A method developed by Lee and Yang (*13*) involves a study of the zeros of the polynomial representing the grand canonical partition function. In general, the use of series expansions is of more interest in the study of three-dimensional cases (*17, 18*), since an exact analytic expression is available for two-dimensional cases as we shall see in the next chapter.

REFERENCES

1. Hill, T. L., *J. Phys. Chem.* **57**, 324 (1953).
2. Hill, T. L., *J. Chem. Phys.* **23**, 812 (1955).
3. Katsura, S., *J. Chem. Phys.* **22**, 1277 (1954).
4. Kramers, H. A., and Wannier, G. H., *Phys. Rev.* **60**, 252 (1941).
5. Onsager, L., *Phys. Rev.* **65**, 117 (1944).
6. Kaufman, B., *Phys. Rev.* **76**, 1232 (1949).
7. Kac, M., and Ward, J. C., *Phys. Rev.* **88**, 1332 (1952).
8. Hurst, C. A., and Green, H. S., *J. Chem. Phys.* **33**, 1059 (1960).

9. Newell, G. F., and Montroll, E. W., *Rev. Mod. Phys.* **25**, 353 (1953).
10. Hill, T. L., "Statistical Mechanics," Chapter VII. McGraw-Hill, New York, 1956.
11. Mayer, J. E., and Mayer, M. G., "Statistical Mechanics." Wiley, New York, 1940.
12. Fuchs, K., *Proc. Roy. Soc. (London), Ser. A* **179**, 340 (1942); **181**, 411 (1943).
13. Lee, T. D., and Yang, C. N., *Phys. Rev.* **87**, 410 (1952).
14. Domb, C., *Proc. Roy. Soc. (London), Ser. A* **199**, 199 (1949).
15. Kirkwood J. G., *J. Chem. Phys.* **6**, 70 (1938).
16. Opechowski, W., *Physica* **4**, 181 (1937).
17. Domb, C., and Sykes, M. F., *J. Math. Phys.* **2**, 63 (1961).
18. Baker, G. A., Jr., *Phys. Rev.* **124**, 768 (1961).
19. Hill, T. L., *J. Chem. Phys.* **17**, 762 (1949).
20. Gordon, R., *J. Chem. Phys.* **48**, 1408 (1968).
21. Runnells, L. K., and Combs, L. L., *J. Chem. Phys.* **45**, 2482 (1966).
22. Orban, J., and Bellemans, A., *J. Chem. Phys.* **49**, 363 (1968).

IV

LOCALIZED ADSORPTION—DEPENDENT SYSTEMS
(Continued)

4.1 Introduction

In this chapter, an analytic solution for a two-dimensional lattice gas at $\theta = \frac{1}{2}$ will be considered. The first such solution was obtained by Onsager in 1944 (1), who introduced an extremely complex matrix method. A new method was introduced in 1961 and developed subsequently (2–7). It was first applied to the dimer problem which is the simplest representation of diatomic molecules adsorbed on a lattice and consists of rigid dimers each of which fills two vertical or horizontal nearest-neighbor sites of a lattice of B sites with no overlapping. The solution involves the use of Pfaffians which are triangular arrays with rules for numerical evaluation similar to those for determinants. The useful feature of the Pfaffian in connection with the dimer problem is that its expansion consists of a series of terms each of which is in one-to-one correspondence to a configuration of a lattice of rigid dimers. The final numerical evaluation of the Pfaffian is made by making use of a theorem which states that the square of the Pfaffian is equal to the determinant of the skew-symmetric matrix to which the given triangular array of coefficients of the Pfaffian extends. The crux of the problem of determining the partition function for the dimer problem is the finding of the total number of ways of distributing horizontal and vertical dimers on the lattice. Evaluation of the Pfaffian leads to this solution. Application of the method to the Ising problem requires some modifications. This is the same Ising problem with which we dealt in Section 3.5. The object is the same: to count the different ways of constructing closed polygons on a lattice. The use of Pfaffians allows an analytic solution to be obtained in place of the series solution. In order to use Pfaffians in a meaningful way, the lattice terminals have to be modified. In Section 3.5, we showed that in the construction of closed polygons, each lattice terminal is connected by zero, two, or four bonds. Thus, contrary to the dimer problem, overlapping of bonds occurs and in each admissible configuration a lattice terminal may appear more than once. To use Pfaffians, the terminals of the lattice are modified to eliminate overlapping. This is done by the simple expedient of introducing a group of four terminals at each lattice point and establishing appropriate rules for their coupling. The solution for the partition function in closed form then follows along the same general principles employed in the dimer problem. Although the solution of the problem at $\theta = \frac{1}{2}$ is important, there still remains the problem of obtaining solutions at more and less than half-coverage.

4.2 The Dimer Problem

The simplest model of a system containing diatomic molecules is a lattice gas of rigid dimers, each of which fills two vertical or horizontal nearest-neighbor sites of a lattice of B sites with no overlapping. The closed solution to this problem involves the same mathematical techniques as that of the Ising model, but is somewhat simpler. We shall use the dimer problem to introduce the solution of the rectangular Ising problem. The first announcement of the solution of the dimer problem was made by Temperley and Fisher (*2*). Kasteleyn's independent study (*3*) followed quickly. Additional studies have been made by both Fisher (*4, 14*) and Kasteleyn (*5*).

Referring to Fig. 4.1, we picture a configuration on a plane quadratic lattice containing l horizontal and m vertical rigid dimers with the bonding energy of each dimer equal to w or w', respectively. The problem in which the lattice contains isolated lattice points or "monomers" as well as dimers has not been solved in closed form. Therefore, we restrict our attention to a lattice which is completely occupied by dimers. The total number of dimers is $l + m = B/2$.

The total energy of a configuration is $lw + mw'$. Let the number of configurations corresponding to l horizontal and m vertical dimers be $g(l, m, B)$. The configurational partition function is

$$Z = \sum_l \sum_m g(l, m, B) x^l y^m, \tag{1}$$

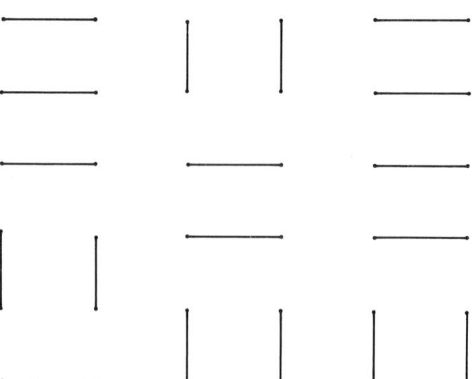

FIG. 4.1. A rectangular configuration of dimers.

where $x = \exp(-w/kT)$ and $y = \exp(-w'/kT)$, and the summations extend over all values of l and m consistent with $l + m = B/2$. The objective of the problem is to calculate Z in closed form, which is tantamount to calculating g. We consider an $r \times s$ plane quadratic lattice, each row containing r lattice terminals and each column s lattice terminals. A configuration of dimers will be referred to as

$$(p_1, p_2), (p_3, p_4), (p_5, p_6), \ldots, (p_{rs-1}, p_{rs}), \tag{2}$$

where each pair of p is a pair of numbers which denotes the lattice terminals occupied by a dimer. The total number of lattice terminals is $rs = B$. At least one of the two numbers r and s must be even, since B must be even for all lattice terminals to be covered by dimers. We adopt the convention of ordering the designation of a configuration given by Kasteleyn (3) as follows:

$$p_1 < p_2; \, p_3 < p_4; \, \ldots; \, p_{rs-1} < p_{rs}, \qquad p_1 < p_3 < p_5 < p_{rs-1}. \tag{3}$$

Also, we order the lattice terminals row after row by selecting the p-numbering system as follows:

$$(a, b) \leftrightarrow p = (b - 1)r + a, \tag{4}$$

where (a, b) is a pair of Cartesian coordinates. For a 3×2 plane quadratic lattice, we have the following numbering and dimer configurations:

$$
\begin{array}{ll}
\overset{\cdot}{4}\ \overset{\cdot}{5}\ \overset{\cdot}{6} & (1, 4), (2, 5), (3, 6) \\
 & (1, 2), (3, 6), (4, 5) \\
\overset{\cdot}{1}\ \overset{\cdot}{2}\ \overset{\cdot}{3} & (1, 4), (2, 3), (5, 6).
\end{array}
\tag{5}
$$

The representation of configurations is now unique.

It is desired to construct a mathematical form consisting of a series of terms each of which is in one-to-one correspondence to a configuration and possesses the "weight" $x^l y^m$ of this configuration. Conditions (4) strongly suggest that this form should be a Pfaffian. A Pfaffian of order M is a triangular array of elements (p, q) such that $0 < p < q \leq M$, and M is even:

$$Pf\ D = \lvert D \rvert = \lvert p, q \rvert = \begin{vmatrix} (1, 2) & (1, 3) & (1, 4) & \cdots & (1, M) \\ & (2, 3) & (2, 4) & \cdots & (2, M) \\ & & (3, 4) & \cdots & (3, M) \\ & & & \vdots & \\ & & & & (M - 1, M) \end{vmatrix} . \quad (6)$$

Each element (p, q) is a number and the Pfaffian may be expanded by its first row in the same way as a determinant, except that the minor of an element (p, q) is obtained from $\lvert D \rvert$ by striking out *both* the pth row and column and the qth row and column. The complete expansion found by iterating this relation is

$$\lvert D \rvert = \sum_j (-1)^j (k_1, k_2)(k_3, k_4) \cdots (k_{M-1}, k_M), \quad (7)$$

where the sum is over the $(M - 1)\,(M - 3) \times \cdots \times 5 \times 3 \times 1$ permutations $(k_1, k_2, \ldots, k_M)$ which satisfy

$$k_1 < k_2,\ k_3 < k_4, \ldots, k_{M-1} < k_M, \\ k_1 < k_3 < k_5 < \cdots < k_{M-1}, \quad (8)$$

and j is the parity of the permutation just as it is in a determinant. For example, when $M = 4$ we have

$$\lvert D \rvert = (1, 2)\,(3, 4) - (1, 3)\,(2, 4) + (1, 4)\,(2, 3). \quad (9)$$

Even permutations of the first term are taken as positive and odd permutations as negative.

In applying the Pfaffian to the dimer problem, we let $M = B$, an even number. The members p and q of any element (p, q) of the Pfaffian are then associated with the pth and qth terminals of the lattice. Further, any term of the Pfaffian as in Eq. (9) corresponds to a particular way of grouping the terminals in pairs. There is thus a one-to-one correspondence between the terms of the Pfaffian and the different ways of coupling the terminals of the lattice. But not all of these ways will correspond to the rules for placing horizontal and vertical dimers on the lattice. In a plane quadratic lattice, for example, such a term as $(1, 4)\,(2, 3)$ in Eq. (9), which involves diagonal couplings, will not correspond. However, all terms which do not correspond to admissible configurations can be made to vanish by setting $(p, q) = 0$, when coupling between p and q is not allowed. When this is done, there is, in view of the conditions in Eqs. (3) and (8), a one-to-one correspondence between the nonvanishing terms of the Pfaffian and admissible configurations of dimers on the lattice. If we can give to each nonvanishing term of the Pfaffian a value equal to the weight of the corresponding configuration as expressed in Eq. (1), then

$$Z = \backslash D| \qquad (10)$$

In an effort to accomplish this, we set each element of the Pfaffian corresponding to a horizontal dimer equal to x and each element corresponding to a vertical dimer equal to y. But the signs of the terms in the Pfaffian expansion are not always positive as they should be to correspond to Eq. (1). For example, see the middle term in Eq. (9). Green and Hurst (6) have shown that if iy, instead of y, is used in the Pfaffian as the weight of the vertical dimers, the difficulty is surmounted. This is easily checked in cases where B is small, but the general proof requires detailed argument for which the reader is referred to Green and Hurst. Temperley and Fisher (2) and Kasteleyn (3) give other ways of ensuring correct signs.

To evaluate $\backslash D|$, we make use of an important theorem which states that the square of a Pfaffian is equal to the determinant of the skew-symmetric matrix to which the given triangular array of coefficients of the Pfaffian extends.

$$Z^2 = \backslash D|^2 = |D| \qquad (11)$$

The skew-symmetric matrix D for a large, plane quadratic lattice is written

$$
D = \begin{bmatrix}
0 & x & 0 & 0 & \cdots & & iy & 0 & \cdots & & 0 \\
-x & 0 & x & 0 & 0 & \cdots & & 0 & iy & 0 & \cdots & 0 \\
0 & -x & 0 & x & 0 & 0 & \cdots & & 0 & iy & \cdots & 0 \\
\vdots & \vdots & \vdots & \vdots & \vdots & \vdots & & & \vdots & \vdots & \vdots & \\
-iy & 0 & \cdots & & & -x & 0 & x & 0 & \cdots & & 0 \\
0 & -iy & 0 & \cdots & & 0 & -x & 0 & x & 0 & \cdots & 0 \\
0 & 0 & -iy & \cdots & & 0 & 0 & -x & 0 & x & \cdots & 0 \\
\vdots & \vdots & \vdots & & & \vdots & \vdots & \vdots & \vdots & \vdots & & \vdots
\end{bmatrix} \qquad (12)
$$

and the Pfaffian from which it is produced consists of the elements lying above the principal diagonal.

Before proceeding with the evaluation of $|D|$, we introduce a modification which simplifies the procedure. Edge effects are troublesome for finite lattices. One way of circumventing the difficulty is to wind the lattice on a torus. In Fig. 4.2, a lattice with B terminals is wound helically on a torus. Note that additional nearest-neighbor pairs of lattice terminals become available. For example, lattice terminals B and 1 are now horizontal nearest-neighbors and lattice terminals $B - (r - 1)$ and 1 are vertical nearest-neighbors.

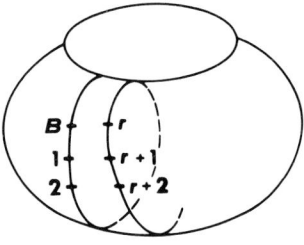

FIG. 4.2. Lattice helically wound on a torus. [After Green and Hurst, "Order–Disorder Phenomena," Wiley (Interscience), New York, 1964.]

Temperley and Fisher (2) and Kasteleyn (3) have evaluated a finite $|D|$ for the plane quadratic lattice, but $|D|$ is simpler to evaluate with the irregularities of edge effects removed as shown by Green and Hurst (6). As $B \to \infty$, the asymptotic values of $|D|$ determined for the plane quadratic lattice and from the lattice wound on a torus become indistinguishable (6). Therefore, edge effects do not influence the thermodynamic properties in a large lattice.

We illustrate the skew-symmetric matrix for a 4×4 lattice wound on a torus in Eq. (13) (see p. 96).

We note that D is not only a skew-symmetric matrix but also has the form

$$\begin{bmatrix} a & b & c & \cdots & & g & h \\ -h & a & b & c & \cdots & \vdots & g \\ -g & -h & a & b & c & \cdots & \vdots \\ \vdots & \vdots & \vdots & \vdots & \vdots & & \\ -b & -c & -d & \cdots & & & a \end{bmatrix}, \tag{14}$$

which is related to the cyclic matrix. A B-dimensional matrix of this form has the eigenvectors

$$\Psi_n = \begin{bmatrix} \exp[(2n-1)\pi i/B] \\ \exp[2(2n-1)\pi i/B] \\ \vdots \\ \exp[B(2n-1)\pi i/B] \end{bmatrix}, \tag{15}$$

There are B eigenvectors, one for each of the values $n = 1, 2, \ldots, B$. From the matrix equation

$$D\Psi_n = \lambda_n \Psi_n, \tag{16}$$

where λ_n is an eigenvalue we find

$$x\omega_n^2 + iy\omega_n^{r+1} + iy\omega_n^{(B-r+1)} + x\omega_n^B = \lambda_n \omega_n \tag{17}$$

or

$$x\omega_n + x\omega_n^{B-1} + iy\omega_n^r + iy\omega_n^{B-r} = \lambda_n$$

$$\omega_n = \exp[(2n-1)\pi i/B].$$

From the fact that

$$\omega_n^j = -\omega_n^{B+j}, \tag{18}$$

$$
D =
\begin{bmatrix}
x & 0 & 0 & iy & 0 & 0 & 0 & 0 & 0 & 0 & 0 & iy & 0 & 0 & x & 0 \\
0 & 0 & iy & 0 & 0 & 0 & 0 & 0 & 0 & 0 & iy & 0 & 0 & x & 0 & -x \\
0 & iy & 0 & 0 & 0 & 0 & 0 & 0 & iy & 0 & 0 & x & 0 & -x & 0 & 0 \\
iy & 0 & 0 & 0 & 0 & 0 & 0 & 0 & iy & 0 & 0 & x & 0 & -x & 0 & 0 \\
0 & 0 & 0 & 0 & 0 & 0 & 0 & iy & 0 & 0 & x & 0 & -x & 0 & 0 & -iy \\
0 & 0 & 0 & 0 & 0 & 0 & iy & 0 & 0 & x & 0 & -x & 0 & 0 & -iy & 0 \\
0 & 0 & 0 & 0 & 0 & iy & 0 & 0 & x & 0 & -x & 0 & 0 & -iy & 0 & 0 \\
0 & 0 & 0 & 0 & iy & 0 & 0 & x & 0 & -x & 0 & 0 & -iy & 0 & 0 & 0 \\
0 & 0 & 0 & iy & 0 & 0 & x & 0 & -x & 0 & 0 & -iy & 0 & 0 & 0 & 0 \\
0 & 0 & iy & 0 & 0 & x & 0 & -x & 0 & 0 & -iy & 0 & 0 & 0 & 0 & 0 \\
0 & iy & 0 & 0 & x & 0 & -x & 0 & 0 & -iy & 0 & 0 & 0 & 0 & 0 & 0 \\
iy & 0 & 0 & x & 0 & -x & 0 & 0 & -iy & 0 & 0 & 0 & 0 & 0 & 0 & 0 \\
0 & 0 & x & 0 & -x & 0 & 0 & -iy & 0 & 0 & 0 & 0 & 0 & 0 & 0 & -iy \\
0 & x & 0 & -x & 0 & 0 & -iy & 0 & 0 & 0 & 0 & 0 & 0 & 0 & -iy & 0 \\
x & 0 & -x & 0 & 0 & -iy & 0 & 0 & 0 & 0 & 0 & 0 & 0 & -iy & 0 & 0 \\
0 & -x & 0 & 0 & -iy & 0 & 0 & 0 & 0 & 0 & 0 & 0 & -iy & 0 & 0 & -x
\end{bmatrix}
\tag{13}
$$

it follows that Eq. (17) may be written

$$x(\omega_n - \omega_n^{-1}) + iy(\omega_n^r - \omega_n^{-r}) = \lambda_n. \tag{19}$$

Since D is the product of its eigenvalues, we have

$$\prod_{n=1}^{B} [x(\omega_n - \omega_n^{-1}) + iy(\omega_n^r - \omega_n^{-r})] = |D|. \tag{20}$$

Equation (20) gives the exact value of the determinant. When B is very large, an excellent approximation can be made by taking the logarithm of Eq. (20)

$$\ln|D| = \sum_{n=1}^{B} \ln[x(\omega_n - \omega_n^{-1}) + iy(\omega_n^r - \omega_n^{-r})] \tag{21}$$

and converting the sum to an integral. To make the conversion, we follow the procedure of Green and Hurst (6). Let $n = (b-1)s + a$ where $1 \leqslant a \leqslant s$, $1 \leqslant b \leqslant r$, which is equivalent to changing the numbering system of lattice terminals from $n = 1, 2, \ldots, B$ to the Cartesian coordinate system (a, b) as in Eq. (4). Then set

$$\varphi_a = (2a - 1)\pi/s, \qquad \theta_{ab} = 2(b-1)\pi/r + \varphi_a/r, \tag{22}$$

so that

$$\omega_n = \exp(i\theta_{ab}), \qquad \omega_n^r = \exp(i\varphi_a). \tag{23}$$

The summation over $\sum_{n=1}^{B}$ is equivalent to $\sum_{a=1}^{s} \sum_{b=1}^{r}$. In order to convert the double sum to an integral, we note that for large r and large s, φ_a and θ_{ab} go from 0 to 2π as a and b go from 1 to s and r, respectively. Also

$$\varphi_{a+1} - \varphi_a = 2\pi/s \to d\varphi_a \qquad \text{as} \quad s \to \infty. \tag{24}$$

Thus

$$\ln|D| = \sum_{b=1}^{r} \sum_{a=1}^{s} \ln[x(e^{i\theta_{ab}} - e^{-i\theta_{ab}}) + iy(e^{i\varphi_a} - e^{-i\varphi_a})]$$

$$= \sum_{b=1}^{r} (s/2\pi) \int_0^{2\pi} \ln[x(e^{i\theta_{ab}} - e^{-i\theta_{ab}}) + iy(e^{i\varphi_a} - e^{-i\varphi_a})] \, d\varphi_a. \tag{25}$$

In a similar manner, the remaining summation may be replaced by an integral. After removing indices we obtain

$$\ln|D| = (B/4\pi^2) \int_0^{2\pi} \int_0^{2\pi} \ln[x(e^{i\theta} - e^{-i\theta}) + iy(e^{i\varphi} - e^{-i\varphi})] \, d\varphi \, d\theta, \tag{26}$$

where we have used the fact that $(r/2\pi)(s/2\pi) = B/4\pi^2$. Equation (26) can be put into trigonometric form (6) and becomes after division by 2,

$$\tfrac{1}{2}\ln|D| = \ln Z = B/4\pi^2 \int_0^\pi \int_0^\pi \ln[2x^2(1 - \cos\theta) + 2y^2(1 - \cos\varphi)]\, d\varphi\, d\theta.$$

$$(27)$$

This expression for the configurational partition function shows that the only singularities of the integral occur when $x = y = 0$. Thus, in spite of the similarity of the dimer problem to the Ising problem, a critical point does not exist.

Fisher (4) obtained an expression equivalent to Eq. (27)

$$\ln Z = B\pi^{-1} \int_0^{x/y} v^{-1} \tan^{-1} v\, dv + \tfrac{1}{2}\ln y.$$

$$(28)$$

If we let $x = y = 1$ in Eq. (1), we obtain

$$Z = \sum_l \sum_m g(l, m, B),$$

$$(29)$$

which is the total number of ways of distributing horizontal and vertical dimers on the lattice. From the asymptotic formula in Eq. (28), we find that

$$\ln Z = B\pi^{-1} \int_0^1 v^{-1} \tan^{-1} v\, dv = GB/\pi,$$

$$(30)$$

where $G = 0.9159656$ is Catalan's constant. Therefore $Z = \exp(BG/\pi)$ is the total number of ways of distributing horizontal and vertical dimers when B is large.

4.3   Solution to the Ising Problem of Green and Hurst

Green and Hurst (6, 7) were the first to apply Pfaffians to the solution of the plane quadratic lattice with nearest-neighbor interactions. This is the same problem discussed in Chapter III, except that our present objective is to obtain a closed solution. We start with Eq. (65) of Chapter III which counts the different ways of constructing closed polygons with l horizontal and m vertical bonds on a lattice. We shall relate the counting of closed polygons to the different ways of coupling nearest-neighbor terminals according to an established set of rules. In the construction of closed polygons, we showed in Chapter III that each lattice terminal is connected to neighboring terminals by zero, two, or four bonds. Thus, contrary to the dimer problem, overlapping of bonds occurs. In each admissible configuration, a lattice terminal may appear more than once. Obviously, if Pfaffians are to be used in the solution of the problem, the lattice terminals must be modified in some way so that overlapping is eliminated, and then a set of rules established for their coupling. In accordance with Green and Hurst, we introduce, in the immediate neighborhood of each lattice point, a group of four terminals. The lattice points are numbered as defined by Eq. (4) and illustrated by Green and Hurst (6) for the dimer problem. The pth lattice point will have the four terminals corresponding to the four points of the compass:

$$
\begin{array}{ccc}
 & \cdot N_p & \\
W_p\cdot & & \cdot E_p \\
 & \cdot S_p &
\end{array}
\tag{31}
$$

Closed polygons are drawn on such a lattice as shown in Fig. 4.3.

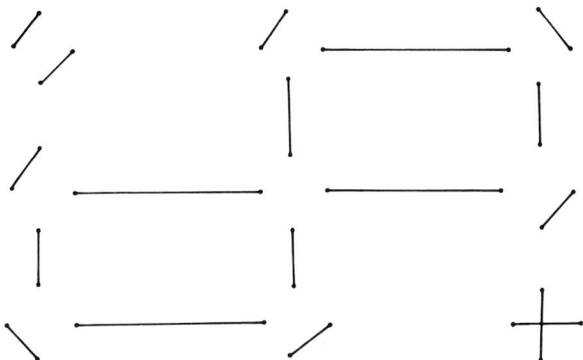

FIG. 4.3. Closed polygon formed by connecting terminals (6). [After Green and Hurst, "Order–Disorder Phenomena," Wiley (Interscience), New York, 1964.]

The east terminal E_j of the jth lattice point may be connected to the west terminal W_{j+1} of the $(j + 1)$th lattice point, and the north terminal N_j may be connected to the south terminal S_{j+r} of the $(j + r)$th lattice point. These are the only connections between different lattice points. The points which are not connected by bonds belonging to polygons are connected in pairs as shown in Fig. 4.3.

The rules for connecting terminals are set forth as follows:

(1) Each terminal must be coupled to another terminal.
(2) E_j may be coupled to W_{j+1}.
(3) N_j may be coupled to S_{j+r}.
(4) E_j may be coupled to N_j, W_j, or S_j.
(5) N_j may be coupled to W_j or S_j.
(6) W_j may be coupled to S_j.

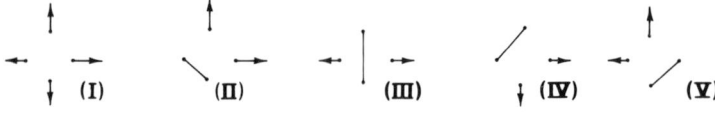

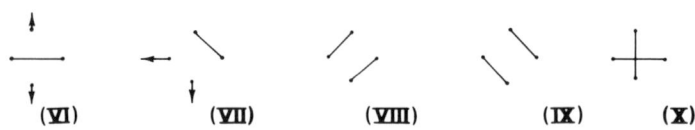

FIG. 4.4. Modes of coupling terminals at lattice points (6). [After Green and Hurst, "Order–Disorder Phenomena," Wiley (Interscience), New York, 1964.]

These are the only couplings which will be considered admissible. Thus, there are ten different ways of coupling the four terminals of a lattice point, as illustrated in Fig. 4.4.

Mode (i) in Fig. 4.4 corresponds to bonds of a polygon drawn in all four directions; modes (ii)–(vii) correspond to bonds of a polygon drawn in two directions, and modes (viii)–(x) correspond to the three different ways of coupling terminals of a lattice point which is not the vertex of a polygon.

A correspondence exists between the number of ways of constructing polygons on the lattice and the number of ways of coupling terminals according to the rules (1)–(6) set forth above. However, the correspondence is not one-to-one. Since there are three ways of coupling the terminals of a lattice point which is not the vertex of a polygon, there are more ways of coupling terminals than constructing polygons. If k lattice points are not involved in polygons, then there are $(N_c + N_p)^k$ total configurations of these points, where $N_c = 1$ is the number of crossed couplings [like (x)] and $N_p = 2$ is the number of parallel couplings [like (viii) and (ix)]. Thus, each configuration of polygons on $B - k$ lattice points is associated with 3^k configuration of points not involved in polygons. From the binomial theorem,

$$(N_c + N_p)^k = \sum_{n=0}^{k} k! N_c^{(k-n)} N_p^{n}/(k - n)! n!, \tag{32}$$

where the summand is the number of configurations for k points having $(k - n)$ crossed couplings and n parallel couplings. Summing all the terms of Eq. (32) with an odd number of crossed couplings [$(k - n)$ odd], we find that the sum is one less than the sum of the terms with an even number of crossed couplings. If we assign a negative value to the number of configurations with an odd number of crossed couplings and a positive value to those with an even number, then the number of ways of constructing polygons on the lattice and the number of ways of coupling terminals will be in one-to-one correspondence. We let $\mathcal{G}(C, l, m, B)$ be the number of configurations with C crossed couplings, l horizontal and m vertical bonds on B sites. Then

$$\mathcal{G}(l, m, B) = \sum_C (-1)^C \mathcal{G}(C, l, m, B) \tag{33}$$

where $\mathcal{G}(l, m, B)$ is the number of closed polygons with l horizontal and m vertical bonds which can be drawn on B lattice sites [Eq. (65), Chapter III]. Therefore, we may recast Eq. (65) as follows:

$$Z = 2^B (\cosh w/4kT \cosh w'/4kT)^B \sum_l \sum_m \sum_C (-1)^C \mathcal{G}(C, l, m, B) x^l y^m$$

$$= \Xi \exp(-Bw/4kT) \exp(-Bw'/4kT), \tag{34}$$

where we have put $c = 4$ for a quadratic lattice.

Having obtained a one-to-one correspondence between the ways of constructing polygons on the lattice and the ways of coupling terminals, we proceed to calculate the number of ways of coupling terminals by means of Pfaffians. The number of lattice terminals is $4B$. They are numbered from 1 to $4B$ so that E_j, N_j, W_j, and S_j, the terminals associated with the jth lattice point, are numbers $4j - 3$, $4j - 2$, $4j - 1$, and $4j$, respectively. Just as in the dimer problem, any element (p, q) of the Pfaffian, $\backslash D|$, is associated with the pth and qth terminals of the lattice. Any term in the expansion of $\backslash D|$, Eq. (6), corresponds to a particular way of grouping the terminals in pairs and satisfies the ordering rules given in Eq. (8). A one-to-one correspondence exists between all possible ways of coupling terminals and terms of the expanded Pfaffian. Not all of these ways are allowed in accordance with the rules (1)–(6). But we can eliminate inadmissible terms by setting (p, q) equal to zero when coupling between p and q is not permitted by the rules. There is now a one-to-one correspondence between nonvanishing terms of the Pfaffian and admissible configurations of the lattice given by rules (1)–(6).

The problem remains of giving the elements of the Pfaffian the proper values so that the expansion of the Pfaffian will give exactly the triple sum Z' in Eq. (34),

$$Z' = \sum_l \sum_m \sum_C (-1)^C \mathcal{G}(C, l, m, B) x^l y^m = \backslash D|. \tag{35}$$

The procedure is straightforward, but lengthy; we summarize the selections of values for nonvanishing elements as given by Green and Hurst (6) in Table 4.1a. To these must be added the selections in Table 4.1b because the lattice is to be wound on a torus, as in the dimer problem, to eliminate troublesome edge-effects in the evaluation of the Pfaffian. Although the vast majority of polygons are given their correct weight by the Pfaffian, Green and Hurst (6) show that some get an incorrect, negative weight. However, these

TABLE 4.1

p	q	Type of coupling	Value
(a)			
$4j-3$	$4j+3$	E_j with W_{j+1}	$-x$
$4j-2$	$4(j+r)$	N_j with S_{j+r}	$-y$
$4j-3$	$4j-2$	E_j with N_j	1
$4j-2$	$4j-1$	N_j with W_j	1
$4j-1$	$4j$	W_j with S_j	1
$4j-3$	$4j$	E_j with S_j	1
$4j-3$	$4j-1$	E_j with W_j	-1
$4j-2$	$4j$	N_j with S_j	-1
(b)			
3	$4B-3$	W_1 with E_B	$-x$
$4j$	$4(B+j-r)-2$	S_j with N_{B+j-r},	$-y$
		$(1 \le j \le r)$	

incorrect values are introduced only by elements associated with the winding of the lattice on a torus. Thus, their effect disappears for a large lattice; they have no influence on thermodynamic properties.

We give the skew-symmetric matrix D associated with the Pfaffian representing a 3×3 lattice ($B = 9$). The matrix is $4B = 36$-dimensional.

$$D = \begin{bmatrix}
U & -X & 0 & -Y & 0 & 0 & -Y' & 0 & -X' \\
X' & U & -X & 0 & -Y & 0 & 0 & -Y' & 0 \\
0 & X' & U & -X & 0 & -Y & 0 & 0 & -Y' \\
Y' & 0 & X' & U & -X & 0 & -Y & 0 & 0 \\
0 & Y' & 0 & X' & U & -X & 0 & -Y & 0 \\
0 & 0 & Y' & 0 & X' & U & -X & 0 & -Y \\
Y & 0 & 0 & Y' & 0 & X' & U & -X & 0 \\
0 & Y & 0 & 0 & Y' & 0 & X' & U & -X \\
X & 0 & Y & 0 & 0 & Y' & 0 & X' & U
\end{bmatrix}, \quad (36)$$

where

$$U = \begin{bmatrix}
0 & 1 & -1 & 1 \\
-1 & 0 & 1 & -1 \\
1 & -1 & 0 & 1 \\
-1 & 1 & -1 & 0
\end{bmatrix}, \quad X = \begin{bmatrix}
0 & 0 & x & 0 \\
0 & 0 & 0 & 0 \\
0 & 0 & 0 & 0 \\
0 & 0 & 0 & 0
\end{bmatrix},$$

$$(37)$$

$$Y = \begin{bmatrix}
0 & 0 & 0 & 0 \\
0 & 0 & 0 & y \\
0 & 0 & 0 & 0 \\
0 & 0 & 0 & 0
\end{bmatrix},$$

and X' and Y' are the transposed matrices of X and Y respectively (reflection in the diagonal). Note that D is a block matrix that has the characteristics expressed by Eq. (14), just as the matrix for the dimer problem, except that the elements are now 4×4 matrices. Substituting the elements of these 4×4 matrices into D, we find that the rows and columns are labeled E_1, N_1, W_1, S_1, E_2, N_2, W_2, S_2, ..., E_9, N_9, W_9, S_9 in that order. The eigenvectors of D are:

$$\Psi_n = \begin{bmatrix} \psi_n(E_1) \\ \psi_n(N_1) \\ \psi_n(W_1) \\ \psi_n(S_1) \\ \vdots \\ \psi_n(E_9) \\ \psi_n(N_9) \\ \psi_n(W_9) \\ \psi_n(S_9) \end{bmatrix}. \tag{38}$$

There are $4B = 36$ eigenvectors, one for each of the values $n = 1, 2, \ldots, 4B$. From the matrix equation (16), we have:

$$U \begin{bmatrix} \psi_n(E_1) \\ \psi_n(N_1) \\ \psi_n(W_1) \\ \psi_n(S_1) \end{bmatrix} - X \begin{bmatrix} \psi_n(E_2) \\ \psi_n(N_2) \\ \psi_n(W_2) \\ \psi_n(S_2) \end{bmatrix} - Y \begin{bmatrix} \psi_n(E_4) \\ \psi_n(N_4) \\ \psi_n(W_4) \\ \psi_n(S_4) \end{bmatrix} - Y' \begin{bmatrix} \psi_n(E_7) \\ \psi_n(N_7) \\ \psi_n(W_7) \\ \psi_n(S_7) \end{bmatrix}$$

$$- X' \begin{bmatrix} \psi_n(E_9) \\ \psi_n(N_9) \\ \psi_n(W_9) \\ \psi_n(S_9) \end{bmatrix} = \lambda_n \begin{bmatrix} \psi_n(E_1) \\ \psi_n(N_1) \\ \psi_n(W_1) \\ \psi_n(S_1) \end{bmatrix}. \tag{39}$$

We make the substitution:

$$\Psi_{n, j} = \begin{bmatrix} \psi_n(E_j) \\ \psi_n(N_j) \\ \psi_n(W_j) \\ \psi_n(S_j) \end{bmatrix}. \tag{40}$$

In general, for a lattice of arbitrary size and for operation on the eigenvector by any matrix row, we can write:

$$U\Psi_{n, j} - X\Psi_{n, j+1} + X'\Psi_{n, j-1} - Y\Psi_{n, j+r} + Y'\Psi_{n, j-r} = \lambda_n \Psi_{n, j}, \tag{41}$$

where we have defined

$$\Psi_{n, j} = -\Psi_{n, B+j}. \tag{42}$$

For example, in the 3×3 lattice where $r = 3$, we have for $j = 1$, $\Psi_{n, j-r} = \Psi_{n, -2} = -\Psi_{n, B-2} = -\Psi_{n, 7}$, which may be checked in Eq. (39). Eq. (42) is satisfied by the substitution:

$$\Psi_{n, j} = \omega_p^{j} \Psi_n, \qquad \omega_p^{B} = -1, \tag{43}$$

for ω_p is obviously any of the Bth roots of -1 and

$$\omega_p^{B+j} = -\omega_p^{j}, \qquad p = 1, 2, \ldots, B,$$
$$\omega_p = \exp[(2p - 1)\pi i/B]. \tag{44}$$

With the substitution of Eq. (43) into Eq. (41), we obtain:

$$(U - X\omega_p + X'\omega_p^{-1} - Y\omega_p^{r} + Y'\omega_p^{-r})\Psi_n = \lambda_n \Psi_n. \tag{45}$$

Equation (45) states that Ψ_n is an eigenvector of the matrix $D\omega_p = (U - X\omega_p + X'\omega_p^{-1} - Y\omega_p^{r} + Y'\omega_p^{-r})$, which may be written:

$$D\omega_p = \begin{bmatrix} 0 & 1 & -1 - \omega_p x & 1 \\ -1 & 0 & 1 & -1 - \omega_p^{r} y \\ 1 + \omega_p^{-1} x & -1 & 0 & 1 \\ -1 & 1 + \omega_p^{-r} y & -1 & 0 \end{bmatrix}. \tag{46}$$

It also shows that there are four eigenvalues λ_n for each root ω_p and that these are eigenvalues of $D\omega_p$ as well as D. The product of the eigenvalues is $\prod_{n=1}^{4B} \lambda_n = |D|$. But the product of the four eigenvalues of $D\omega_p$ is $|D\omega_p|$, and therefore,

$$\prod_{n=1}^{4B} \lambda_n = \prod_{p=1}^{B} |D\omega_p| = |D|$$

$$= \prod_{p=1}^{B} (1 + x^2)(1 + y^2) - (\omega_p + \omega_p^{-1})x(1 - y^2) - (\omega_p^{r} + \omega_p^{-r})y(1 - x^2). \tag{47}$$

Using the substitutions given under the dimer problem in Eq. (22) and also the procedure for changing summations to integrals in Eqs. (24) and (25), we obtain

$$\ln|D| = (B/4\pi^2) \int_0^{2\pi} \int_0^{2\pi} \ln[(1 + x^2)(1 + y^2) - 2x(\cos \theta) \times (1 - y^2) - 2y(\cos \varphi)(1 - x^2)] \, d\theta \, d\varphi. \tag{48}$$

where we have used $e^{i\varphi} + e^{-i\varphi} = 2 \cos \varphi$.

One of the integrations in Eq. (48) can be carried out with the aid of the identity

$$\int_0^{2\pi} \ln(2 \cosh a - 2 \cos \varphi) \, d\varphi = 2\pi a. \tag{49}$$

The identity is proved by substituting $2 \cosh a = e^a - e^{-a}$ and $2 \cos \varphi = e^{i\varphi} + e^{-i\varphi}$ in Eq. (49). Then

$$\int_0^{2\pi} \ln(e^a + e^{-a} - e^{i\varphi} - e^{-i\varphi}) \, d\varphi$$

$$= \int_0^{2\pi} [\ln e^a + \ln(1 - e^{-a}e^{i\varphi}) + \ln(1 - e^{-a}e^{-i\varphi})] \, d\varphi. \qquad (50)$$

But

$$\int_0^{2\pi} [\ln(1 - e^{-a}e^{i\varphi}) + \ln(1 - e^{-a}e^{-i\varphi})] \, d\varphi = 0 \qquad (51)$$

Therefore,

$$\int_0^{2\pi} \ln(2 \cosh a - 2 \cos \varphi) \, d\varphi = \int_0^{2\pi} \ln e^a \, d\varphi = 2\pi a. \qquad (52)$$

We let

$$\cosh a(\theta) = [(1 + x^2)(1 + y^2) - (2 \cos \theta)x(1 - y^2)]/2y(1 - x^2). \qquad (53)$$

Substituting this into Eq. (48), we obtain

$$\ln|D| = B \ln y(1 - x^2) + (B/2\pi) \int_0^{2\pi} a(\theta) \, d\theta, \qquad (54)$$

where

$$a(\theta) = \cosh^{-1}[(1 + x^2)(1 + y^2) - (2 \cos \theta)x(1 - y^2)]2y(1 - x^2). \qquad (55)$$

From Eq. (35), we have

$$\ln Z' = \ln|D| = (1/2) \ln|D| = (B/2) \ln y(1 - x^2) + (B/4\pi) \int_0^{2\pi} a(\theta) \, d\theta. \qquad (56)$$

From Eq. (34), we find that

$$(1/B) \ln \Xi = (w + w')/4kT + \ln[2 \cosh(w/4kT) \cosh(w'/4kT)]$$
$$+ (1/2B) \ln|D| = \pi/kT. \qquad (57)$$

Equation (57) gives the two-dimensional pressure as a function of temperature at constant density (lattice points half covered).

The integral in Eq. (54) does not belong to a common type. But the thermodynamic functions involving derivatives of these integrals can be expressed in terms of elliptic integrals. These integrals have been analyzed for arbitrary w and w' by Onsager (*1*) and also by Green and Hurst (*6*). Letting

$w = w'(x = y)$, the analysis is considerably simplified. The expression for $\cosh a(\theta)$, Eq. (53), with this simplification becomes

$$\cosh a(\theta) = [(1 + x^2)^2 - 2(\cos \theta)x(1 - x^2)]/2x(1 - x^2)$$
$$= \cosh 2K \coth 2K - \cos \theta, \qquad K = (-w/4kT). \qquad (58)$$

With these changes, $\ln Z$ from Eqs. (34), (35), (54), and (56) becomes

$$(1/B) \ln Z = \ln(2 \cosh^2 K) + \tfrac{1}{2} \ln(\sinh K/\cosh^3 K) + 1/4\pi \int_0^{2\pi} a(\theta) \, d\theta,$$
$$(59)$$

where $\sinh K/\cosh^3 K = y(1 - x^2)$.

The configurational energy E_{sc} can now be found from

$$E_{sc} = kT^2 \, \partial \ln Z/\partial T = (Bw/4)\partial \ln Z/\partial K$$
$$= (Bw/4) \coth 2K[1 + (2/\pi)(\cosh^2 2K)(2 \tanh^2 2K - 1)K_1(k_1)], \qquad (60)$$

where

$$k_1 = \sinh^2 2K, \qquad (61)$$

and $K_1(k_1)$ is the complete elliptical integral of the first kind,

$$K_1(k_1) = \int_0^{\pi/2} (1 - k_1{}^2 \sin^2 \varphi)^{-1/2} \, d\varphi. \qquad (62)$$

In Section 3.6, it was predicted that the critical point for this system was $|\sinh 2K_c| = 1$. Thus $\cosh^2 2K_c = 2$; and Eq. (61) shows that $k_1 = 1$ at the critical point. If we expand the integrand of Eq. (62) about $T = T_c$ and $\varphi = \pi/2$, we find that $K_1(k_1)$ behaves as $\ln|T - T_c|$ near $T = T_c$. However, the coefficient of $K_1(k_1)$ in Eq. (60) vanishes linearly with $T - T_c$ near the critical point, since $2 \tanh^2 2K_c = 1$. Therefore, E_{sc} is continuous at the critical point, and there is no latent heat at that point.

If we differentiate Eq. (60) again with respect to temperature, we obtain the configurational heat capacity C_{sc}. From the term $(T - T_c) \ln|T - T_c|$ mentioned above we see that C_{sc} is proportional to $\ln|T - T_c|$. Therefore, the configurational heat capacity has a logarithmic singularity at the critical point. We recall that the one-dimensional case discussed in Section 3.3 had no singularity in the heat capacity at the critical point.

Green and Hurst (6) have applied their method to other lattices such as the triangular and the hexagonal. The triangular lattice problem was first solved independently by Wannier (9), Newell (8), Temperley (10) and Houtappel (11). They used the algebraic method of Onsager (1). Potts (12) solved the triangular lattice problem by applying the intuitive method of Kac and Ward (13). Among the problems yet unsolved in closed form are lattices with crossed bonds, second and higher interactions, more and less than half coverage, more than two dimensions, and the dimer problem with less than complete coverage of the lattice by dimers. Hurst (*15*) has shown that the Pfaffian method breaks down when the lattice is nonplanar.

REFERENCES

1. Onsager, L., *Phys. Rev.* **65**, 117 (1944).
2. Temperley H. N. V., and Fisher, M. E., *Phil. Mag.* **6**, 1061 (1961).
3. Kasteleyn, P. W., *Physics* **27**, 1209 (1961).
4. Fisher, M. E., *Phys. Rev.* **124**, 1664 (1961).
5. Kasteleyn, P. W., *J. Math. Phys.* **4**, 287 (1963).
6. Green, H. S., and Hurst, C. A., "Order–Disorder Phenomena," p. 253, Wiley (Interscience), New York, 1964.
7. Hurst, C. A., and Green, H. S., *J. Chem. Phys.* **33**, 1059 (1960).
8. Newell, G. F., *Phys. Rev.* **79**, 876 (1950).
9. Wannier, G. H., *Phys. Rev.* **79**, 357 (1950).
10. Temperley, H. N. V., *Proc. Roy. Soc. (London), Ser. A* **202**, 202 (1950).
11. Houtappel, R. M. F., *Physica* **16**, 425 (1950).
12. Potts, R. B., *Proc. Phys. Soc. (London), Ser. A* **68**, 145 (1955).
13. Kac, M., and Ward, J. C., *Phys. Rev.* **88**, 1332 (1952).
14. Fisher, M. E., *J. Math. Phys.* **7**, 1776 (1966).
15. Hurst, C. A., *J. Math. Phys.* **7**, 81 (1966).

V

NONLOCALIZED ADSORPTION

5.1 Introduction

In the last three chapters, we have dealt with localized adsorption where molecules are adsorbed as a result of interaction with discrete surface positions of minimum potential energy. We discussed specifically the cases of immobile and partially mobile adsorption. In the former case, potential minima are presumed to be so deep and the rate of change of potential in any lateral direction away from the minima so rapid that lateral movement of adsorbate is essentially prohibited (see Fig. 2.1a, p. 18). In the latter case, the constraints causing immobility are sufficiently relaxed that some, but not complete, mobility can occur (see Fig. 2.1d). We pointed out that localized adsorption approaches the condition of complete mobility as the depths of the potential minima approach kT (see Fig. 2.1b, c). Thus localized adsorption approaches nonlocalized adsorption as the potential minima disappear and the surface potential becomes uniform (or changes gradually with no minima). In this chapter, we shall discuss nonlocalized adsorption, which, in general, is mobile adsorption. However, in some cases, the mathematical path to complete mobility proceeds through relaxation of those constraints which cause immobility, so that in principle such equations have general applicability. The extreme of complete immobility or complete mobility rarely occurs on real surfaces, although low temperatures strongly favor immobile adsorption and high temperatures mobile adsorption.

There are two general approaches to the theoretical study of dilute mobile layers. In one approach, the adsorbed phase is considered separately, usually as a two-dimensional gas. We express the relationship between the adsorbed phase and the gas phase by equating the chemical potentials of the two phases at equilibrium. This approach was also used in localized adsorption. In the other approach, we relate the properties of the adsorbed phase and gas phase to each other directly by considering all the gas in the adsorption container; this is usually referred to as the gas-surface virial expansion. Both approaches are applicable only for dilute adsorbed layers, because the more dense the adsorbed layer the more terms in the expansions of partition functions, equations of state, etc., must be used, and these become progressively more difficult to calculate. For higher coverages, approximate two-dimensional equations of state similar to three-dimensional equations, such as the van der Waals equation, are used. Another approach is the Lennard-Jones–Devonshire theory for the three-dimensional condensation problem which was adapted by Devonshire to the two-dimensional case. It belongs to the category of cell theories of liquids. Still another approach is the application of Eyring's significant structure theory. All these approaches will be discussed in this chapter. The effect of surface heterogeneity for both dilute and concentrated adsorbed phases will also be discussed. Finally, a theory of multilayer, mobile adsorption will be discussed. This theory, though approximate, is more realistic than the BET theory of multilayer, localized adsorption.

5.2 The Two-Dimensional Adsorbed Gas

The simplest situation of mobile adsorption in which the adsorbed phase is considered separately is that of the dilute, perfect, two-dimensional gas. The adsorbed molecule vibrates perpendicular to the surface, as it does in immobile adsorption (Section 2.2), but, in the x and y directions parallel to the surface, the molecule is assumed to move without constraint. A monatomic adsorbate under these conditions will have one vibrational and two translational degrees of freedom. As in Eqs. (78) and (79) of Chapter II, we write the canonical partition function of such a molecule as follows:

$$Q_s = q_s^{N_s}/N_s!, \qquad q_s = q_{xy}q_z \exp(-U_0/kT), \qquad q_{xy} = (2\pi mkT)\alpha/h^2, \quad (1)$$

where q_{xy} is the translational partition function of a molecule with two degrees of translational freedom, and U_0 is the uniform adsorption potential. By well-known principles of statistical thermodynamics, we find the two-dimensional equation of state

$$\varphi/kT = (\partial \ln Q_s/\partial \alpha)_{N_s, T} = N_s/\alpha. \qquad (2)$$

Equation (2) is the two-dimensional counterpart of the equation of state for a three-dimensional perfect gas. Using the methods of Section 2.2, we find

$$\mu_s/kT = -(\partial \ln Q_s/\partial N_s)_{\alpha,\,T} = \ln N_s/q_s = \mu_G/kT = [\mu^0(T)/kT] + \ln p, \quad (3)$$

which gives the adsorption isotherm

$$N_s/\alpha = a(T)p, \qquad a = (2\pi mT/h^2)q_z \exp(-U_0/kT) \exp(\mu^0/kT). \quad (4)$$

Note that for a dilute adsorbed phase, where total molecular cross-section of adsorbed molecules is small compared to total surface area, the amount adsorbed is directly proportional to the pressure. Such behavior is often referred to as Henry's Law.

The problem may be extended to include an imperfect, two-dimensional adsorbed gas. The general procedure is to start with an appropriate form of the grand canonical partition function containing the classical configuration integral. It is then possible to expand the partition function and by standard methods to convert it into a series expansion of the two-dimensional pressure in terms of the powers of the density of adsorbed gas, the equation of state for an imperfect, mobile, adsorbed gas. The coefficients of the powers of the density contain the various intermolecular pair potentials which are evaluated in terms of the hard sphere model and the more realistic Lennard-Jones potential. We express the grand canonical partition function in the form

$$\Xi_s = \exp(\varphi\alpha/kT) = \sum_{N_s \geqslant 0} Q_{N_s} \lambda_s^{N_s} = \sum_{N_s \geqslant 0} (Z_{N_s} z_s^{N_s})/N_s!$$

$$= 1 + Q_{1s}\lambda_s + \cdots = 1 + Z_{1s}z_s + \cdots, \quad (5)$$

where

$$Q_{Ns}(\alpha, T) = Q_s(N_s, \alpha, T), \qquad \lambda_s = \exp(\mu/kT),$$

$$z_s = Q_{1s}\lambda_s/Z_{1s}, \qquad Q_{1s} = q_s. \quad (6)$$

We want z_s not only proportional to λ_s, but also defined so that $z_s \to \rho_s = N_s/\alpha$ as $\rho_s \to 0$. From

$$\varphi\alpha/kT = \ln \Xi_s = Z_{1s}z_s + \cdots = N_s + \cdots, \quad (7)$$

it is evident that we must define $Z_{1s} = \alpha$. Therefore,

$$z_s = Q_{1s}\lambda_s/\alpha, \qquad z_s \to \rho_s \quad \text{as} \quad \rho_s \to 0. \quad (8)$$

From Eq. (5), we find that Z_{N_s} must be defined by

$$Z_{N_s} = (\alpha/Q_{1s})^{N_s} N_s! Q_{N_s}. \quad (9)$$

With these definitions of z_s and Z_{N_s}, we have not committed ourselves to a classical system or to any other type of system. The appropriate values of Q_{N_s} for the system under consideration are used. In the particular case of a two-dimensional adsorbed gas, it will include a term for the adsorption potential, translational motion, etc. However, the particular form of the grand canonical partition function which we have developed is most useful in dealing with classical systems. It can be readily shown (*1*) that for a classical system, Z_{N_s} becomes the configurational integral, which is written,

$$Z_{N_s} = \int_\alpha \exp[-U(\mathbf{r}_1, \mathbf{r}_2, \ldots, \mathbf{r}_{N_s})]\, d\mathbf{r}_1\, d\mathbf{r}_2 \cdots d\mathbf{r}_{N_s}, \tag{10}$$

where U gives the total potential energy of molecular interactions in the adsorbed phase for configurations of N_s molecules specified by the vectors $\mathbf{r}_1, \mathbf{r}_2, \ldots, \mathbf{r}_{N_s}$. When these vectors start at the origin of a rectangular coordinate system, then vector $\mathbf{r}_j(j = 1, 2, \ldots, N_s)$ locates the jth adsorbed molecule with coordinates x_j, y_j. In the absence of intermolecular forces, $U = 0$, and $Z_{N_s} = \alpha^{N_s}$, so that Eq. (5) reverts to the equation for a perfect two-dimensional gas. We assume that the intermolecular potential energy is pairwise additive so that we may write

$$U = \sum_{1 \leqslant i \leqslant j \leqslant N_s} u_s(\mathbf{r}_i, \mathbf{r}_j), \tag{11}$$

where $u_s(\mathbf{r}_i, \mathbf{r}_j)$ is the intermolecular pair potential between the ith and jth molecules located by the vectors $\mathbf{r}_i$ and $\mathbf{r}_j$.

Now

$$\varphi\alpha/kT = \ln \Xi_s = \ln \sum_{N_s \geqslant 0} (Z_{Ns}\, z_s^{N_s}/N_s\,!). \tag{12}$$

We expand the right-hand side of Eq. (12), divide by α and obtain an expansion for φ in powers of z_s,

$$(1/\alpha) \ln \Xi_s = \varphi/kT = \sum_{j \geqslant 1} b_j(T) z_s^{\,j}, \tag{13}$$

where

$$1!\,\alpha b_1 = Z_{1s} = \alpha, \qquad b_1 = 1,$$
$$2!\,\alpha b_2 = Z_{2s} - Z_{1s}^2, \tag{14}$$
$$3!\,\alpha b_3 = Z_{3s} - 3Z_{1s} Z_{2s} + 2Z_{1s}^3.$$

The b_n are called the cluster integrals and were first introduced by Mayer and Mayer (*2*).

We now use

$$(kT/\alpha)(\partial \ln \Xi_s/\partial \mu_s)_{\alpha,\,T} = (z_s/\alpha)(\partial \ln \Xi_s/\partial z_s)_{T,\,\alpha} = N_s/\alpha = \rho_s, \tag{15}$$

to give

$$\rho_s = \sum_{j \geqslant 1} j b_j(T) z_s^{\,j}. \tag{16}$$

By standard methods, Eq. (16) may be inverted to give $z_s(\rho_s)$ which may then be substituted in Eq. (13) to obtain an equation of state for the imperfect two-dimensional gas. The result is

$$\varphi/kT = \rho_s + B_{2s}(T)\rho_s{}^2 + B_{3s}(T)\rho_s{}^3 + \cdots, \tag{17}$$

where from Eqs. (13)–(17),

$$B_{2s} = -b_2 = -\tfrac{1}{2}\alpha(Z_{2s} - Z_{1s}^2) = -\tfrac{1}{2}\alpha \iint_\alpha \left\{ \exp[-u_s(\mathbf{r}_1, \mathbf{r}_2)/kT] - 1 \right\} d\mathbf{r}_1 \, d\mathbf{r}_2$$

$$B_{3s} = 4b_2{}^2 - 2b_3 = -(1/3\alpha) \iiint_\alpha f_{12} f_{13} f_{23} \, d\mathbf{r}_1 \, d\mathbf{r}_2 \, d\mathbf{r}_3, \tag{18}$$

$$f_{ij} = \exp[-u_s(\mathbf{r}_i, \mathbf{r}_j)/kT] - 1.$$

Equation (17) is known as the virial expansion and the B_{ns} are called the virial coefficients. When the intermolecular pair potentials are all zero, Eq. (17) reverts to the perfect gas Eq. (2).

In the three-dimensional gas problem, the intermolecular pair potential $u_s(\mathbf{r}_i, \mathbf{r}_j)$ is usually taken to be $u_s(r_{ij})$ where $r_{ij} = |\mathbf{r}_i - \mathbf{r}_j|$, that is, the potential depends only on the distance of separation of the ith and jth molecules. Molecules are assumed spherical so that no orientation effects enter. When a surface is present, as in the case of a two-dimensional adsorbed gas, the interaction energy appears to be about 20% lower than the gas phase value (3), indicating additional repulsion between molecules as a result of the presence of a surface. The gas–surface interaction is in general about ten times greater than the gas–gas interaction and therefore should not be appreciably affected by lateral interactions of adsorbed molecules.

The intermolecular pair potential $u_s(r_{ij})$ goes rapidly to zero for intermolecular distances greater than 15 or 20 Å, and therefore the integrand in Eq. (18) is nonzero only when the area elements $d\mathbf{r}_1 = dx_1 \, dy_1$, and $d\mathbf{r}_2 = dx_2 \, dy_2$ are close to each other. We change variables in Eq. (18) from $\mathbf{r}_1 = x_1, y_1$ and $\mathbf{r}_2 = x_2, y_2$ to $\mathbf{r}_1$ and $\mathbf{r}_{12} = \mathbf{r}_2 - \mathbf{r}_1 = x_2 - x_1, y_2 - y_1$. But integration over $\mathbf{r}_2 - \mathbf{r}_1$ is independent of the location of $\mathbf{r}_1$, except for insignificant edge effects. Thus, for the purpose of integrating over $\mathbf{r}_{12}$, we can fix molecule 1 at the origin ($x_1 = y_1 = 0$) and write B_{2s}, Eq. (18), in the form

$$B_{2s} = -(1/2\alpha) \int_\alpha d\mathbf{r}_1 \int_\alpha \{\exp[-u_s(r_{12})/kT] - 1\} \, d\mathbf{r}_{12}$$

$$= -(1/2\alpha) \int_\alpha dx_1 \, dy_1 \int_\alpha \{\exp[-u_s(r_{12})/kT] - 1\} \, dx_2 \, dy_2$$

$$= -(1/2) \int_0^\infty \{\exp[-u_s(r)/kT] - 1\} 2\pi r \, dr, \tag{19}$$

where we have changed to polar coordinates $r_{12} = r$, and set the upper limit $r = \infty$, since the integral contributes nothing after about the first 20 Å.

Equation (19) for B_{2s} and equations for higher virial coefficients become useful only when explicit expressions for $u_s(r)$ are available. Various forms have been used. The simplest form is for a hypothetical gas of hard spheres,

$$u_s(r) = +\infty, \quad r < a; \quad u_s(r) = 0, \quad r \geqslant a, \tag{20}$$

which gives from Eq. (19)

$$B_{2s} = \frac{\pi a^2}{2}, \tag{21}$$

and after a long calculation from Eq. (15),

$$B_{3s} = (5/8)B_{2s}^2 = (5/32)\pi^2 a^4. \tag{22}$$

A much more realistic function is the Lennard-Jones potential. This function stems in part from quantum mechanical calculations and in part from empirical considerations. A quantum mechanical calculation has shown that the attractive force between two spherically symmetrical, chemically saturated molecules falls off as $1/r^7$ for sufficiently large r. Therefore the potential of this force varies as $1/r^6$. This is the van der Waals attractive force, sometimes also called the London dispersion force. When r is small, the two molecules repel each other because of the internuclear repulsion and overlap of electron shells. No quantum mechanical calculations of this repulsive force have been made yet. A useful approximation is to assume that the repulsive potential varies as $1/r^{12}$. Combining the attractive and repulsive potentials gives the so-called Lennard-Jones 6–12 potential

$$u_s(r) = 4\epsilon_{2s}[-(r_0/r)^6 + (r_0/r)^{12}] = -2\epsilon_{2s}(r^*/r)^6 + \epsilon_{2s}(r^*/r)^{12}, \tag{23}$$

where $u_s = 0$ at $r = r_0$, and u_s has a minimum ($u_s = -\epsilon_{2s}$) at $r = r^*$. Fig. 5.1 shows the behavior of $u_s(r)$. With this potential, the second virial coefficient becomes

$$B_{2s} = -\pi r_0^2 \int_0^\infty \{\exp[(4\epsilon_{2s}/kT)(y^{-6} - y^{-12})] - 1\} y \, dy, \quad y = r/r_0, \tag{24}$$

which states that B_{2s}/r_0^2 is a universal function of kT/ϵ_{2s} for all molecules with an intermolecular potential of the Lennard-Jones form. If experimental values of B_{2s} as a function of T are fitted to the theoretical curves of B_{2s}/r_0^2 versus kT/ϵ_{2s}, values for ϵ_{2s} and r_0 are obtained. Similarly, B_{3s}/r_0^4 is a universal function of kT/ϵ_{2s}.

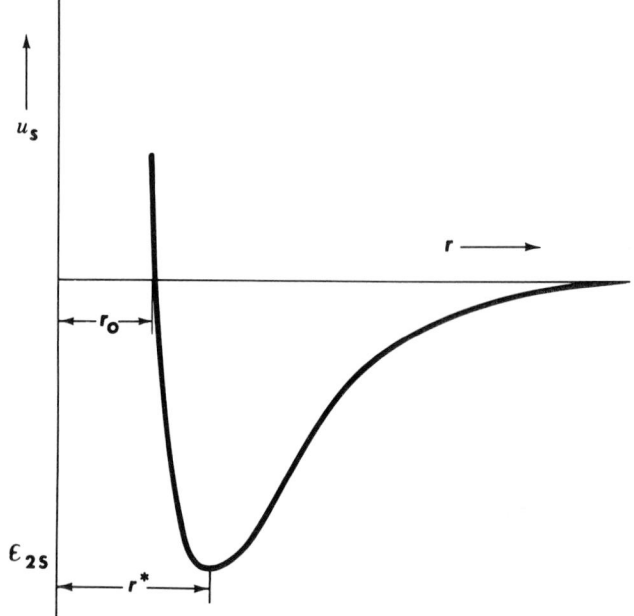

Fɪɢ. 5.1. Lennard-Jones 6–12 intermolecular pair potential.

5.3 Gas–Surface Virial Expansion[1]

The procedure is similar to that in the previous section except that one starts with the grand canonical partition function of the entire system and a potential energy of interaction of the system which includes the interaction of molecules with themselves and with the surface. An adsorption isotherm is obtained in the form of a virial expansion of powers of (p/kT). The virial coefficients are evaluated by use of Lennard-Jones and similar potentials. Adsorption energies calculated using a 9–3 potential energy function compare favorably with values determined from experimental isotherms as shown below.

We consider a gas in contact with a solid adsorbent at fixed p and T. The volume available to the gas in the adsorption container is V and the total number of molecules in the gas phase and in the adsorbed phase is N. The energy of interaction between the ith gas molecule and the surface will be denoted by $u_s(\mathbf{r}_i)$, where $\mathbf{r}_i$ is the vector which specifies the position of the ith molecule with respect to the surface. The fact that the dimensions of a solid adsorbent often change with changes in amount of gas adsorbed indicates

[1] See Steele (4).

that $u_s(\mathbf{r}_i)$ does not accurately describe the interaction of a molecule with a surface under all conditions. But the theoretical treatment of these changes is difficult. The interaction energies of gaseous molecules will be assumed to be pairwise additive as before. In this case, we do not specify the exact dimensions of the adsorbed phase; it may be two dimensional or it may be an ill-defined layer exhibiting an increasing density near the surface. As stated previously, the presence of a solid may have an appreciable effect on the pairwise interaction near the surface. Approximate estimates indicate that, compared to gaseous molecules, the attractive energy appears to be 20% greater when the vector $\mathbf{r}_{ij}$ is perpendicular to the surface, and 20% less when it is parallel. As usual, molecules will be assumed to be spherically symmetrical. The potential energy $U(\mathbf{r}_1, \mathbf{r}_2, \ldots, \mathbf{r}_N)$ of N molecules in a configuration specified by the vectors $\mathbf{r}_1, \mathbf{r}_2, \ldots, \mathbf{r}_N$, and which interact with each other and with the surface of the adsorbent, is

$$U(\mathbf{r}_1, \mathbf{r}_2, \ldots, \mathbf{r}_N) = \sum_{i=1}^{N} u_s(\mathbf{r}_i) + \sum_{1 \leqslant i \leqslant j \leqslant N} u(\mathbf{r}_i, \mathbf{r}_j). \tag{25}$$

The classical grand canonical partition function of the entire system, adsorbed phase and gas phase, is

$$\Xi = \sum_{N \geqslant 0} Z_N z^N / N!, \tag{26}$$

where

$$z = Q_1 \lambda / V, \qquad Z_N = \int_V \exp[-U(\mathbf{r}_1, \mathbf{r}_2, \ldots, \mathbf{r}_N)/kT]\, d\mathbf{r}_1\, d\mathbf{r}_2 \cdots d\mathbf{r}_N. \tag{27}$$

For simplicity, we shall assume a mobile monatomic adsorbate. Thus

$$Q_1 = q = V/\Lambda^3, \qquad z = \lambda/\Lambda^3, \qquad \Lambda = h/(2\pi m k T)^{1/2}. \tag{28}$$

The term of the partition function corresponding to the adsorption potential is incorporated in Z_N as shown in Eqs. (25) and (27) and extends over all N molecules in the system, which is tantamount to assuming that the adsorbed phase consists of all the molecules in the container. Both Q_1 and $\lambda = \exp(\mu/kT)$ remain constant throughout all parts of the system. Therefore, z also remains constant and may be related to experiment by observing the properties of the gas far from the surface of the adsorbent. If we assume that the gas phase is ideal, then from Eqs. (15) and (16) of Chapter II,

$$\lambda = \lambda_G = p\Lambda^3/kT, \tag{29}$$

and z becomes

$$z = z_G = p/kT. \tag{30}$$

The partition function for the entire system may be separated into two functions, one for the adsorbed phase $\Xi^{(s)}$ and one for the gas phase Ξ_G, with the usual relationship

$$\Xi = \Xi^{(s)}\Xi_G. \tag{31}$$

Therefore,

$$\Xi^{(s)} = \Xi/\Xi_G = \frac{\sum_{N \geq 0} Z_N z^N/N!}{\sum_{N_G \geq 0} Z_{N_G} z^{N_G}/N_G!}. \tag{32}$$

In order to define more precisely what is meant by the gas phase and the adsorbed phase when this separation may be diffuse, we make use of surface excess functions. For example, we define the quantity of adsorbed gas N_s as

$$N_s = N - N_G, \tag{33}$$

where N_G is the number of gas molecules which would be in the volume V if no adsorption occurred. Now

$$N - N_G = (\partial \ln \Xi/\partial \ln z)_{T,V} - (\partial \ln \Xi_G/\partial \ln z)_{T,V}$$
$$= (\partial \ln \Xi^{(s)}/\partial \ln z) = N_s. \tag{34}$$

Similarly, a surface excess entropy may be defined by

$$S - S_G = kT(\partial \ln \Xi/\partial T)_{V,\mu} + k \ln \Xi - kT(\partial \ln \Xi_G/\partial T)_{V,\mu} - k \ln \Xi_G$$
$$= kT(\partial \ln \Xi^{(s)}/\partial T)_{V,\mu} + k \ln \Xi^{(s)} = S_s. \tag{35}$$

Note that we distinguish between Ξ_s of Section 5.2, which refers specifically to a two-dimensional adsorbed layer, and $\Xi^{(s)}$ of this section, which is more general.

For a perfect gas, we set $z = p/kT$, and the configurational integral for the gas phase $Z_{N_G} = V^{N_G}$, since $u(\mathbf{r}_i, \mathbf{r}_j) = 0$ for all i and j in the gas phase. We now write Eq. (32)

$$\Xi^{(s)} = \frac{1 + Z_1(p/kT) + (Z_2/2!)(p/kT)^2 + \cdots}{1 + V(p/kT) + (V^2/2!)(p/kT)^2 + \cdots}$$
$$\equiv \sum_{N \geq 0} [Z_N^{(s)}(p/kT)^N/N!], \tag{36}$$

which defines $Z_N^{(s)}$. In order to determine explicitly the values of the $Z_N^{(s)}$, we expand the middle term of Eq. (36) to get

$$\Xi^{(s)} = 1 + (Z_1 - V)(p/kT) + [(Z_2 - 2Z_1 V + V^2)/2!](p/kT)^2 + \cdots. \tag{37}$$

Equation (37) has been applied to a number of studies of the properties of gases in contact with high-area solids at relatively high temperatures (3–14). Under these conditions, the adsorbed phase is dilute and is pictured physically as a diffuse layer of increased density over the gas phase due to the attractive forces of the surface, rather than as a two-dimensional layer. At low density, we have approximately

$$\lambda_s = \frac{p_s(\mathbf{r})\Lambda^3}{kT \exp[-u_s(\mathbf{r})/kT]} = \lambda_G = \frac{p_G \Lambda^3}{kT} \tag{38}$$

or

$$\rho_s(\mathbf{r}) = \rho_G \exp[-u_s(\mathbf{r})/kT],$$

which states that the density as a function of position with respect to the surface can be determined by multiplying the density, ρ_G, at infinite distance from the surface by a Boltzmann probability factor. At infinite separation from the surface $u_s(\mathbf{r})$ is zero; it becomes increasingly negative as one approaches the surface until it passes through a minimum, and finally becomes positive. From Eq. (38), it is evident that $\rho_s(\mathbf{r})$ passes through a maximum. The maximum will be higher and sharper the lower the temperature. Thus at higher temperatures, the gas must deviate from its bulk properties over distances of several angstroms. It is because the volume of the adsorbed phase is not a well-defined quantity at higher temperatures that Eq. (37) provides a good starting point for a theoretical treatment.

Expanding Eq. (37) as the logarithm, we have

$$\ln \Xi^{(s)} = (Z_1 - V)(p/kT) + [(Z_2 - Z_1^2)/2!](p/kT)^2$$
$$+ [(Z_3 - 3Z_2 Z_1 + 2Z_1^3)/3!](p/kT)^3 + \cdots. \tag{39}$$

Taking the derivative with respect to $\ln z = \ln(p/kT)$, we find from Eq. (34)

$$(\partial \ln \Xi^{(s)}/\partial \ln z)_{T,V} = N_s = N - N_G$$
$$= B_{AS}(p/kT) + C_{AAS}(p/kT)^2 + D_{AAAS}(p/kT)^3 + \cdots, \tag{40}$$

where the coefficients of $(p/kT)^n$ are called the gas–surface virial coefficients. From Eqs. (25), (27), and (39), explicit equations for them are found to be

$$B_{AS} = \int_V \{\exp[-u_s(\mathbf{r}_1)/kT] - 1\} \, d\mathbf{r}_1 \tag{41}$$

$$C_{AAS} = \int_V \exp\{-[u_s(\mathbf{r}_1) + u_s(\mathbf{r}_2)]/kT\} f_{12} \, d\mathbf{r}_1 \, d\mathbf{r}_2 \tag{42}$$

$$D_{AAAS} = (1/2) \iiint_V \exp\{-[u_s(\mathbf{r}_1) + u_s(\mathbf{r}_2) + u_s(\mathbf{r}_3)]/kT\}$$
$$\times (f_{12} f_{13} f_{23} + 3f_{12} f_{13}) \, d\mathbf{r}_1 \, d\mathbf{r}_2 \, d\mathbf{r}_3, \tag{43}$$

where

$$f_{ij} = \exp[-u(\mathbf{r}_i, \mathbf{r}_j)/kT] - 1. \tag{44}$$

Experimentally, B_{AS} can be determined from isotherms—that is, from a plot of excess volume $V_{ex} \equiv N_s kT/p$ versus p. Equation (40) shows that

$$\lim_{p \to 0} V_{ex} = B_{AS}. \tag{45}$$

Also Eq. (40) shows that C_{AAS} can be determined from

$$\lim_{p \to 0} dV_{ex}/dp = C_{AAS}. \tag{46}$$

In order to compute virial coefficients, it is necessary to obtain explicit expressions for $u_s(\mathbf{r}_i)$, the interaction potential of a molecule with the solid and for $u(\mathbf{r}_i, \mathbf{r}_j)$ the pairwise interaction potential of gas molecules with each other. For computation of B_{AS}, only an expression for $u_s(\mathbf{r}_i)$, $i = 1, \ldots, N$, is required. The Cartesian coordinates corresponding to $\mathbf{r}_i$ are τ_i, z_i, where z_i is the perpendicular distance between the ith gas molecule and the solid surface, and $\tau_i = x_i$, y_i are coordinates in a plane parallel to the surface. If surface atoms of the solid are considered to be located at discrete intervals and the pairwise interactions of them with a gas molecule are summed, then $u_s(\tau_i, z_i)$ will be a periodic function of τ_i. On the other hand, if the solid surface is considered to be a continuum and the summation is replaced by integration, then u_s will be independent of τ_i. Most of the work reported to date involves integration. The general form resulting from integration of an inverse power pair potential of attractive and repulsive terms over a structureless semi-infinite solid is

$$u_s(z) = \epsilon_s[n/(m-n)](m/n)^{m/(m-n)}[(z_0/z)^m - (z_0/z)^n], \tag{47}$$

where $u_s(z)$ has a minimum value $u_s(z^*) = -\epsilon_s$ and $u_s(z) = 0$ at $z = z_0$. If the equation which is integrated is the Lennard-Jones 6–12 pair potential, then the result is Eq. (47) with $m = 9$, $n = 3$.

This is shown briefly as follows (15). Suppose that the solid surface is the xy plane with z increasing away from the surface. Let ρ be the number of atoms of the solid per unit volume. The energy of interactions of a gaseous molecule above the solid at x, y, z with the atoms of the solid in $dx'\,dy'\,dz'$ at x', y', z' is

$$u_s(r)\rho \, dx' \, dy' \, dz'$$
$$r^2 = (x - x')^2 + (y - y')^2 + (z - z')^2, \tag{48}$$

where $u_s(r)$ is the Lennard-Jones 6–12 single pair potential given by Eq. (23). Integration over $-\infty < x' < +\infty$, $-\infty < y' < +\infty$, $-\infty < z' < 0$ gives

$$u_s(z) = (4\epsilon r_0^{12}\pi\rho/45z^9) - (2\epsilon r_0^6\pi\rho/3z^3). \tag{49}$$

Both r_0 and ϵ (single-pair value) may be replaced by z_0 and ϵ_s in Eq. (49) using the definitions of z_0 and ϵ_s following Eq. (47) to obtain

$$u_s(z) = (3\sqrt{3}/2)\epsilon_s[(z_0/z)^9 - (z_0/z)^3]. \tag{50}$$

When a potential of the form of Eq. (47) is substituted for $u_s(\mathbf{r}_1)$ in Eq. (41), we have

$$B_{AS}/\alpha z_0 = \int_0^\infty \{\exp[-u_s(z^0)/kT] - 1\}\, dz^0, \tag{51}$$

where z^0 is the reduced distance z/z_0 and α is the area resulting from integration over τ. Values of $B_{AS}/\alpha z_0$ may be plotted against ϵ_s/kT for various potential functions $u_s(z)$. In general, B_{AS} is not very sensitive to the form of the potential function. If experimental values of B_{AS} as a function of T measured according to Eq. (45) are fitted to theoretical curves of $B_{AS}/\alpha z_0$ versus ϵ_s/kT, values for ϵ_s/k and αz_0 are obtained. In Table 5.1, we give values of ϵ_s and αz_0 determined for various materials using the 9–3 potential function, Eq. (50).

TABLE ·5.1

PARAMETERS FOR GAS–SURFACE INTERACTIONS BASED ON A
9–3 POTENTIAL FUNCTION[a]

Gas–solid system	ϵ_s, cal/mole	αz_0, mm³/gm
H_2–porous glass (117 m²/gm)	1860	12.7
Ne–porous glass (117 m²/gm)	1300	11.4
Ar–porous glass (117 m²/gm)	3180	8.7
O_2–porous glass (117 m²/gm)	3460	6.7
N_2–porous glass (117 m²/gm)	3600	5.3
Ar–alumina (181 m²/gm)	2360	14
Kr–alumina (181 m²/gm)	3040	16
H_2–Saran charcoal (2080 m²/gm)	1440	201
Ne–Saran charcoal (2080 m²/gm)	1080	161
Ar–Saran charcoal (2080 m²/gm)	3100	166
N_2–Saran charcoal (2080 m²/gm)	3120	169
CH_4–Saran charcoal (2080 m²/gm)	3900	180
Ne–Saran charcoal (800 m²/gm)	1120	117
Ar–Saran charcoal (800 m²/gm)	3480	96

[a] See Steele (*4*).

If an independent method were available for determining z_0, then surface areas could be obtained from the experimental values of αz_0. Several attempts to calculate z_0 have been made, but they must be considered as very rough approximations (*6, 11, 12*). The gas–surface attractive potential at a distance

may be identified with the formula for the distant London forces of attraction
of two isolated systems,

$$u_s(r) = -C/r^6, \tag{52}$$

where C is given by the Kirkwood–Mueller formula [Eq. (56)]. Starting with
the Lennard-Jones 6–12 potential of Eq. (23), we may make the association

$$C = 4\epsilon r_0{}^6, \tag{53}$$

so that

$$u_s(r) = C[(r_0{}^6/r^{12}) - 1/r^6)]. \tag{54}$$

If we integrate Eq. (54) over a semi-infinite solid, we get an equation comparable to Eq. (49) in which ϵ and r_0 may be replaced by ϵ_s and z_0 just as in
Eq. (50). The net result is

$$\epsilon_s = (1/9\sqrt{3})(\pi C \rho / z_0{}^3), \tag{55}$$

where C is related to the atomic polarizability α and the diamagnetic susceptibility χ by the Kirkwood–Mueller formula

$$C = \frac{6mc^2\alpha_1\alpha_2}{\alpha_1/\chi_1 + \alpha_2/\chi_2}, \tag{56}$$

with m the mass of an electron and c the velocity of light. With values of ϵ_s
determined by curve fitting of experimental values of B_{AS} as a function of T
to theoretical plots of $B_{AS}/\alpha z_0$ versus ϵ_s/kT and C determined by Eq. (56),
values of z_0 may be calculated from Eq. (55). Table 5.2 gives the areas calculated by use of the Kirkwood–Mueller formula to determine C and the BET
areas. In the same table, values of ϵ_s determined from the best fit to theoretical
curves are compared to isosteric heats determined from adsorption isotherms
(8).

TABLE 5.2

AREAS AND ADSORPTION ENERGIES BASED ON THE 9–3 POTENTIAL
FUNCTION GAS–SOLID SYSTEM

Gas–solid system	ϵ_s, from best fit cal/mole	ϵ_s, from isotherms cal/mole	Area from K-M formula, m²/gm	BET area m²/gm
Helium on carbon black	536.5		235	337
Argon on carbon black	3818	3742 3855	162	337
Argon on P-33 graphitized carbon black	2714	2180	8.1	14
Krypton on alumina	3034	3051	105	181

To calculate C_{AAS}, both $u_s(z)$ and $u(r_{ij})$ are needed, and computations of this coefficient are less extensive. We shall not review them in detail. Freeman and Halsey (7) and Freeman (8) have used the following potential functions:

(1) ∞–3 for $u_s(z)$ and hard sphere for $u(r_{ij})$;
(2) 9–3 for $u_s(z)$ and hard sphere for $u(r_{ij})$;
(3) 9–3 for $u_s(z)$, a fixed value of $\epsilon_s/kT = 6$ and a (12–6) law for $u(r_{ij})$.

Only a few values of D_{AAAS} have been measured (3) and compared with the hard-sphere virial coefficient. Sams et al. (3) have related the virial coefficients for the imperfect two-dimensional gas of Section 5.2 to the gas–surface coefficients of this section. Using the Gibbs adsorption isotherm,

$$d\varphi = (N_s kT/\alpha) \, d \ln p,$$

they obtained the relationships

$$B_{2s}/\alpha = -C_{AAS}/2B_{AS}^2, \tag{57}$$

$$B_{3s}/\alpha^2 = -[(2D_{AAAS}/3B_{AS}^3) - (C_{AAS}^2/3B_{AS}^4)]. \tag{58}$$

The formal approach of this section is sufficiently general to treat low-coverage adsorption at any temperature. However, as the temperature is decreased, the probability of finding an adsorbed molecule near a potential minimum increases greatly. Steele (4) has shown how the equations of this section can be modified to represent a generalized two-dimensional gas constrained by a probability factor, which involves the effect of potential barriers, to motion in the two-dimensional plane of the solid surface. The equations developed are valid for arbitrary $u_s(\mathbf{r})$ and may be applied to localized as well as nonlocalized adsorption. The most frequent applications, however, have been to systems in which the τ dependence of $u_s(\mathbf{r})$ is negligible. Steele shows that when the local potential barriers are strong, the equations revert to those developed in Chapter III for immobile adsorption. Additional discussions of the nature of forces between adsorbing molecules and between adsorbed molecules and the surface will be found in Chapter VI.

5.4 High-Coverage Adsorption

The virial expansion can be applied to dilute adsorbed gases. As the density of an adsorbed gas increases, higher virial coefficients are required. These higher coefficients are difficult to compute, for the nth virial coefficient involves an n-body problem. Furthermore, the assumption of pairwise additivity of intermolecular potentials, which enters for virial coefficients concerned with three or more molecules, causes discrepancy with experiment. When condensation occurs, the virial expansion approach breaks down altogether as singularities appear in the function $p(\theta)$. Among the methods which have been employed at higher coverages are (1) approximate two-dimensional equations of state, (2) the Lennard-Jones–Devonshire (*16*) cell model, and (3) the Eyring "significant structures" theory (*17*). First we shall discuss the use of approximate two-dimensional equations of state.

The most commonly used equation of state is the two-dimensional analog of the van der Waals equation. By making certain simplifying assumptions, this equation may be derived by the methods of statistical mechanics (*18*). We assume that each molecule moves independently in a uniform potential field provided by the other molecules, all of which are distributed in a random manner. We further assume that the potential energy between a pair of adsorbed molecules is

$$u_s(r) = +\infty, \quad r < r^*; \quad u_s(r) = -\epsilon_{2s}(r^*/r)^6, \quad r \geqslant r^*. \quad (59)$$

Here we have used the "hard-sphere core" and the usual r^{-6} attractive term for large r; the minimum occurs at $u_s(r^*) = -\epsilon_{2s}$. The partition function will have the form $Q_s = q_s^{N_s}/N_s!$. The individual partition function q_s takes on a particular form as a result of intermolecular forces. First, because of the hard-sphere core, not all of the area is available for the motion of a molecule. We call the free area α_f. Second, a Boltzmann factor $\exp(-\Psi/2kT)$ is inserted to take care of the intermolecular potential field. The energy $\Psi(N_s/\alpha)$ is the potential energy of interaction of a molecule with all others in the system; and the factor of two in the exponent of the Boltzmann factor is introduced to prevent overcounting of the pairwise interactions, since each pairwise interaction is caused and shared by two molecules. The partition function is written

$$q_s = \frac{\alpha_f \exp(-\Psi/2kT)q_z \exp(-U_0 kT)}{\Lambda^2}, \quad \Lambda = \frac{h}{(2\pi mkT)^{1/2}}, \quad (60)$$

where q_z is the vibrational partition function of the adsorbed molecule in the z direction, perpendicular to the surface, and $\exp(-U_0/kT)$ is the Boltzmann factor for the adsorption potential. For simplicity we assume a monatomic adsorbate so that there are no rotations or internal vibrations. Explicit forms for α_f, and Ψ are easily developed. To obtain an expression for α_f, we note that a particular molecule wandering about the adsorption area has denied to it an area πr^{*2} for each other molecule. Again we divide by two because the excluded area arises from a pairwise interaction and only half of the effect belongs to each molecule. Thus, we write

$$\alpha_f = \alpha - N_s b_s, \quad b_s = \pi r^{*2}/2. \quad (61)$$

As the density increases, excluded areas will overlap, and the additivity of excluded areas, as implied by $N_s b_s$, will not hold. As an approximation, α_f will be used at all densities.

To arrive at an expression for Ψ, we note that in the neighborhood of a particular molecule the density of other molecules will be zero between $r = 0$, the location of the center of the specified molecule, and $r = r^*$, because of the hard core. Between $r = r^*$ and $r = \infty$, the density will be constant at N_s/α because of the assumption that the distribution is random. In the region $r + dr$ for $r \geq r^*$, there are $(N_s/\alpha) 2\pi r\, dr$ other adsorbed molecules, and the potential energy of interaction of each of these with the molecule at $r = 0$ is given by Eq. (59). Thus

$$\Psi = - \int_{r^*}^{\infty} \epsilon_{2s}(r^*/r)^6 (N_s/\alpha) 2\pi r\, dr = -2N_s a_s/\alpha, \qquad a_s = \pi \epsilon_{2s} r^{*2}/4. \quad (62)$$

We may now write the canonical partition function

$$Q_s = q_s^{N_s}/N_s! = \frac{(\alpha - N_s b_s)^{N_s} \exp(N_s^2 a_s/\alpha kT) q_z^{N_s} \exp(-N_s U_0/kT)}{\Lambda^{2N_s} N_s!} \quad (63)$$

or

$$-A_s/kT = \ln Q_s = N_s \ln[(\alpha - N_s b_s)/\Lambda^2] + (N_s^2 a_s/\alpha kT) + N_s \ln q_z$$
$$- (N_s U_0/kT) - N_s \ln N_s + N_s. \quad (64)$$

Taking the derivative with respect to α, we obtain

$$[\varphi + (N_s^2 a_s/\alpha^2)](\alpha - N_s b_s) = N_s kT, \quad (65)$$

which is the two-dimensional counterpart of the three-dimensional van der Waals equation

$$[p + (N_G^2 a_G/V^2)](V - N_G b_G) = N_G kT,$$
$$a_G = 2\pi \epsilon r^{*3}/3, \qquad b_G = 2\pi r^{*3}/3. \quad (66)$$

If we take the derivative of Eq. (64) with respect to N_s, and set it equal to $(\mu_G/kT) = (\mu^0/kT) + \ln p$, we obtain

$$-\mu/kT = \partial \ln Q_s/\partial N_s = \ln[(\alpha - N_s b_s)/\Lambda^2] + \ln q_z - [N_s b_s/(\alpha - N_s b_s)]$$
$$+ (2N_s a_s/\alpha kT) - (U_0/kT) - \ln N_s = -[(\mu^0/kT) + \ln p], \quad (67)$$

or

$$p = \left(\frac{2\pi mkT}{h^2}\right)^{1/2} \frac{kT}{q_z b_s} \exp\left(\frac{U_0}{kT}\right) \frac{\rho_s b_s}{1 - \rho_s b_s} \exp\left(\frac{\rho_s b_s}{1 - \rho_s b_s} - \frac{2a_s \rho_s}{kT}\right), \quad (68)$$

where we have used $\rho_s = N_s/\alpha$ and $(\mu^0/kT) = -\ln(kT/\Lambda^3)$. Equation (68) is the adsorption isotherm corresponding to the two-dimensional van der Waals equation of state. If the two-dimensional van der Waals equation (65) is expanded about $N_s/\alpha = \rho_s$, we get the virial expansion

$$\varphi/kT = \rho_s + [b_s - (a_s/kT)]\rho_s^2 + b_s^2\rho_s^3 + b_s^3\rho_s^4 + \cdots. \tag{69}$$

The second virial coefficient calculated from Eq. (19) using the potential given in Eq. (59) agrees in the high temperature limit [replace $\exp(-u_s/kT)$ by $(1 - u_s/kT)$ for $r \geqslant r^*$] with $B_{2s} = b_s - a_s/kT$ in Eq. (69). This comes about because the random molecular distribution assumed in calculating φ is approached at high temperatures and the correction in α_f is exact if the density is sufficiently low that only pair interactions need to be taken into account.

In order to determine the critical conditions, we take the first and second derivatives of φ with respect to α/N_s in Eq. (65), the two-dimensional equation of state, and set them equal to zero. We can then solve for

$$\begin{aligned}\alpha_c/N_s &= 3b_s, &\text{or}\quad (\rho_s)_c &= 1/3b_s, \\ T_{2c} &= 8a_s/27kb_s, &\varphi_c &= a_s/27b_s^2,\end{aligned} \tag{70}$$

or

$$a_s = (27/64)(k^2T_{2c}^2/\varphi_c), \qquad b_s = kT_{2c}/8\varphi_c \tag{71}$$

From the values of a_s and b_s in Eqs. (61) and (62) and the values of a_G and b_G in Eq. (66), we can relate the two-dimensional and three-dimensional constants

$$a_s = (\pi a_G/4b_G(3b_G/2\pi)^{2/3}, \qquad b_s = (\pi/2)(3b_G/2\pi)^{2/3}. \tag{72}$$

From Eqs. (71) and (72), we find the relations between the two- and three-dimensional critical constants.

$$T_{2c} = 0.5T_{3c}, \tag{73}$$

$$\alpha_c/N_s = 1.38(V_c/N_G)^{2/3}, \tag{74}$$

$$\varphi_c = 0.361p_c(V_c/N_G)^{1/3}. \tag{75}$$

The two-dimensional equation of state, Eq. (65), will have the familiar van der Waals loops associated with condensation of a two-dimensional gas to a two-dimensional liquid when $T < T_{2c}$. Similarly, the adsorption isotherm is expected to show a sudden jump in the amount adsorbed when $T < T_{2c}$. Experimental evidence for sudden jumps in adsorption isotherms is available from several sources (19–29). In Fig. 5.2, the data of Fisher and McMillan (25) for krypton on sodium bromide are shown.

Davis and Pierce (64) have demonstrated for uniform surfaces that first— and higher-layer jumps in adsorption isotherms may occur on the same substrate.

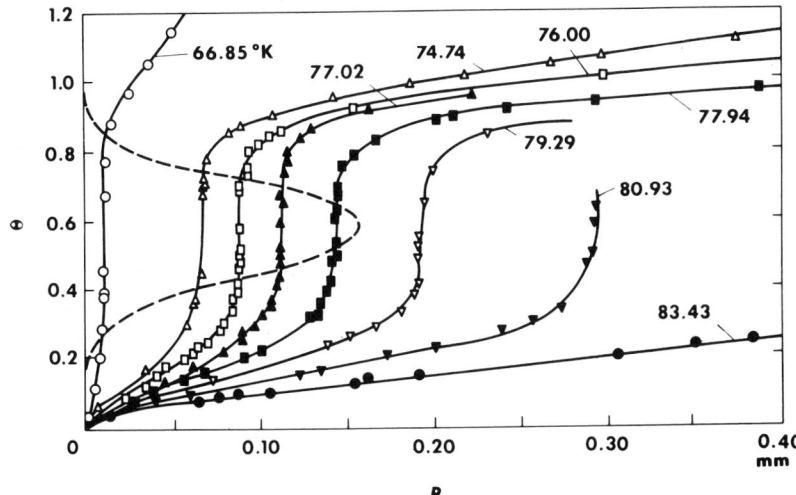

FIG. 5.2. Adsorption isotherms of krypton on sodium bromide (*25*). [After Fisher and McMillan, *J. Am. Chem. Soc.* **79**, 2969 (1957). By permission of The American Chemical Society.]

TABLE 5.3

VALUES OF T_{2c}/T_{3c} FOR VARIOUS SYSTEMS

Gas	Adsorbent	T_{2c}, °K	T_{3c}, °K	T_{2c}/T_{3c}	Reference
C_2H_6	NaCl	131	305.5	0.43	(*19–21, 28*)
CH_4	NaCl	90	191.1	0.47	(*21*)
B_2H_6	NaCl	133	289.7	0.46	(*22*)
Xe	NaCl	104	289.7	0.36	(*21*)
Kr	NaBr	78	209.4	0.37	(*25*)
$n\text{-}C_7H_{16}$	Ag	⟨286–298⟩	540.0	⟨0.53–0.55⟩	(*27*)
$n\text{-}C_7H_{16}$	Fe_2O_3	302	540.0	0.56	(*27*)
$n\text{-}C_7H_{16}$	Graphite	302	540.0	0.56	(*27*)

In Table 5.3, two-dimensional and three-dimensional critical temperatures T_{2c} and T_{3c}, are compared for various systems. It will be observed the ratios, T_{2c}/T_{3c}, correspond roughly to the theoretical value 0.5 derived by Hill (*30*) from the van der Waals equation of state, Eq. (73). The other critical constants of Eqs. (74) and (75) are not in good agreement. The agreement with theory for the last three rows of data of Table 5.3 on *n*-heptane is better than we would anticipate in view of the complicated shape of the heptane molecule.

Another approach to the high-density, mobile, adsorbed state with lateral interactions is the Lennard-Jones–Devonshire theory (*31*) for the three-dimensional condensation problem which was adapted by Devonshire (*16*) to the two-dimensional case. We shall give only a brief sketch of the theory.

The Lennard-Jones–Devonshire theory belongs to the category of cell theories of liquids. In general in these theories, the volume V is divided into a lattice of N cells, one molecule in each cell. In the cell model of a crystal, each molecule is assumed to be confined to its cell in the potential field of its neighbors, and its motion is assumed to be independent of the motion of other molecules. In the other extreme of a cell model—a dilute gas—it is presumed that a molecule of the gas wanders over the entire volume V. The additional motion of the gas molecules over those of a crystal gives an additional entropy $N_s k$ which is called the "communal entropy." In the case of a cell model of a liquid, the ability of a molecule to escape from its cell is considered to be intermediate between the two extreme cases. The exact picture of the communal entropy of a liquid is still obscure. In the Lennard-Jones–Devonshire model, each molecule moves in its cell in the potential field of the neighboring molecules which are assumed fixed at the centers of their cells. The simplifying assumption is made that the c (usually taken as 12) nearest-neighbor molecules are smeared over the surface of a sphere of radius a, where a is the distance between centers of the nearest-neighbor cells. The motion of the central molecule is determined by its interaction with the smeared shell of its nearest-neighbors, the interaction energy being a function of the distance of the central molecule from the center of its cell. A term is included in the partition function to take care of the possibility that a molecule may eventually escape from its cell. Devonshire (16) has worked out the two-dimensional analog of this theory and has calculated critical constants. He found the following relationships between two- and three-dimensional critical constants:

$$T_{2c} = 0.53 \ T_{3c}, \tag{76}$$

$$\alpha_c/N_s = 1.03 \ (V_c/N_G)^{2/3}, \tag{77}$$

$$\varphi_c = 0.79 \ p_c(V_c/N_G)^{1/3}, \tag{78}$$

which may be compared to those obtained from the van der Waals theory, Eqs. (73)–(75). Observe that the critical temperature ratio T_{2c}/T_{3c} shows good agreement between the two theories. The other two critical constants are in poor agreement. Since we have shown in Table 5.3 that a ratio of $T_{2c}/T_{3c} = 0.5$ is in reasonable agreement with experiment, we may assume that 0.5 is a true value. No firm conclusions can be drawn at this time about the other constants.

Application of Eyring's significant structure theory to a mobile adsorbed phase has been developed by McAlpin and Pierotti (17). The adsorbed phase is described as a two-dimensional fluid on an energetically homogeneous solid surface. The adsorbed molecules are assumed to vibrate normal to the surface. Interaction with the solid is presumed to fall off rapidly enough so that only a monolayer exists at low temperatures. As in the case of a three-dimensional liquid, three significant structures are proposed: (1) a filled lattice-like structure in which each molecule is restrained by its neighbors to occupy an equilibrium site like a molecule in a crystal lattice, (2) a structure with vacancies which therefore has configurational degeneracy, and (3) a structure with sufficient vacancies that some molecules may move over a portion of the surface like gas molecules. Each of these structures contributes to the overall partition function in proportion to its pressure under a particular condition. Each adsorbed molecule has definite probabilities of being in each of these structures, and the probabilities are dependent on the mole fraction of vacancies. When the number of vacancies approaches zero, the contribution is entirely from the filled lattice-like structure. When the mole fraction of vacancies approaches unity, the gas-like structure will represent the total contribution. The filled lattice-like structure is described by the partition function of a molecule in a two-dimensional Einstein crystal. In their second paper (*17*), McAlpin and Pierotti replaced the Einstein partition function by the two-dimensional cell model of Devonshire, while retaining the general features of the significant structure theory. This eliminated difficulties in evaluating Einstein characteristic temperatures. They derived equations for the isotherm, isosteric heat, and critical properties, and obtained good agreement for the systems of argon and krypton on graphite. Because of the lack of mathematical rigor in the foundations of the significant structure theory, both good correlations and anomalies are difficult to interpret physically.

5.5 Heterogeneous Surfaces

When heterogeneity of adsorption sites is introduced, two distinct cases must be considered: (1) sites of each energy are grouped in patches, (2) sites of different energy are distributed randomly on the surface. In the first case, the patches are assumed sufficiently large that they can be considered to act as mutually independent bits of surface area. If the concentrations of adsorbed molecules are large enough that lateral molecular interactions must be taken into account, then the patches must be large compared to the range of these interactions. At equilibrium, each patch will have the same chemical potential, and the surface concentrations and spreading pressures to achieve this equality will be different. Application of the Gibbs equation to the experimental isotherm will yield an average spreading pressure. In the second case, where sites of different energy are arranged randomly, we can no longer consider the surface as a collection of independent phases. The concentration of mobile, adsorbed molecules will fluctuate rapidly as the molecules move from point to point on the surface.

Tompkins (*32*) has considered these two cases for both dilute and concentrated adsorbed phases. For a dilute, adsorbed phase on a surface with sites of each energy grouped in independent patches, we write the two-dimensional canonical partition function of the ith patch

$$Q_{si} = q_{si}^{N_{si}}/N_{si}! = (1/N_{si}!)[(2\pi mkT/h^2)\alpha_i]^{N_{si}}q_{zi}^{N_{si}} \exp[-(N_{si}U_{0i})/kT], \quad (79)$$

where α_i is the area of the ith patch, q_{zi} is the molecular partition function for vibration normal to the surface, and U_{0i} is the potential energy of adsorption of a molecule. For a polyatomic adsorbate, rotational and internal vibrational terms must be included. Taking the derivative of $\ln Q_{si}$ with respect to N_{si}, we find the chemical potential, μ_{si},

$$\mu_{si}/kT = \ln(N_{si}/q_{si}) = \mu_{1s}/kT = \mu_{2s}/kT = \cdots = \mu_G/kT = \mu^0/kT + \ln p,$$

$$N_{si} = q_{si}\exp(\mu^0/kT)\,p, \qquad\qquad\qquad\qquad\qquad (80)$$

and summing over i and dividing by the total area α,

$$N_s/\alpha = \rho_s = \frac{\sum q_{si}\exp(\mu^0/kT)p}{\alpha}. \qquad\qquad (81)$$

Since q_{si} is independent of p, we have a linear isotherm as though the surface were uniform [see Eq. (4)].

With the sites of different energy assumed to be randomly distributed, we must consider the surface as a whole rather than as a collection of independent patches. The same result, Eq. (81), is obtained for the isotherm except that each molecule is now assumed to have access to the entire surface area α instead of being restricted to α_i. Thus we come up against the same question discussed in Section 5.4, namely, the area available to the molecule for use in the partition function. In either case—random or patches—it is most likely at low coverages that the molecule has available to it the entire surface area. At any rate, in both cases, the isotherm turns out to be linear just as it does in the case of energetically homogeneous surfaces. The question of area does not arise for immobile adsorption in the absence of interactions.

Tompkins (32) also developed the case of a mobile layer at high density for patchwise and random distribution of site energies. For the patchwise distribution the canonical partition function of the ith patch is

$$Q_{si} = q_{si}^{N_{si}}/N_{si}! = 1/N_{si}![(2\pi mkT/h^2)\alpha_{fi}]^{N_{si}}q_z^{N_{si}} \exp[-(N_{si}U_{0i})/kT],$$

$$\ln Q_s = \sum_i \ln Q_{si}, \qquad\qquad\qquad\qquad\qquad\qquad (82)$$

where α_{fi} is the free area available to a molecule on the ith patch. Lateral interactions are not included. As a crude approximation, α_{fi} is taken as

$$\alpha_{fi} = \alpha_i - N_{si}b_s, \qquad\qquad\qquad\qquad (83)$$

which is the expression used in the van der Waals approximation of the previous section, Eq. (61). Taking the derivative of $\ln Q_{si}$ in Eq. (82) with respect to N_{si}, we obtain the chemical potential μ_i, which we set equal to $\mu_G = \mu^0 + kT$ ln p. The isotherm equation thus obtained is

$$p/p_{0i} = \rho_{si}/(1 - \rho_{si} b_s) \exp[\rho_{si} b_s/(1 - \rho_{si} b_s)],$$

$$p_{0i} = \frac{(2\pi mkT)^{1/2} kT \exp(U_{0i}/kT)}{hq_{zi}} = p_0 \exp\frac{U_{0i}}{kT}, \tag{84}$$

$$\rho_{si} = N_{si}/\alpha_i,$$

which cannot be solved analytically for ρ_{si}, only graphically or numerically. At low pressures, $\rho_{si} \to 0$, we have the simplified equation

$$p/p_{0i} = \rho_{si}/(1 - \rho_{si} b_s),$$

$$\rho_{si} = p/(p_{0i} + pb_s). \tag{85}$$

Defining $\theta_i = N_{si}/B_i$, where B_i is the number of molecules on the ith patch in a completed monolayer, we find that $\rho_{si} = \theta_i/\alpha_s$ where $\alpha_s = \alpha_i/B_i$ is the cross-sectional area of an adsorbed molecule and is assumed independent of i. Also from Section 5.4, Eq. (61), $b_s = \pi r^{*2}/2$, where in the hard-sphere approximation, r^* is twice the radius of an adsorbed molecule, so that $b_s = 2\alpha_s$ (since total area is usually determined by adsorption). With these substitutions, Eq. (85) becomes

$$\theta_i = \frac{\alpha_s p}{p_{0i} + 2\alpha_s p} = \frac{p}{(p_{0i}/\alpha_s) + 2p}. \tag{86}$$

The factor of 2 before the pressure in Eq. (86) is the result of the assumption $b_s = 2\alpha_s$ which is true for dilute layers. As explained in Section 5.4, the factor is no longer correct at higher densities because of overlapping. A factor of $b_s = \alpha_s$ is then more nearly correct. The general arguments of this section are not affected by this fact.

Now the fraction of the total surface covered, N_s/B, is

$$N_s/B = \theta = (1/B) \sum N_{si} = (1/B) \sum B_i \theta_i. \tag{87}$$

To replace summation by integration, we proceed as in Chapter II following Eq. (101) by letting $Bf(U_0) \, dU_0$ be the number of molecules in a completed monolayer with the range of adsorption energies U_0 to $U_0 + dU_0$. Thus, at low coverage,

$$\theta = \int \frac{p}{(p_0/\alpha_s) \exp(U_0/kT) + 2p} f(U_0) \, dU_0, \tag{88}$$

where we assume that $p_0(q_{zi})$ from Eq. (84) is independent of U_0. If the distribution function $f(U_0)$ is assumed to be exponential,

$$f(U_0) = (1/B)[dB(U_0)/dU_0] = a_0 \exp(-a_1 U_0),$$

where $B(U_0)$ is the number of molecules in a completed monolayer with adsorption energies between 0 and U_0, then it is easily shown that the Freundlich isotherm is obtained. In Chapter II, we showed that the Freundlich isotherm was also obtained for immobile adsorption with an exponential distribution of adsorption energies. This is a reflection of the fact that at low coverages, the partition functions for the two models are approximately the same.

Tompkins (32) found that the general isotherm, Eq. (84) (in terms of θ_i),

$$\frac{p}{p_{0i}/\alpha_s} = \frac{\theta_i}{1 - 2\theta_i} \exp \frac{2\theta_i}{1 - 2\theta_i} \tag{89}$$

is in good agreement with the equation

$$\frac{p}{p_{0i}/\alpha_s} = \frac{1 \cdot 3\theta_i}{1 - 3\theta_i} \tag{90}$$

except at very low and very high values of θ_i. Using an exponential distribution of adsorption site energies in Eq. (88) along with Eq. (90), we arrive again at a Freundlich isotherm after integration.

In the corresponding case for random distribution of sites of different energies, Tompkins (32) uses the same partition function, $\ln Q_s = \sum_i \ln Q_{si}$, Eq. (82), with α_{fi}, the free area available to an adsorbed molecule, replaced by

$$\alpha_f = \alpha - \sum_i N_{si} b_s, \tag{91}$$

so that a molecule has access to the entire area except for the area $\sum_i N_{si} b_s$ covered by molecules already adsorbed. Proceeding as before, we obtain the isotherm

$$p/p_0' = [\theta_i/a_i(1 - 2\theta)] \exp[2\theta_i/(1 - 2\theta)],$$

$$\theta_i = N_{si}/B, \qquad a_i = (2\pi mkT/h^2)q_{zi} \exp(-U_{0i}/kT),$$

$$p_0' = (2\pi mkT)^{3/2} kT/h^3 \alpha_s, \qquad \alpha_s = \alpha/B, \tag{92}$$

$$\theta = \sum \theta_i = N_s/B.$$

Note that θ_i is defined as the fraction of the *total* surface in this case. At low pressures, θ_i, $\theta \to 0$, we have

$$p/p_0' = \theta_i/a_i(1 - 2\theta), \tag{93}$$

from which we find

$$\theta_1/a_1 = \theta_2/a_2 = \cdots, \tag{94}$$

or

$$\theta = \sum_i \theta_i = (\theta_1/a_1)(a_1 + a_2 + \cdots) = (p/p_0')(1 - 2\theta)(\sum_i a_i).$$

Since $\sum_i a_i$ is a function of T but not of p, we have at constant T,

$$p/p_0{}' = k\theta/(1 - 2\theta), \qquad k = \text{constant}. \tag{95}$$

Equation (95) is of the Langmuir type and is independent of the distribution function. Unfortunately, the case for random topographical distribution with an exponential distribution of energies of sites is not available for comparison with the patchwise model.

If the energy distribution of a surface is not continuous, vertical steps in the isotherms would be expected, representing phase transitions. Isotherms with more than one vertical step have actually been reported many times. The steps have been attributed to surface nonuniformities and to a multiplicity of polymorphic transitions in the adsorbed phase. Studies by Ross and Boyd (*33*) and Ross and Winkler (*34*) have revealed different phase transitions for ethane on different crystal faces of sodium chloride and calcium fluoride, thus strongly supporting surface heterogeneity as a cause of multistep isotherms.

Ross and Oliver (*35, 36*) have done considerable work on adsorption isotherms for mobile adsorption on heterogeneous surfaces. Considering that sites are distributed in patches of equal energies, they used an adsorption equation of the general form

$$\theta = \int \varphi(U_0)\psi(p, U_0)\, dU_0, \tag{96}$$

where $\varphi(U_0)$ is the distribution function for adsorption energies and $\psi(p, U_0)$ is an adsorption isotherm. For $\varphi(U_0)$ they employed a number of distributions including the Gaussian function, and for $\psi(p, U_0)$ the adsorption isotherm associated with the two-dimensional van der Waals equation of state [Eq. (68)]. Such models were computed numerically. It was found that experimental isotherms of argon and nitrogen on various carbon blacks, a synthetic zeolite, etc. agreed well over a temperature range with models employing a Gaussian distribution function. Harris (*65–67*) discusses in detail the errors involved in determining the distribution function $\varphi(U_0)$ by approximating the "correct" adsorption isotherm $\psi(p, U_0)$ by a step-function.

5.6 Multilayer Adsorption

In Chapter II, Section 3, the BET model of multilayer adsorption was discussed. In that model, adsorbed molecules are arranged in immobile, vertical piles with vertical interactions, but not lateral interactions, allowed. A variant of this is the Hüttig (*37*) model in which lateral interactions, but not vertical interactions, are allowed (*38*). Abandoning the lattice theory approach, Wheeler (*39*), Ono (*40*), and later Freeman and Halsey (*7*) employed a more general approach which treats physical adsorption systems as examples of fluids in the external force field of the surface of an adsorbent. The results of Wheeler, Ono, and Freeman and Halsey are expressed essentially in terms of the cluster integral expansions of Ursell (41) and Mayer (42), as in Sections 5.2 and 5.3. They are of little practical use at high densities because of the difficulty in calculating the higher terms of the expansions. Several subsequent papers (43–48) have advanced the formalism of the theory, using the method of distribution functions. Distribution functions give the probabilities of observing the different configurations of sets of n molecules out of the total number N, and are affected by external force fields. Steele and Ross (43), for example, have given formal equations for the number density distribution function that apply at all densities and from which, in principle, adsorption isotherms can be calculated. The formal approach initiated by Wheeler and Ono constitutes the most rigorous approach to the problem of physical adsorption, though numerical calculations based on the formal equations cannot yet be carried out. The newer theories, employing distribution functions, are based on the theories of uniform liquids of Kirkwood (49), Mayer (42), and Born and Green (50). Presumably the liquid-state theories are rigorous, but numerical results can be obtained only with the use of approximations. The difficulties encountered in dealing with uniform fluids are greatly magnified when an external force field is introduced, as in the theory of physical adsorption. The current theories set up formal equations for multilayer adsorption on plane surfaces possessing a uniform potential. Pairwise interactions between gas molecules in all possible configurations and also the pairwise interactions of gas molecules with the solid as a function of distance z from the surface of the solid are taken into account in the configurational integral. Since the severe mathematical difficulties of the rigorous formal approach have not yet been solved, approximate theories play an important role at present.

The most important approximate theory is that due to Frenkel (51), Halsey (52), Hill (53), and McMillan and Teller (54) [see Eq. (65), Chapter II], who considered the problem of a spherically symmetrical molecule adsorbed on a nonporous, nonpolar adsorbent. They limit the theory approximately to $\theta > 2$, for in the third and higher layers the effects of detailed surface structure are smoothed out, and the molecular environment may be considered to be essentially that of the bulk liquid. The model bears some resemblance to that employed by Polanyi (55) in his potential theory. We consider the development given by Hill (53).

Let the system consist of a slab of liquid of uniform thickness h adsorbed on the plane surface of a semi-infinite ($-\infty < x, y < \infty, z \leqslant 0$) solid of density ρ located in the region $z \gtrless 0$. The adsorbed film is assumed to have the same uniform density as bulk liquid, $\rho_L = \Gamma/h$, where Γ is the number of molecules adsorbed per square centimeter. This system will be compared to a system in which the solid is replaced by a semi-infinite liquid at $z \gtrless 0$. In view of what was said above about the molecular environment at $\theta < 2$, it is assumed that the adsorption behavior is determined predominantly by the potential energy field in the slab when it is adsorbed on the semi-infinite solid compared with the field when it is supported on the semi-infinite bulk liquid. The difference in chemical potential of the slab in the two situations is

$$\mu_s - \mu_L = kT \ln p/p_L = -kT \partial \ln(Q_s - Q_L)/\partial N_s = \partial(A_s - A_L)/\partial N_s. \quad (97)$$

The subscript "L" refers to properties of the slab when it is supported on, and is assumed to have the same properties as, the bulk liquid, and p_L is the vapor pressure of bulk liquid. The subscript "s" refers to properties of the slab when it is adsorbed on the solid, and p is the vapor pressure in equilibrium with the adsorbed liquid. The only contribution to the Helmholtz free energy, as mentioned above, is assumed to be the potential energy,

$$A_s - A_L = U_s - U_L,$$

where U_s and U_L are the potential energies of N_s molecules in the slab and in bulk liquid. Entropy and kinetic energy terms of A_s and A_L are assumed equal to those of bulk liquid and thus cancel. We may write

$$-kT \ln (p/p_L) = \partial U_s/\partial N_s - U_L/N_s \quad (98)$$

The energy $(\partial U_s/\partial N_s)$ is the change dU_s in U_s due solely to the adsorption of dN_s additional molecules, for the density of the slab is assumed to be uniform and constant, independent of h. It is therefore the sum of the energies of interaction of a liquid molecule at $z = h$ with

(1) all the molecules in the liquid slab $0 \leqslant z \leqslant h$,

(potential energy $= u_{slab}$);

(2) all the molecules of the adsorbent $z \gtrless 0$.

[potential energy $= u_s(h)$].

The energy U_L/N_s can be considered to be the sum of the energies of interaction of a liquid molecule at $z = h$ with

(1) all the molecules in the liquid slab $0 \leqslant z \leqslant h$,

(potential energy $= u_{slab}$);

(2) all the molecules of the semi-infinite liquid, $z \gtrless 0$.

[potential energy $= u_L(h)$].

The use of the semi-infinite liquid instead of the infinite eliminates the double counting of each interaction. Thus we may write Eq. (98)

$$kT \ln (p/p_L) = u_s(h) - u_L(h). \quad (99)$$

Now $u_s(h)$, the interaction energy of a molecule at $z = h$ with all the molecules of the adsorbent, $z \leqslant 0$, may be written as the integrated Lennard-Jones
potential given in Eq. (49). By replacing ϵ, ρ, and r_0 in Eq. (49) by ϵ_L, ρ_L,
and r_{0L}, we obtain a corresponding expression for $u_L(h)$. The short range
repulsive term in Eq. (49) may be dropped since we are considering only $\theta > 2$.
Eq. (99) becomes

$$kT \ln(p/p_L) = -2[(\epsilon r_0{}^6 \pi \rho/3h^3) - (\epsilon_L r_{0L}^6 \pi \rho_L/3h^3)], \tag{100}$$

or substituting $h = \Gamma/\rho_L$, we obtain

$$\ln (p/p_L) = -\kappa/\Gamma^3, \tag{101}$$

$$\kappa = 2[(\epsilon r_0{}^6 \pi \rho \rho_L{}^3/3kT) - (\epsilon_L r_{0L}^6 \pi \rho_L{}^4/3kT)]. \tag{102}$$

From Eq. (23) and Fig. 5.1, we see that the substitution $r_0 = 2^{-1/6} r^*$ can be
made, where, in the single-pair Lennard-Jones potential, $u_s(r_0) = 0$ and
$u_s(r^*) = -\epsilon_{2s}$.

Steele and Ross (43) showed that the formal approach using distribution
functions leads to Eq. (101) when certain simplifications are introduced.
Halsey (52) deduced the isotherm

$$\ln (p/p_L) = -\kappa/\Gamma^s \tag{103}$$

by a semiquantitative approach. The constants κ and s are not given in terms
of molecular properties. Halsey (52) found excellent agreement with Eq. (103)
using $s = 2.67$ for nitrogen on anatase from the data of Jura and Harkins (56).
In general, however, wide variations in s have been found even for spherically
symmetrical molecules. For example, Halsey (52), working with the data of
Boyd and Livingston (57), shows that s has a value of about 6 for adsorption
of propyl alcohol on anatase, and at least 15 for the same vapor on $BaSO_4$,
while for the adsorption of water on graphite the value is about unity. Jura
and Harkins (58) have proposed an empirical equation of the form

$$\ln (p/p_L) = B - A/\Gamma^2, \tag{104}$$

which has fairly wide applicability suggesting that $s = 2$ is characteristic of
quite a few systems.

It is not surprising that the simple form of Eq. (101) has limited applicability. At the expense of increasing complexity, refinements have been proposed. Barrer and Robins (59), have employed the van der Waals equation of
state for the gas phase and adsorbed film, treating them as one fluid. The
density of the fluid decreases with increasing distance from the surface and
the function of distance can be calculated. The equations were solved graphically for specific van der Waals constants and for various values of the
energy of interaction of the first layer with the surface. Further refinements
have been introduced by Hill (60–63), but the complexity of the calculations

limit their usefulness. Kim and Oh (*68*) have used a model comprising a localized first layer and nonlocalized higher layers. Barrer and Lee (*69, 70*) showed that the sorption isotherms of paraffinic hydrocarbons on zeolites could not be described over the complete range in degree of filling of the intracrystalline pores in terms of any single model isotherm. Localized sorption equations applied over considerable ranges, but in general an isotherm based on an osmotic virial equation of state was more successful. These refinements represent further steps in the direction of the Wheeler–Ono approach.

REFERENCES

1. Hill, T. L., "Introduction to Statistical Thermodynamics," Chapter 15-2. Addison-Wesley, Reading, Massachusetts, 1960.
2. Mayer, J. E., and Mayer, M. G., "Statistical Mechanics." Wiley, New York, 1940.
3. Sams, J. R., Jr., Constabaris, G., and Halsey, G. D., Jr., *J. Chem. Phys.* **36**, 1334 (1962).
4. Steele, W. A., "The Solid-Gas Interface" (E. A. Flood, ed.), Vol. 1, Chapter 10. Dekker, New York, 1967.
5. Steele, W A., and Halsey, G. D., Jr., *J. Chem. Phys.* **22**, 979 (1954).
6. Steele, W. A., and Halsey, G. D., Jr., *J., Phys. Chem.* **59**, 57 (1955).
7. Freeman, M. P., and Halsey, G. D., Jr., *J. Phys. Chem.* **59**, 181 (1955).
8. Freeman, M. P., *J. Phys. Chem.* **62**, 723, 729 (1958).
9. Kwan, T., Freeman, M. P., and Halsey, G. D., Jr., *J. Phys. Chem.* **59**, 600 (1955).
10. Constabaris, G., and Halsey, G. D., Jr., *J. Chem. Phys.* **27**, 1433 (1957).
11. Sams, J. R., Constabaris, G., and Halsey, G. D., Jr., *J. Phys. Chem.* **64**, 1689 (1960).
12. Constabaris, G., Sams, J. R., Jr., and Halsey, G. D., Jr., *J. Phys. Chem.* **65**, 367 (1961).
13. Sams, J. R., Jr., *J. Chem. Phys.* **37**, 1883 (1962).
14. Sams, J. R., Jr., and Yaris, R., *J. Phys. Chem.* **67**, 1931 (1963).
15. Hill, T. L., *J. Chem. Phys.* **16**, 181 (1948).
16. Devonshire, A. F., *Proc. Roy. Soc. (London), Ser. A* **163**, 132 (1937).
17. McAlpin, J. J., and Pierotti, R. A., *J. Chem. Phys.* **41**, 68 (1964); **42**, 1842 (1955).
18. Hill, T. L., "Introduction to Statistical Thermodynamics," Chapter 16-1. Addison-Wesley, Reading, Massachusetts, 1960.
19. Clark, H., and Ross, S., *J. Am. Chem. Soc.* **75**, 6081 (1953).
20. Ross, S., and Winkler, W., *J. Am. Chem. Soc.* **76**, 2637 (1954).
21. Ross, S., and Clark, H., *J. Am. Chem. Soc.* **76**, 4291 (1954).
22. Ross, S., and Clark, H., *J. Am. Chem. Soc.* **76**, 4297 (1954).
23. Ross, S., and Winkler, W., *J. Colloid Sci.* **10**, 330 (1955).
24. Clark, H., *J. Phys. Chem.* **59**, 1068 (1955).
25. Fisher, B. B., and McMillan, W. G., *J. Am. Chem. Soc.* **79**, 2969 (1957).
26. Jura, G., Loeser, E. H., Basford, P. R., and Harkins, W. D., *J. Chem. Phys.* **13**, 535 (1945).
27. Jura, G., Loeser, E. H., Basford, P. R., and Harkins, W. D., *J. Chem. Phys.* **14**, 117 (1946).
28. Jura, G., Harkins, W. D., and Loeser, E. H., *J. Chem. Phys.* **14**, 344 (1946).
29. Ross, S., and Boyd, G. E., New Observations on Two-Dimensional Condensation Phenomena. MDDC Rept. No. 864 (1947).

30. Hill, T. L., *J. Chem. Phys.* **14**, 441 (1946).
31. Lennard-Jones, J. E., and Devonshire, A. F., *Proc. Roy. Soc. (London), Ser. A* **163**, 53 (1937); **165**, 1 (1938).
32. Tompkins, F. C., *Trans. Faraday Soc.* **46**, 569 (1950).
33. Ross, J., and Boyd, G. E., New Observations on Two-Dimensional Condensation Phenomena, MDDC Rept. No. 864 (1947).
34. Ross, S., and Winkler, W., *J. Am. Soc. Chem.* **76**, 2637 (1954).
35. Ross, S., and Olivier, J. P., "On Physical Adsorption," Chapter 4, 5. Wiley (Interscience), New York, 1964.
36. Ross, S., and Olivier, J. P., *J. Phys, Chem.* **65**, 608 (1961).
37. Hüttig, G. F., *Monatsh.* **78**, 177 (1948).
38. Barrer, R. M., *J. Chem. Soc.* 1874 (1951).
39. Wheeler, A., Am. Chem. Soc. Natl. Meeting, Atlantic City, New Jersey, September (1949).
40. Ono, S., *J. Chem. Phys.* **18**, 397 (1950).
41. Ursell, H. D., *Proc. Cambridge Phil Soc.* **23**, 685, (1927).
42. Mayer, J. E., *J. Chem. Phys.* **5**, 67 (1937).
43. Steele, W. A., and Ross, M., *J. Chem. Phys.* **33**, 464 (1960).
44. Hill, T. L., and Saito, N., *J. Chem. Phys.* **34**, 1543 (1961).
45. Lebowitz, J. L., and Percus, J. K., *J. Math. Phys.* **4**, 116 (1963).
46. Stillinger, F. H., and Buff, F. P., *J. Chem. Phys.* **37**, 1 (1962).
47. Hart, E. W., *Phys. Rev.* **113**, 412 (1959).
48. Lebowitz, J. L., and Percus, J. K., *Phys. Rev.* **122**, 1675 (1961).
49. Kirkwood, J. G., *J. Chem. Phys.* **3**, 300 (1935).
50. Born, M., and Green, H. S., *Proc. Roy. Soc. (London), Ser. A* **188**, 10 (1946); **189**, 103 (1947).
51. Frenkel, J., "Kinetic Theory of Liquids." Oxford Univ. Press, London and New York, 1946.
52. Halsey, G., *J. Chem. Phys.* **16**, 1931 (1948).
53. Hill, T. L., *J. Chem. Phys.* **17**, 590, 668 (1949).
54. McMillan, W. G., and Teller, E., *J. Chem. Phys.* **19**, 25 (1951); *J. Phys. Colloid Chem.* **55**, 17 (1951).
55. Polanyi, M., *Verhandl. Deut. Physik. Ges.* **15**, 55 (1916).
56. Jura, G., and Harkins, W. D., *J. Am. Chem. Soc.* **66**, 1356 (1944).
57. Boyd, G. E., and Livingston, H. K., *J. Am. Chem. Soc.* **64**, 2383 (1942).
58. Jura, G., and Harkins, W. D., *J. Chem. Phys.* **11**, 430 (1943).
59. Barrer, R. M., and Robins, A. B., *Trans. Faraday Soc.* **47**, 773 (1951).
60. Hill, T. L., *J. Chem. Phys.* **19**, 261 (1951).
61. Hill, T. L., *J. Chem. Phys.* **19**, 1203 (1951).
62. Hill, T. L., *J. Chem. Phys.* **20**, 141 (1952).
63. Hill, T. L., *J. Phys. Chem.* **56**, 526 (1952).
64. Davis, B. W., and Pierce, C., *J. Phys. Chem.* **70**, 1051 (1966).
65. Harris, L. B., *Surface Sci.* **10**, 129 (1968).
66. Harris, L. B., *Surface Sci.* **13**, 377 (1969).
67. Harris, L. B., *Surface Sci.* **15**, 182 (1969).
68. Kim, S. K., and Oh, B. K., *Thin Solid Films*, **2**, 445 (1968).
69. Barrer, R. M., and Lee, J. A., *Surface Sci.* **12**, 341 (1968).
70. Barrer, R. M., and Lee, J. A., *Surface Sci.* **12**, 354 (1968).

CHAPTER

VI

PHYSICAL FORCES OF ADSORPTION

6.1 The Forces of Adsorption

In this chapter and the two following ones, we shall look more closely at
the nature of the forces acting in the adsorbed state. Forces exist between
molecules and the solid surface and also between the adsorbed molecules
themselves. All intermolecular forces have their origin in the electromagnetic
interactions of nuclei and electrons. The distribution of these forces can be
determined, in principle, from their mutual interactions and the procedures
of quantum mechanics, leading to a complete knowledge of the quantum-
mechanical states of the system. Such procedures are, however, too complex
for application even to the simplest of adsorption systems. Approximations
and simplifications must be introduced. One simplification is to classify forces
into those associated with physical adsorption and with chemisorption. Physical adsorption results from the action of van der Waals forces which are
considered to be made up of London dispersion forces and classical electrostatic forces. With these forces, there is no transfer or sharing of electrons between a molecule and the solid. As a molecule approaches the solid, electrons
may take up a new equilibrium distribution, but they maintain their respective
associations in the interacting species. Chemisorption arises from the transfer
or sharing of electrons between the adsorbate and adsorbent. The chemisorptive bond has all the characteristics of a chemical bond. Physical adsorption

is usually weaker than chemisorption and experimentally determined heats of adsorption are often used to distinguish between the two types of adsorption. But there are systems in which the distinction is not clear.

In this chapter, we consider physical adsorption. We shall discuss (1) the London dispersion forces and repulsive forces between a single pair of molecules, (2) the application of these forces to the case of a molecule interacting with a surface, (3) the electrostatic interactions between a molecule and a surface, and (4) the interactions between two molecules under the influence of a solid surface. Although in principle these forces act simultaneously, it is convenient to treat them separately. In some systems, all but one or two forces may be ignored. Dispersion and repulsive forces are probably always present, and for nonpolar molecules interacting with covalent or metallic solids they are the predominant forces. If the gas molecule has a permanent dipole moment and/or higher moments, an attracting charge distribution will be induced in the solid. Such interactions are significant only when polar molecules interact with metals. If the solid possesses an external field additional attractive interactions with the gas molecule will be set up. Strong interactions of this type occur only in the case of ionic crystals.

6.2   Dispersion and Repulsive Forces between Single Molecules

In Eq. (23) of Chapter V, dispersion and repulsive forces between two molecules were introduced. We now wish to pursue the subject further. In 1930, London (1, 2), was the first to recognize the existence of dispersion forces between pairs of atoms or molecules. Until then, there was no satisfactory explanation of the forces which were known to exist between nonionic, nonpolar molecules. According to the quantum mechanical theory developed by London, electrons in atoms and molecules are in continuous motion even when they are in their ground states. Thus they possess rapidly fluctuating dipole moments, and the fleeting dipole moment in one atom or molecule perturbs a neighboring one inducing a moment in it. The temporary moment in the first molecule and the moment induced in the neighboring one lead to an attractive force between them. London showed that these forces are large enough to account for the forces observed in gases whose molecules do not have permanent dipole moments. He obtained an expression for the interaction potential

$$u_d(r) = -(C_1/r^6) - (C_2/r^8) - (C_3/r^{10}) - \cdots, \tag{1}$$

where the first term on the right represents dipole–dipole interaction; the second, dipole–quadrupole; the third, quadrupole–quadrupole; etc. Relations defining the constants C_i have been developed (3–5). In numerical calculations, the second and higher terms have often been neglected. However, inspection of data for gases given by Margenau (5) shows that dipole–quadrupole interactions are negligible only when $r \geqslant 8$ Å. In physical adsorption, the adsorbate–adsorbent distance is usually less than 8 Å, so that dipole–quadrupole interaction is not usually negligible in this case. Quadrupole–quadrupole interactions and higher terms are in general very small compared to the others.

Depending on the trial wave function and other approximations used, various expressions for the constant C_1 have been obtained. For simple atoms in their ground state approximated by simple isotropic harmonic oscillators with single frequencies v_1 and v_2, London (1) obtained the expression

$$C_{1L} = \tfrac{3}{2}\alpha_1\alpha_2[hv_1v_2/(v_1 + v_2)], \tag{2}$$

where α_1 and α_2 are the polarizabilities of the two atoms. Since hv_1 and hv_2 are often approximately equal to the ionization energies of the atoms, I_1 and I_2, we may write

$$C_{1L} = \tfrac{3}{2}\alpha_1\alpha_2[I_1I_2/(I_1 + I_2)]. \tag{3}$$

Using a variational technique, Slater and Kirkwood (6) derived the expression

$$C_{1SK} = \frac{3eh}{4\pi m^{1/2}} \frac{\alpha_1\alpha_2}{(\alpha_1/N_1)^{1/2} + (\alpha_2/N_2)^{1/2}}, \tag{4}$$

where e and m are the charge and mass of an electron, α_1 and α_2 the polarizabilities of the atoms, and N_1 and N_2 the number of electrons in the outer shells of the atoms. Equation (4) always gives somewhat higher values than Eq. (2) or (3).

A third expression is known as the Kirkwood–Mueller equation (7), and has been applied to the interaction of molecules with solid surfaces (see Section 5.3),

$$C_{1KM} = 6mc^2 \frac{\alpha_1\alpha_2}{(\alpha_1/\chi_1) + (\alpha_2/\chi_2)}, \tag{5}$$

where m is the mass of the electron, c the velocity of light, and χ_1 and χ_2 the magnetic susceptibilities of the atoms.

The constants C_2 and C_3 of Eq. (1) may be determined in principle by methods similar to those employed in determining C_1. For simple molecules and for the harmonic oscillator approximation, a few results have been obtained. Kiselev (8), in a study of the adsorption of various gases on graphite, used expressions for C_2 and C_3 analogous to C_{1KM} of Eq. (5). He found that the dipole–quadrupole term (C_2) contributed about 10% and the quadrupole–quadrupole term (C_3) about 1% to the total dispersion energy.

Important in adsorption studies, as we shall see, is the property of additivity possessed by dispersion forces. By additivity, we mean, for example, that the total dispersion force for three atoms is given by the sum of three pairs of direct perturbations of the atoms on each other (as given by London's equations). Strictly, we should add to this sum the perturbations of these perturbations due to the presence of the third atom. But the perturbations themselves are small, and the perturbations of the perturbations are considered negligible. In fact, Margenau (5) has shown that this is generally the consequence of temporary dipole–dipole (dispersion) interactions, but not of permanent multipole interactions. Extending the effect to dipole–quadrupole interactions, Axilrod and Teller (9) showed that the property of additivity no longer applies, though the contribution of this effect may be small.

Two interacting molecules are also subject to a mutual repulsion force which becomes increasingly significant the more the electron clouds of the two molecules interpenetrate. Under conditions of appreciable penetration of electron clouds, the calculation of the interaction of two molecules demands an entirely different approach. Both the Pauli principle and electron exchange must be taken into account. Chemical bonds may form if electrons on one molecule have the same velocity as some electrons on the other molecule, as well as antiparallel spins. If the spins are parallel, strong repulsive forces arise. Quantum mechanical calculations, involving first-order perturbation theory, made by Born and Mayer (10), indicate that the repulsion potential for inert gases is exponential in form

$$u_r(r) = Be^{-ar}, \tag{6}$$

where B and a are appropriate constants. An empirical equation is often assumed for mathematical convenience, and without theoretical justification,

$$u_r(r) = C_4/r^m. \tag{7}$$

Values of B and a can be determined empirically. Values of m between 9 and 14 usually fit the experimental data, with 12 being the most common value. Although additivity of the repulsive term is usually assumed, there is no justification for this assumption except mathematical convenience.

Dispersive and repulsive terms are often combined to give the total interaction potential. The procedure is completely unjustified on a theoretical basis, since the dispersive terms are calculated assuming only weak perturbations and the repulsive terms assuming strong perturbations. For distances at which appreciable overlapping of electron clouds occurs, the dispersive terms given above are meaningless. However, the combination gives reasonably satisfactory results from a practical point of view for potentials between two inert gas molecules and between various nonpolar molecules and nonionic surfaces, where no appreciable overlapping of electron clouds occurs. Combining the dispersive terms (dipole–dipole and dipole–quadrupole only) of Eq. (1) with the repulsive term of Eq. (7), setting m equal to 12, we obtain

$$u(r) = -(C_1/r^6) - (C_2/r^8) + (C_4/r^{12}). \tag{8}$$

As in Chapter V, at the equilibrium distance r^*, the forces of attraction and repulsion just balance, so that $[\partial u(r)/\partial r]_{r=r^*} = 0$, and $u(r^*) = -\epsilon$, the minimum potential energy ($\epsilon > 0$). Using these conditions, we can express C_4 in terms of C_1 and C_2,

$$C_4 = \tfrac{1}{2} C_1 r^{*6} + \tfrac{2}{3} C_2 r^{*4}, \tag{9}$$

giving

$$u(r) = -(C_1/r^6) - (C_2/r^8) + \tfrac{1}{2}(C_1 r^{*6}/r^{12}) + \tfrac{2}{3}(C_2 r^{*4}/r^{12}), \tag{10}$$

and

$$u(r^*) = -\epsilon = -\tfrac{1}{2}(C_1/r^{*6}) - \tfrac{1}{3}(C_2/r^{*8}). \tag{11}$$

Solving Eq. (11) for C_2 and substituting in Eq. (10), we obtain

$$u(r) = \epsilon[3(r^*/r)^8 - 2(r^*/r)^{12}] - \tfrac{1}{2}(C_1/r^{*6})[2(r^*/r)^6 - 3(r^*/r)^8 + (r^*/r)^{12}]. \tag{12}$$

When $C_2 = 0$, Eq. (12) reduces to

$$u(r) = -2\epsilon(r^*/r)^6 + \epsilon(r^*/r)^{12}, \qquad u(r^*) = -\epsilon, \tag{13}$$

which is similar to Eq. (23) of Chapter V, and is known as the Lennard-Jones potential.

An analogous potential, called the Buckingham potential, is obtained by combining the dispersion term (dipole–dipole only) with the exponential repulsive term, Eq. (6),

$$u(r) = [ar^*\epsilon/(ar^* - 6)]((6/ar^*) \exp\{ar^*[1 - (r/r^*)]\} - (r^*/r)^6). \tag{14}$$

As previously stated, the dipole–dipole term (C_1) in the dispersion potential is often the only one retained in numerical calculations, giving either the Lennard-Jones, Eq. (13), or the Buckingham potential, Eq. (14), when combined with a repulsive potential. This is done for mathematical convenience, in spite of the fact that the dipole–quadrupole term is usually significant. For this reason, calculated interaction potentials are often somewhat low. It has been stated many times (4, 11–15) that the repulsion term (C_4/r^{12}) approximately cancels the dipole–quadrupole dispersion term (C_2/r^8). Thus, if the term r^{-8} is neglected then the term r^{-12} should be discarded also, as has been done in many cases.

Complications arise in applying the potential equations discussed above when the molecules, though nonpolar, are polyatomic or asymmetric, or both.

One way of dealing with this problem is to consider each atom as a force center and to sum, for each orientation of the molecules, all the intermolecular pairwise interactions of the atoms in the two molecules. The procedure is justified by the property of additivity. However, such an approach can be only qualitative, because the spherical symmetry of the atoms about force centers is destroyed by distortion of the electrons of the atoms. Several examples of the application of this method to the problems of physical adsorption will be discussed in the next section.

Kihara (*16*) has made an important contribution to the study of molecular interactions with his molecular core model. The method is applicable to polyatomic, asymmetrical molecules as well as simple ones. Within the molecule, a hard core of given shape is assumed to exist. Kihara defines the intermolecular separation, ρ, as the shortest distance between the outer surfaces of the cores. The intermolecular potential is assumed to have the form of the Lennard-Jones function with ρ in place of r, the distance between centers,

$$u(\rho) = -2\epsilon(\rho^*/\rho)^6 + \epsilon(\rho^*/\rho)^{12}, \tag{15}$$

where ρ^* and ϵ are defined similarly to the corresponding terms, r^* and ϵ, in Eq. (13). The size and shape of the core is calculated with the aid of interatomic distances and bond angles within the molecule; it is always assumed that the core is convex. For a spherically symmetrical model, the Kihara function approaches the Lennard-Jones function as the diameter of the core is allowed to shrink to zero.

TABLE 6.1

CONSTANTS IN THE LENNARD-JONES AND KIHARA POTENTIAL MODELS

Molecule	Model	ρ^* or r^*, Å	ϵ/k, °K	Molecular shape	Reference
Argon	Kihara	3.482	138.0	Spherical	(*18*)
	Lennard-Jones	3.870	121.1		
Krypton	Kihara	3.735	196.1		
	Lennard-Jones	4.144	174.3		
Xenon	Kihara	4.078	263.6		
	Lennard-Jones	4.562	225.3		
Hydrogen	Kihara	2.81	39.4	Nonspherical	(*16*)
Nitrogen	Kihara	3.47	124.0		
Carbon dioxide	Kihara	3.36	309.0		
Methane	Kihara	3.15	226.0		
Carbon tetra fluoride	Kihara	2.55	368.0		
Ethylene	Kihara	4.2	256.0		
Benzene	Kihara	3.6	740.0		

Considerable work has been done on the determination of empirical constants for Eqs. (13)–(15). Physical properties of matter are calculated with the aid of these equations as illustrated in Section 5.3 and compared with measured values. Among the physical properties frequently selected are the virial coefficients (see Section 5.3), transport properties of gases (17), thermal diffusion and diffusion (17), and viscosity (18). In Table 6.1, constants determined for various symmetrical and asymmetrical gas molecules are given. The constants for the rare gases, argon, krypton, and xenon were determined for both the Kihara and the Lennard-Jones potentials by simultaneously fitting both second virial and viscosity coefficient data. The constants for the nonspherical molecules were determined for the Kihara potentials by fitting second virial coefficient data.

6.3 Dispersion and Repulsive Forces—Lattice Summation

London dispersion forces are present in all systems. They are often the major attractive forces in systems of polar molecules adsorbed on the surfaces of covalent solids and spherically symmetrical inert atoms adsorbed on the surfaces of ionic solids; though electrostatic forces may play a significant role. Attractive interactions between nonpolar molecules and metals are considered as dispersion forces; but due to the nonlocalized character of the conduction electrons in metals, these interactions are not amenable to the approximations of this section. Prime examples of attractive interactions consisting almost exclusively of dispersion forces are found in systems of nonpolar molecules adsorbed on the surfaces of covalent solids. And the particular solid which has received the most attention in this class is graphitized thermal carbon black which is made up of basal graphite planes.

The starting point for adsorption calculations involving lattice summations is the equations of Section 6.2 for the interactions between single molecules. Various combinations of dispersion and repulsive potentials have been employed. Sometimes (3, 19) as explained in Section 6.2, the dipole–dipole term of Eq. (8) is used for the total interaction potential, assuming that the dipole–quadrupole attraction term effectively cancels the repulsive term as in the case of potentials between single molecules. The Lennard-Jones potential, Eq. (13), finds frequent application to the adsorption problem. The Buckingham potential, Eq. (14), has also been used. Poshkus (20) has used the dipole–dipole and dipole–quadrupole dispersion terms plus the exponential repulsive term. Kiselev and co-workers (8, 21) have presented even more elaborate calculations including the quadrupole–quadrupole dispersion term.

The various equations mentioned above are summed over all distances r between the location of the adsorbed particle and the lattice points of the solid.

The summations of the attractive and repulsive potentials may be carried out by directly summing interacting pairs or by integration as in Section 5.3. As an approximation, both procedures assume the property of additivity discussed in the preceding section. We shall look first at the physical implications of integration. The real lattice is replaced by a semi-infinite continuum with a mathematically plane bounding surface. The density ρ of this continuum is taken to be equal to the number of lattice atoms per unit volume of the real crystal, but is assumed to hold in every infinitesimal volume of the solid. As a consequence of replacing the real crystal by a continuum, the potential energy of interaction of an atom with the surface is independent of the location of the atom on the surface. Performing the integration of the Lennard-Jones potential in terms of C_1 and C_4 in place of ϵ and r^* [Eq. (13)], we obtain

$$u_s(z) = (\rho\pi C_4/45z^9) - (\rho\pi C_1/6z^3), \tag{16}$$

which is identical to Eq. (50) of Chapter V when the relations $u_s(z_0) = 0$ and $u_s(z^*) = -\epsilon_s$, the potential minimum, are used, and where z, as before, is the vertical distance from the adsorbed atom to the plane through the nuclei of the surface atoms. Integrating the Buckingham potential in terms of C_1, B, and a instead of ϵ, r^*, and a [Eq. (14)], we obtain

$$u_s(z) = (2\rho\pi B/a^3)(za + 2)\exp(-az) - (\rho\pi C_1/6z^3). \tag{17}$$

The parameters in Eqs. (16) and (17) may be determined, as mentioned in Section 6.2, by comparing with measured physical properties. Data obtained by using virial coefficients and the Lennard-Jones potential are given in Tables 5.1 and 5.2. Values of the equilibrium distance z^* of the adsorbed molecule from the surface are often assumed on the basis of some reasonable argument such that $(\partial u_s/\partial z) = 0$ at $z = z^*$, and values of the parameter C_1 are determined from the equations of Section 6.2. Calculations involving the integrated forms of the Lennard-Jones and Buckingham potentials, as well as more complex integrated potentials, may not be very reliable, for the process of integration wipes out the detailed structure of the solid surface making the adsorption potential independent of the location of the adsorbed atom on the surface.

More accurate results are obtained by direct summation of interacting pairs,

$$u_s = \sum_j u_j(r_j), \tag{18}$$

where $u_j(r_j)$ is the interacting potential with the jth atom of the solid at a distance r_j from the adsorbed atom. Usually, interacting pairs are summed directly for atoms of the solid nearer to the adsorbed molecule, and the remaining contribution from more distant atoms is obtained by integration. Enough terms are ordinarily summed so that only the attractive dispersion potential must be integrated. If it is assumed that only the dipole–dipole term of the dispersion potential contributes to the direct summation and the integral, then Eq. (18) becomes

$$u_s = \sum_{j=1}^{N_0} u_r(r_j) - \sum_{j=1}^{N_0} (C_1/r_j^6) - (2\pi\rho C_1/3R_0^3)[1 - \tfrac{3}{4}(z/R_0)], \qquad (19)$$

where N_0 is the number of nearest atoms of the solid whose pair interactions with the adsorbed atom are directly summed; R_0 the radius, measured from the adsorbed atom, of a spherical shell containing N_0 atoms of the solid; ρ the density of atoms of the solid, and $u_r(r_j)$ the pairwise repulsive energy between the adsorbed atom and the atoms of the solid.

Orr (22) has made a detailed calculation for the adsorption of argon atoms on the 100 face of the ionic crystal, potassium chloride, using Eq. (19). In this particular case of a nonpolar molecule interacting with an ionic solid, the electrostatic forces of interaction turn out to be no greater than 4% of the total for any location of the adsorbed atom. Therefore, the interactions to be described may be attributed almost solely to dispersion and repulsive forces. Orr used the Kirkwood–Mueller formula, Eq. (5), to determine C_1. For the repulsion term, he used the exponential function, Eq. (6). Parameter a in this equation was taken as 2.90 Å^{-1}, the value which Born and Mayer (23) gave for alkali halides. Parameter B was evaluated from (24)

$$B_{12} = (B_{11}B_{22})^{1/2}, \qquad (20)$$

where B_{12} stands for the parameter between a pair of unlike molecules 1 and 2, and B_{11} and B_{22} were obtained for this case from Huggins and Mayer (25) and Herzfeld (26). For several values of z, Orr calculated the interaction energy of adsorption on sites over the center of the lattice cell, above the midpoint of the lattice edge, above a potassium ion, and above a chloride ion. He summed the dispersion term over the nearest 250 sites and obtained the remaining contribution by integration. The repulsive potential was obtained by summing over the nearest 14 to 16 sites. After fitting the results for several values of z and the various locations of the adsorbed atom to an interpolation formula, he constructed a plot of u_s versus z for each of the locations. Potential curves with minima were obtained, and the potential minimum $u_s(z^*) = -\epsilon_s$ was found to depend on the adsorption site considered. Such information cannot be obtained from integrated potential functions. Young (27) carried out similar claculations for argon adsorbed on the two kinds of 111 faces of potassium chloride—all cations or all anions exposed—and obtained similar results. Table 6.2 gives the results of Orr for ϵ_s, the potential minimum; ϵ_0, the zero-point energy calculated from the curvature of the potential energy curves at the minimum; and equilibrium distances z^*.

TABLE 6.2

ADSORPTION ENERGIES AND EQUILIBRIUM DISTANCES FOR THE
ADSORPTION OF Ar ON THE 100 PLANE OF KCl AS A FUNCTION
OF THE LOCATION OF THE ADSORBATE[a]

Type of site	z^*, Å	ϵ_s kcal/mole	ϵ_0, kcal/mole
Above center of lattice cell	3.22	1.62	0.03
Above midpoint of lattice edge	3.44	1.34	0.03
Above a K^+ ion	3.48	1.45	0.03
Above a Cl^- ion	3.74	1.27	0.04

[a] See Orr (22).

Similar calculations of u_s have been made by Crowell and Young (28) using Eq. (19) for an argon atom adsorbed on graphite. They determined C_1 from the Kirkwood–Mueller formula, Eq. (5), and C_4 for the r^{-12} repulsive potential Eq. (7) was adjusted to make the equilibrium distance z^*, directly above a carbon atom equal to 3.60 Å, the arithmetic mean between the interlaminar spacing of graphite and the equilibrium nuclear separation of a pair of argon atoms. The value of N_0 was taken to be 100. Three different locations of the adsorbed argon atom were studied, each at several values of z: above the center of a lattice hexagon, above the midpoint between two carbon atoms, and above a carbon atom. The results for z^*, the equilibrium distance of the adsorbed atom from the surface; $u_s(z^*)$, the potential energy minimum; and ϵ_0, the zero-point energy are given as a function of location of the adsorbed atom in Table 6.3. Note that these data show that the adsorption potential of graphite, contrary to that of potassium chloride, is essentially independent of the location of the adsorbed molecule on the surface, and

TABLE 6.3

ADSORPTION ENERGIES AND EQUILIBRIUM DISTANCES FOR THE ADSORPTION OF
Ar ON GRAPHITE AS A FUNCTION OF LOCATION OF THE ADSORBATE[a]

Type of site	z^*, Å	ϵ_s kcal/mole	ϵ_0, kcal/mole	Energy adsorption of $\epsilon_s - \epsilon_0$, kcal/mole
Above the center of a lattice hexagon	3.55	1.78	0.057	1.72
Above the midpoint between two carbon atoms	3.60	1.77	0.064	1.71
Above a carbon atom	3.60	1.76	0.063	1.70

[a] See Crowell and Young (28).

depends only on z, the perpendicular distance from the surface. The surface appears to be uniform in the sense that there are no preferred sites of lower adsorption potential.

The insensitivity of u_s to the structure of the basal planes of graphite suggests that a less laborious method than direct summation might be appropriate for representing the interactions of inert atoms with the graphite lattice. Crowell (29) proposed to represent the semi-infinite graphite lattice by a set of planes of uniform density σ of carbon atoms separated by the interlaminar spacing h of graphite. He assumed that adsorption potentials could be approximated by integrating over each plane and then forming the sum of the resulting terms over all planes. He also assumed interaction terms of the form Kr^{-n} where K and n are constants. Thus energies were calculated by evaluating $K\sum r^{-n}$. The summation $\sum r^{-n}$ was found to be

$$\sum r^{-n} = [2\pi\sigma(-1)^n/(n-2)h^{(n-2)}(n-3)!]\psi^{(n-3)}(z/h), \qquad (21)$$

where $\psi^m(z/h)$ are the polygamma functions tabulated by Davis (30). Using Eq. (21) with the Lennard-Jones 6–12 potential, Crowell (31) obtained

$$u_s(z) = -(C_1\pi\sigma/12h^4)\psi^3(z/h) + (C_4\pi\sigma/5h^{10})(z/h)^{-10}, \qquad (22)$$

in which the summation over r^{-12} was approximated by integration over the surface plane only. From $u_s(z^*) = -\epsilon_s$, the minimum of Eq. (22), we find

$$u_s(z^*) = -\epsilon_s = (-C_1\pi\sigma/12h^4)[\psi^3(z^*/h) + (z^*/10h)\psi^4(z^*/h)], \qquad (23)$$

which may be approximated to within a few percent by

$$u_s(z^*) = -\tfrac{1}{3}C_1\pi\sigma z^{*-4}. \qquad (24)$$

The parameters C_1 and z^* were determined in the same manner as Crowell and Young (28) for their calculations by direct summing. In Table 6.4, values of the calculated energies of adsorption, $(\epsilon_s - \epsilon_0) = -[u_s(z^*) + \tfrac{1}{2}hv]$, are

TABLE 6.4

COMPARISON OF CALCULATIONS OF INTERACTION ENERGIES OF
INERT GAS ATOMS WITH GRAPHITE

Gas	z^*, Å	Calculated energy of adsorption, $\epsilon_s - \epsilon_0$, kcal/mole	Experimental values,[a] energy of adsorption
Ne	3.21	0.746	0.729
Ar	3.59	1.800	2.130
Kr	3.69	2.720	2.790
Xe	3.95	3.370	3.700

[a] See Sams et al. (32).

compared with experimental values for Ne, Ar, Kr, and Xe. The value for argon is in reasonable agreement with that calculated in Table 6.3 by direct summation.

Other methods have been used to determine the constants C_1 and C_4 in Eq. (22) and C_1, a, and B in the corresponding expression derived from the Buckingham potential. For example, Crowell and Steele (*33*) used empirically determined constants for interactions between like gas atoms and between carbon atoms in graphite to determine the corresponding parameters in potential energy equations such as Eq. (22). Thus, they hoped to avoid the uncertainties of an essentially arbitrary choice of z^* used earlier. However, the results agree about equally well with experiment as those in Table 6.4. A more critical appraisal of parameter values used in calculations is needed before it can be determined whether agreement between theory and experiment is significant.

The summation method has been extended to polyatomic molecules by Corner (*34*), Hill, (*35*) and the Russian school. Poshkus (*20*), for example, employs the approximation

$$u_s = \sum_i \sum_j u(r_{ij}), \tag{25}$$

where $u(r_{ij})$ is the interaction energy between atom i of an adsorbed molecule and atom j of a solid at a separation r_{ij}. He uses both the dipole–dipole and dipole–quadrupole terms of the attractive dispersion potential and the exponential form of the repulsive term so that $u(r_{ij})$ is given by

$$u(r_{ij}) = -(C_{1ij}/r_{ij}^6) - (C_{2ij}/r_{ij}^8) + B_{ij}\exp(-a_{ij}r_{ij}). \tag{26}$$

Like Crowell (*29*), he assumes that the basal planes of graphite have uniform density and show no periodicity of interaction potentials. Thus, he used the same general procedures as Crowell for determining values of parameters and for obtaining total interaction potentials. Kiselev (*8, 21*), on the other hand, has carried out direct summations over 100 to 250 of the nearest carbon atoms for attractive potentials and over 40 to 50 nearest atoms for repulsive potentials.

6.4   Dispersion and Repulsive Forces—Collective Treatment

The use of summation procedures where solids are involved is open to objections. The close packing of the atoms changes the nature of their electronic envelopes and the electromagnetic field between the interacting atoms varies with the medium between them. A further objection in the case of metals stems from the presence of free electrons, which mask interactions with the individual ion cores. Early attempts to develop a collective approach, where the solid is treated as a single unit, were concerned with interactions of simple molecules with metals. Later, techniques were developed which are applicable to dielectrics.

In 1932, Lennard-Jones (*36*) derived an expression for the interaction of a neutral molecule or atom with metal surfaces (perfectly conducting). He used the classical image method in which it is assumed that the electrons in an adsorbed molecule behave as though they produced images of opposite charge within the metal, and that these images move in definite phase relation with the electrons of the molecule. The equation for the attractive potential energy of interaction which Lennard-Jones derived is

$$u_s = -e^2 \langle r^2 \rangle / 12z^3, \tag{27}$$

where e is the electronic charge, $\langle r^2 \rangle$ is the mean-square position of all the electrons in the molecule with respect to the center of charge, and z is the distance from the center of charge to the surface. According to Bardeen (*37*), Eq. (27) expresses an upper limit to the interaction energy and should be corrected by taking into account the interaction of electrons in the metal. He obtained

$$u_s = -\frac{e^2 \langle r^2 \rangle}{12z^3} \frac{Ce^2/2r_s \Delta}{1 + Ce^2/2r_s \Delta}, \tag{28}$$

where C is a numerical constant approximately equal to 2.6; r_s is the radius of a sphere in the metal containing one conduction electron, and Δ is the difference in energy between the ground state and an excited state of the adsorbed atom, usually taken equal to the ionization potential.

Margenau and Pollard (*38*) reasoned that the electrons in the metal could not maintain proper phase relations with the instantaneous, rapidly changing dipoles of the nonpolar, adsorbed atom. Thus, the metal in this situation behaves more like a dielectric than an ideally polarizable structure. They calculated interaction energies between the adsorbed atom and elementary portions in the metal surface using perturbation theory and summing over the surface. Their expression is

$$u_s = -(e^2 \alpha / 8z^3)[(C/r_s) - (hn_0/\pi m v_0)], \tag{29}$$

where α is the static polarizability of the adsorbed atom, n_0 is the number of conduction electrons, m is the electronic mass, v_0 is the characteristic frequency of adsorption of the adsorbed atom, h is Planck's constant, and C and r_s are the same as in Eq. (28).

Prosen and Sachs (*39*) applied second-order perturbation theory, neglecting electron–electron interactions in the metal. They found that the interaction energy for conductors or semiconductors is inversely proportional to z if electron degeneracy is ignored,

$$u_s = -(\alpha/2)(\pi n_0 e^2/z).\tag{30}$$

If the electron degeneracy is taken in account, the relation is

$$u_s = -(e^2\alpha k_m{}^2/8\pi^2)[\ln(2k_m z)/z^2],\tag{31}$$

where $k_m = (3\pi^2\rho)^{1/3}$, ρ being the electron density in the metal taking degeneracy into account, and is the maximum value of the electron propagation constant in the metal. Their expression is limited to small values of z because of the neglect of electron interactions in the metal.

In more recent studies, the solid surface is considered as a uniform continuous medium with electromagnetic properties described by a frequency-dependent dielectric constant. Casimir and Polder (40) were the first to study the effect of retardation on the London forces. Considering the interaction between a neutral atom and a perfectly conducting plane by means of quantum electrodynamics, they found that the effect of retardation leads to a reduction of interaction energy by a correction factor that decreases monotonically with increasing distance z. When z is small compared with the wavelengths λ, corresponding to transitions between the ground state and the excited states of the adsorbed atom, this factor is unity. For distances large compared with λ, this factor is proportional to z^{-1}. In the latter case, the total interaction energy is

$$u_s = -3hc\alpha/16\pi^2 z^4, \quad \text{for} \quad z \gg \lambda.\tag{32}$$

McLachlan (41) obtained the same result using only the methods of elementary quantum mechanics.

Lifshitz (42) developed a general macroscopic theory of interaction forces between two solid bodies separated by an empty gap. The theory has been extended by Dzyaloshinskii et al. (43–45) to interactions between solid bodies separated by a gap filled with an arbitrary medium using modern field theory. Basically, the theory considers that the interactions between bodies take place through a fluctuating electromagnetic field. They derived an expression for the interaction of a molecule with a solid body for large z ($z \gg \lambda$):

$$u_s = -(3hc\alpha/16\pi^2 z^4)[(\epsilon - 1)/(\epsilon + 1)]\varphi_{ad}(\epsilon),\tag{33}$$

where ϵ is the electrostatic dielectric constant of the solid and $\varphi_{ad}(\epsilon)$ is a complicated function of ϵ which varies between 0.77 at $\epsilon = 1$ and unity at $\epsilon \to \infty$. For a perfectly conducting surface ($\epsilon \to \infty$), Eq. (33) becomes identical with Eq. (32) of Casimir and Polder. Applying the methods of field theory, Mavroyannis (46) considered a model in which each electron of the adsorbed atom interacts, according to the method of images, with its image located at the same distance below the surface of a solid as the electron is above. Using the S-matrix formalism developed by Dzyaloshinskii (47), he gives the following general expression for the interaction of a neutral atom with a dielectric surface, assuming the atom to be in empty space:

$$u_s = -\frac{1}{3\pi} \sum_n \int_0^\infty \frac{k_{n0}|q_{0n}|^2}{k_{n0}^2 + \xi^2} \left[\frac{\epsilon(i\xi) - 1}{\epsilon(i\xi) + 1}\right] \left[\frac{2}{(2z)} + \frac{4}{\xi(2z)^2} + \frac{4}{\xi(2z)^3}\right] e^{-2z\xi\xi^2} \, d\xi, \tag{34}$$

where ϵ is the dielectric constant (or permeability) of the solid dependent on the imaginary frequency $i\xi$, $\mathbf{q} = -e \sum_i \mathbf{r}_i$ is the electric dipole operator with $\mathbf{r}_i$ the displacement of the ith electron of the adsorbed atom from the center of charge, and $(h/2\pi) ck_{n0} = (h/2\pi)c(k_n - k_0)$ is the difference in energy, $E_{n0} = E_n - E_0 = nh\nu$, between the ground state and the excited states of the atom. In this expression, a complicated and unwieldy term which corrects the energies at large distances z, where retardation effects become important, has been omitted. For $z \gg \lambda$, an expansion of Eq. (34) in terms of λ/z yields the first term

$$u_s \cong -(3hc\alpha/16\pi^2z^4)[(\epsilon_0 - 1)/(\epsilon_0 + 1)], \tag{35}$$

where ϵ_0 is the dielectric constant at $i\xi = 0$. Note that Eq. (35) is the same as Eq. (33) of Dzyaloshinskii *et al.* (43–45) except for a factor $\varphi_{ad}(\epsilon)$ which varies from 0.77 to unity. For metals, or more strictly, for perfect conductors $(\epsilon \to \infty)$, Eq. (35) reduces to Eq. (32) of Casimir and Polder. For $z \ll \lambda$, the first term in an expansion of Eq. (34) in terms of z/λ is

$$u_s = -\frac{1}{6\pi z^3} \sum_n \int_0^\infty \frac{E_{n0}|q_{0n}|^2}{E_{n0}^2 + \xi^2} \frac{\epsilon(i\xi) - 1}{\epsilon(i\xi) + 1} \, d\xi. \tag{36}$$

This equation is similar to the expression obtained by McLachlan (48) emphasizing an image–dipole approach,

$$u_s = -(h/8\pi^2z^3) \int_0^\infty \alpha(i\xi)\{[\epsilon(i\xi) - 1]/[\epsilon(i\xi) + 1]\} \, d\xi, \tag{37}$$

where $\alpha(i\xi)$ is the frequency-dependent polarizability of the adsorbed atom. For a perfect conductor, Eq. (36) reduces to the Lennard-Jones expression, Eq. (27).

Mavroyannis has used Eq. (36) to derive the interaction energies between a neutral atom and metal surfaces. For a perfect conductor, $\epsilon(i\xi) \to \infty$; but for a real metal, the dielectric constant may be expressed

$$\epsilon(i\xi) = 1 + (4\pi n_m e^2/m\xi^2), \tag{38}$$

where n_m is the number of electrons in all the atoms per unit volume of the metal, and e and m are the electron charge and mass, respectively. After substitution of Eq. (38) into Eq. (36), the integration may be performed. Mavroyannis then replaces the summation over the excited states of the adsorbed atom by simpler expressions approximated as suggested in one of his earlier publications (49), and obtains either

$$u_s = -\frac{\alpha^{1/2}n_a^{1/2}}{8z^3}\frac{h\omega_p/2^{1/2}}{2\pi[(n_a/\alpha)^{1/2} + (h/2\pi)(\omega_p/2^{1/2})]}, \tag{39}$$

or

$$u_s = -\frac{\langle q^2 \rangle}{12z^3}\frac{h\omega_p/2^{1/2}}{2\pi[(3n_a/2\langle q^2 \rangle) + (h/2\pi)(\omega_p/2^{1/2})]}, \tag{40}$$

where α and n_a are the polarizability and the number of electrons in the adsorbed atom, respectively, $\omega_p = (4\pi n_m e^2/m)^{1/2}$ is the characteristic resonance frequency of absorption of electrons in the metal, and $(h/2\pi)\omega_p$ may be interpreted as the dielectric energy loss. Experimental values of α are available and $\langle q^2 \rangle$ was taken as $e^2\langle(\sum_i \mathbf{r}_i)^2\rangle$ or $e^2\langle\sum_i \mathbf{r}_i^2\rangle$.

Using Eqs. (39) and (40), Mavroyannis has calculated the energies of interactions of rare gas atoms with metals. These values are compared in Table 6.5 with those obtained from Eq. (27) of Lennard-Jones, Eq. (28) of Bardeen,

TABLE 6.5

COMPARISON OF THEORETICAL AND EXPERIMENTAL VALUES FOR THE
INTERACTION OF INERT GAS ATOMS WITH METALS[a]

System	Experimental cal/mole	Calculated					
		Mavroyannis		Lennard-Jones Eq. (27)	Bardeen Eq. (28)	Margenau and Pollard Eq. (29)	Prosen and Sachs Eq. (31)
		Eq. (39)	Eq. (40)				
Pt–He (z = 2.70 Å)	265	325	280	890	230	320	350
Pt–Ne (z = 2.98 Å)	330	535	510	2190	370	460	580
Pt–Ar (z = 3.30 Å)	1320	1435	1815	5170	940	1260	2080
Zn–Ar (z = 3.24 Å)	1570	1520	1660	5465	1100	1400	—
Cu–Ar (z = 3.18 Å)	2090	1475	1870	5800	1090	1450	2700
Pt–Kr (z = 3.36 Å)	2110	2170	2010	7120	1240	1610	3060

[a] See Mavroyannis (46).

Eq. (29) of Margenau and Pollard, and Eq. (31) of Prosen and Sachs. Experimental values are also included and were obtained from Chon *et al.* (*50*) along with values calculated from Eqs. (28), (29), and (31). It is clear from the table that the Lennard-Jones expression, Eq. (27), for a perfect conductor gives values which are far too high, while the various approximations taking the dielectric properties of metals into account are in better agreement with experiment.

The recent collective treatments of the interactions of atoms and molecules with solids which we have outlined are, indeed, promising. However, some of the problems that must be faced ultimately are: (1) the treatment of solids as crystalline rather than continuous media, (2) the introduction of repulsive forces which have an appreciable effect at the usual equilibrium distances, and (3) the difficulties engendered by the "roughness" of real surfaces which makes their location uncertain with respect to an adsorbed atom by an amount comparable to the interatomic crystal spacing.

6.5   Classical Electrostatic Interactions

In Section 6.3, we stated that the interactions of nonpolar molecules with ionic solids are partially the result of electrostatic forces. An attractive potential is generated by the polarization of the atom in the electric field **E** of the ionic lattice. Lennard-Jones and Dent (*51*) derived a relation between the field **E** and the electrostatic potential P at any point outside the 100 face of a simple cubic crystal, using a method due to Madelung (*52*). After several simplifying approximations, they obtained

$$P = (d_0/2^{1/2}\pi)E_z, \tag{41}$$

where

$$E_z = -\partial P/\partial z \cong (8\pi ek/d_0^2)\exp(-2^{1/2}\pi z/d_0) \tag{42}$$

is the component of the electric field normal to the 100 lattice plane, $2d_0$ the lattice parameter (edge of a unit cell), and k the charge on the lattice ion. It can be shown that $E_x = E_y = 0$ at positions directly above a lattice ion. For a nonpolar atom in an external electric field, the potential energy of interaction is given by

$$u_s = -\tfrac{1}{2}\alpha E^2, \tag{43}$$

where α is the polarizability of the adsorbed atom. This equation holds providing the field is homogeneous or slowly varying over the region occupied by the molecule. Lenel (*53*), however, has pointed out that in general the field outside ionic crystals varies too rapidly to justify the use of Eq. (43). Thus, it is necessary to subdivide the total volume of the electronic cloud of the adsorbed atom into elements dV and to determine the polarization in each element. The expression becomes

$$u_s = -\int [(P - P_0)\rho/\Delta]\, dV, \tag{44}$$

where P and P_0 are the electrostatic potentials in the volume element dV and at the center of the atom, respectively, ρ the electron density, and Δ, a characteristic energy approximately equal to the ionization potential for the inert atom.

Calculations presented by deBoer (54) show the magnitude of these interaction energies. For adsorption of argon on KCl, he used $\alpha = 1.68 \times 10^{-24}$ cm^3 and $z^* = 3.14$ Å. The electric field above a potassium ion at this distance is $E_z = 1.45 \times 10^5$ esu. The interaction energy u_s obtained by using these values in Eq. (43) is -255 cal/mole, while Eq. (44) gives -450 cal/mole. However, the interaction energy resulting from induced polarization is in general small compared to dispersion energies in such systems as shown in Section 6.4 and by Hayakawa (55).

Although dispersion forces are generally considered to be predominant in the interactions of nonpolar atoms with metals, Mignolet (56–60) found by surface potential measurements that xenon atoms, for example, adsorbed on nickel were strongly polarized with the positive charge outward. Initially, he explained the results on the basis of induced polarization (see above). Later, he invoked the charge-transfer, no-bond theory of Mulliken (61). In this theory, the ground state of the complex is described by a linear combination of the wave function for a no-bond state (A, B) (dispersion forces) and that for a dative state $(A^- - B^+)$ in which an electron has been transferred from the Lewis base to the Lewis acid. Thus, it is evident that this kind of adsorption represents a hybrid of physical and chemical forces. In the case of adsorption of an inert atom on a metal, the metal is the electron acceptor (Lewis acid) and the atom the donor (Lewis base), for the measured surface potentials are generally positive. Matsen et al. (62) give the potential energy of adsorption on metals at zero coverage

$$u_s = +(u_D - u_{NB}) - [(u_D - u_{NB})^2 + 4\beta^2]^{1/2}, \tag{45}$$

where $u_s < 0$ for exothermic adsorption, u_{NB} and u_D are the potential energies of the zero-order no-bond and dative states, respectively, and β the interaction integral, The difference in energy between the dative and no-bond forms are given approximately by

$$u_D - u_{NB} = Ie - \varphi e - e^2/4z^*, \tag{46}$$

where I is the ionization potential of the electron donor (adsorbed atom), φ is the work function of the electron acceptor (metal), and $e^2/4z^*$ is the image energy with $z*$ having its usual meaning. Ionization energies for adsorbates usually lie in the range 9–12 eV and work functions in the range 2–5 eV. Thus the transfer of an electron from the adsorbed atom to the metal will be an endothermic process $(I - \varphi > 0)$. But the potential energy of adsorption u_s can be negative, as required for exothermic adsorption, by virtue of the resonance between the no-bond and dative states. Matsen *et al.* have plotted u_s versus Ie using $\varphi e = 4.7$ eV, $z* = 1$ Å, $e^2/4z* = 3.6$ eV, and $\beta = 1.3$ eV. Exothermic (negative) and monotonically decreasing energies are found in the region $Ie \gtrsim 8.3$, corresponding to $(u_D - u_{NB}) \gtrsim 0$. There are no data available as yet which distinguish unequivocally between the two types of interaction charge-transfer no-bond and induced polarization.

Polar molecules will interact with the external field of a solid. Significant electrical fields occur on ionic solids, and the interaction of a molecule whose dipole moment is μ with such a solid is given by

$$u_s = -\boldsymbol{\mu} \cdot \mathbf{E} = -\mu E \cos \theta = -\mu E_t, \tag{47}$$

where $\mathbf{E}$ is the electrical field of the solid and E_t is the component of the field in the direction t along the axis of the dipole which makes an angle θ with the electric field vector. Since $(\partial P/\partial t) = -E_t$, where P is the potential at any point in the neighborhood of the surface, we may write

$$u_s = \mu(\partial P/\partial t). \tag{48}$$

Equations (47) and (48) hold, provided that the field changes slowly over the volume of a molecule.

If a molecule possesses a quadrupole moment, expressions corresponding to Eqs. (47) and (48) are more complicated. For example, when the charge distribution has an axis of rotational symmetry, the quadrupole is called linear, and its moment may be defined by

$$Q = \tfrac{1}{2} \int \rho(r, \theta)(3 \cos^2 \theta - 1)r^2 \, dV, \tag{49}$$

where $\rho(r, \theta)$ is the charge density at the point (r, θ) with origin at the center of the atom, and the integration is over the entire molecule. In this case, the interaction energy is

$$u_s = \tfrac{1}{2}Q(\partial^2 P/\partial t^2). \tag{50}$$

Interactions of polar atoms or molecules with metals which *per se* have negligible permanent fields are caused chiefly by the induction of a field in the solid by the adsorbate. The interaction energy is determined by considering the dipole and its image. The problem was studied by Lorenz and Landé (*63*) and by Magnus (*64*) for atoms with permanent dipoles. The interaction energy is given by

$$u_s = -(\mu^2/16z^3)(1 + \cos^2 \theta). \tag{51}$$

This contribution must be added to the dispersion energies discussed in Section 6.4. Values obtained from Eq. (51) are quite low, much lower, in fact, than those obtained for the interaction of a dipole with the field of an ionic solid.

Classical electrostatic interactions of covalent solids with polar atoms are negligible, for they possess neither significant permanent fields nor fields induced by polar atoms or molecules.

6.6   Adsorbate–Adsorbate Interactions

Both theoretical and experimental information on the interactions between molecules adsorbed on a solid surface is scarce. A possible method is the use of the two-dimensional analog of the van der Waals equation (see Chapter V). Comparison of two-dimensional and three-dimensional van der Waals parameters could lead to a knowledge of the effects of a surface on the interactions of molecules in its vicinity. However, the van der Waals equation is semiempirical, and it is desirable to have a sounder theoretical basis for such studies. The method of virial coefficients developed by Sams *et al.* (*65*) and refined by Everett (*66*) and by Barker and Everett (*67*) is a more satisfactory starting point. The general procedure is:

(1)   Determine "experimental" values of $B_{2s}(T)/\alpha$, where $B_{2s}(T)$ is the strictly two-dimensional second virial coefficient [Eq. (18), Chapter V] and α is the surface area.

(2)   Fit values of $B_{2s}(T)/\alpha$ to a theoretical plot of $B_{2s}/\pi r^{*2}$ versus ϵ_{2s}/kT, in which one of a number of available equations for the potential energy of interaction is used, where r^* is the distance of separation of two adsorbed molecules at their minimum potential, $-\epsilon_{2s}$ [Eq. (23), Chapter V].

To determine "experimental" values of B_{2s}/α, the two-dimensional virial coefficient, we use the second and third gas–surface virial coefficients, B_{AS} and C_{AAS}, of Eq. (40) of Chapter V. The relationship between these virial coefficients, as given by Sams *et al.* is Eq. (57) of Chapter V, and we repeat it here for convenience,

$$B_{2s}/\alpha = -C_{AAS}/2B_{AS}^2. \tag{52}$$

A more refined expression, taking into account the imperfect nature of the bulk gas, is given by Everett (66) and by Barker and Everett (67),

$$(B_{2s} - a)/\alpha = (-C_{AAS}/2B_{AS}^2) - (B/B_{AS}), \tag{53}$$

where B is the second virial coefficient of bulk gas and a corrects for deviations of the adsorbed phase from exact planarity. As explained in Chapter V, Eqs. (45) and (46), B_{AS} and C_{AAS}, may be determined from the adsorption isotherm.

In the theoretical expression for B_{2s} [Eq. (19), Chapter V]

$$B_{2s} = -\int_0^{\infty} \{\exp[-u_s(r)/kT] - 1\}\pi r\, dr, \tag{54}$$

Sams et al. (65), used the Lennard-Jones 6–12 potential for $u_s(r)$. After integrating, they obtained

$$B_{2s}/\pi r^{*2} = -\sum_{j=0}^{\infty} (1/12j)(kT/4\epsilon_{2s})^{[(3j+1)/6]}\Gamma[(3j-1)/6], \tag{55}$$

which they evaluated for different values of ϵ_{2s}/kT. Fitting the experimental values of $B_{2s}(T)/\alpha$ to the theoretical curve $B_{2s}/\pi r^{*2}$ versus ϵ_{2s}/kT gives values of the parameters ϵ_{2s}/k and $\alpha/\pi r^{*2}$. It is usually assumed that r^* has just the Lennard-Jones (6–12) gas–gas value. The parameters ϵ_{2s} and α are then obtained. For argon on carbon black, Sams et al. (65) found a value of ϵ_{2s} that was 20% lower than the gas-phase value ϵ, indicating additional repulsion between gas atoms under the influence of the surface.

For their calculations, Barker and Everett (67) assumed that the potential for the interaction between adsorbed atoms could be written as

$$u_s(r) = 4\epsilon[-\xi(r_{0\,gas}/r)^6 + (r_{0\,gas}/r)^{12}], \tag{56}$$

where ϵ is the depth of the potential minimum and $r_{0\,gas}$ the value of r at $u_s = 0$ for the *unperturbed interaction in the gas*, and the factor ξ ($\leqslant 1$) takes into account the reduction in the attraction arising from perturbation by the surface. This, of course, is expressible in the usual 6–12 form,

$$u_s(r) = 4\epsilon_{2s}[-(r_0/r)^6 + (r_0/r)^{12}], \tag{57}$$

where $\epsilon_{2s}(= \xi^2\epsilon)$ is the depth of the energy minimum of the *perturbed potential* and $r_0(= \xi^{-1/6}r_{0\,gas})$ is the new value of r at which $u_s(r) = 0$.

Johnson and Klein (68) in view of theoretical calculations by Sinanoglu and Pitzer (69) assumed that the perturbed potential between adsorbed atoms could be expressed by

$$u_s(r) = 4\epsilon[-(r_{0\,gas}/r)^6 + (r_{0\,gas}/r)^{12} + \chi(r_{0\,gas}/r)^3], \tag{58}$$

where again ϵ and $r_{0\,gas}$ are parameters characterizing the *bulk gas interaction* and χ a constant. Fitting the experimental values of B_{2s} to the theoretical curves essentially as described above, they found that the two potentials, Eqs. (57) and (58), agreed with the data of Sams *et al.* for argon on carbon black equally well.

Two completely theoretical approaches have been developed for the inter, actions of two adsorbed molecules. Using third-order perturbation theory-Sinanoglu and Pitzer (*69*) showed that the London dispersion energy is modified not only by the polarization effect, but also by the nonadditivity of dispersion forces in many-atom systems. Their calculations show that the interaction energy between adsorbed atoms differs from that of bulk gas by an additional repulsion which is approximately proportional to r^{-3}. McLachlan (*70*), treating the same problem used a model in which the instantaneous dipole of an adsorbed atom interacts with its image in the surface of the solid (treated as a continuum) and with the image of a second neighboring adsorbed atom. He showed that the additional interaction energy between two adsorbed atoms is proportional to r^{-3} for small distances, but at sufficiently large r becomes proportional to r^{-6}. Despite the empirical agreement with experimental data of the potential energy form predicted by Sinanoglu and Pitzer (*69*), Everett (*66*) has shown that a calculation based on their exact equation with no empirical constants predicts a perturbation far greater than that deduced from experiment. Furthermore, he found that both the theory of Sinanoglu and Pitzer and that of McLachlan indicate that the perturbation of the interaction energy due to the surface should increase with the size of the molecule, whereas the experimental data give a variation in the opposite direction.

REFERENCES

1. London, F., *Z. Physik* **63**, 245 (1930).
2. London, F., *Z. Physik. Chem. B* **11**, 222 (1930).
3. Brunauer, S., "The Adsorption of Gases and Vapors," Princeton Univ. Press, Princeton, New Jersey, 1943.
4. de Boer, J. H., *Advan. Colloid Sci.* **3** (1950).
5. Margenau, H., *Rev. Mod. Phys.* **11**, 1 (1939).
6. Slater, J. C., and Kirkwood, J. G., *Phys. Rev.* **37**, 682 (1931).
7. Müller, A., *Proc. Roy. Soc. (London), Ser. A* **154**, 624 (1936).
8. Kiselev, A. V., *Proc. Intern. Cong. Surface Activity 2nd London* **2**, 168 (1957).
9. Axilrod, B. M., and Teller, E., *J. Chem. Phys.* **20**, 1812 (1952).
10. Born, M., and Mayer, J. E., *Z. Phys.* **75**, 1 (1932).
11. Barrer, R. M., *Proc. Roy. Soc. (London), Ser. A* **161**, 476 (1937).
12. Brunauer, S., Emmett, P. H., and Teller, E., *J. Am. Chem. Soc.* **60**, 309 (1938).
13. de Boer, J. H., *Trans. Faraday Soc.* **32**, 10, (1936).
14. Hellman, H., "Einführung in die Quantenchemie," Deutike, Leipzig, Germany, and Vienna, Austria (1937).

15. Lenel, F. V., *Z. Physik Chem.* **B23**, 379 (1933).
16. Kihara, T., *Advan. Chem. Phys.* **5** (1963).
17. Mason, E. A., Munn, R. J., and Smith, F. J., *Discussions Faraday Soc.* **40**, 97 (1965).
18. Rossi, J. C., and Danon, F., *Discussions Faraday Soc.* **40**, 97 (1965).
19. de Boer, J. H., *Advan. Catalysis* **8**, 17 (1956).
20. Poshkus, D. P., *Discussions Faraday Soc.* **40**, 97 (1965).
21. Avgul, N. W., Isirkyan, A. A., Kiselev, A. V., Lygina, I. A., and Poshkus, D. P., *Izv. Akad. Nauk. SSSR Otd. Khim. Nauk* 1314 (1957).
22. Orr, W. J. C., *Trans. Faraday Soc.* **35**, 1247 (1939).
23. Born, M., and Mayer, J. E., *Z. Physik.* **75**, 1 (1932).
24. Mason, E. A., and Rice, W. E., *J. Chem. Phys.* **22**, 522 (1954).
25. Huggins, M. L., and Mayer, J. E., *J. Chem. Phys.* **1**, 643 (1935).
26. Herzfeld, K. F., *Phys. Rev.* **52**, 374 (1937).
27. Young, D. M., *Trans. Faraday Soc.* **47**, 1228 (1951).
28. Crowell, A. D., and Young, D. M., *Trans. Faraday Soc.* **49**, 1080 (1953).
29. Crowell, A. D., *J. Chem. Phys.* **22**, 1397 (1954).
30. Davis, H. T., " Tables of Higher Mathematical Functions," Principia Press, Bloomington, Indiana, 1933.
31. Crowell, A. D., *J. Chem. Phys.* **26**, 1407 (1957).
32. Sams, J. R., Jr., Constabaris, G., and Halsey, G. D., Jr., *J. Phys. Chem.* **64**, 1689 (1960).
33. Crowell, A. D., and Steele, R. B., *J. Chem. Phys.* **34**, 1347 (1961).
34. Corner, J., *Proc. Roy. Soc. (London), Ser. A* **192**, 275 (1948).
35. Hill, T. L., *J. Chem. Phys.* **16**, 181 (1948).
36. Lennard-Jones, J. E., *Trans. Faraday Soc.* **28**, 334 (1932).
37. Bardeen, J., *Phys. Rev.* **58**, 727 (1940).
38. Margenau, H., and Pollard, W. G., *Phys. Rev.* **60**, 128 (1941).
39. Prosen, E. J. R., and Sachs, R. G., *Phys. Rev.* **61**, 65 (1942).
40. Casimir, H. B. G., and Polder, D., *Phys. Rev.* **73**, 360 (1948).
41. McLachlan, A. D., *Proc. Roy. Soc. (London), Ser. A* **271**, 387 (1963).
42. Lifshitz, E. M., *J. Exptl. Theor. Phys.* **2**, 73 (1956).
43. Dzyaloshinskii, I. E., and Pitaevskii, L. P., *J. Exptl. Theor. Phys.* **9**, 1282 (1959).
44. Dzyaloshinskii, I. E., Lifshitz, E. M., and Pitaevskii, L. P., *J. Exptl. Theor. Phys.* **10**, 161 (1960).
45. Dzyaloshinskii, I. E., Lifshitz, E. M., and Pitaevskii, L. P., *Advan. Phys.* **10**, 165 (1961).
46. Mavroyannis, C., *Mol. Phys.* **6**, 593 (1963).
47. Dzyaloshinskii, I. G., *J. Exptl. Theor. Phys.* **3**, 977 (1957).
48. McLachlan, A. D., *Mol. Phys.* **7**, 381 (1964).
49. Mavroyannis, C., and Stephen, M. J., *Mol. Phys.* **5**, 629 (1962).
50. Chon, H., Fisher, R. A., McCammon, R. D., and Aston, J. G., *J. Chem. Phys.* **36**, 1378 (1962).
51. Lennard-Jones, J. E., and Dent, B. M., *Trans. Faraday Soc.* **24**, 92 (1928).
52. Madelung, E., *Physik Z.* **19**, 524 (1918).
53. Lenel, F. V., *Z. Physik. Chem.* **B23**, 379 (1933).
54. deBoer, J. H., *Advan. Colloid Sci.* **3**, 1 (1950).
55. Hayakawa, T., *Bull. Chem. Soc. Japan*, **30**, 124, 230, 243, 332, 337 (1957).
56. Mignolet, J. C. P., *Discussions Faraday Soc.* **8**, 105 (1950).
57. Mignolet, J. C. P., *J. Chem. Phys.* **21**, 1298 (1953).
58. Mignolet, J. C. P., *Bull. Soc. Chim. Belges.* **64**, 126 (1955).
59. Mignolet, J. C. P., *Rec. Trav. Chim.* **74**, 685, 701 (1955).
60. Mignolet, J. C. P., "Chemisorption " (W. E. Garner, ed.), Butterworth, London and Washington, D.C., 1957.

61. Mulliken, R. S., *J. Am. Chem. Soc.* **74**, 811 (1952).
62. Matsen, F. A., Makrides, A. C., and Hackerman, N., *J. Chem. Phys.* **22**, 1800 (1954).
63. Lorenz, R., and Landé, A., *Z. Anorg. Chem.* **125**, 47 (1922).
64. Magnus, A., *Z. Physik Chem.* **142**, 401 (1922).
65. Sams, J. R., Jr., Constabaris, G., and Halsey, G. D., Jr., *J. Chem. Phys.* **36**, 1334 (1962).
66. Everett, D. H., *Discussion Faraday Soc.* No. 40, 177 (1965).
67. Barker, J. A., and Everett, D. H., *Trans. Faraday Soc.* **58**, 1608 (1962).
68. Johnson, J. D., and Klein, M. L., *Trans. Faraday Soc.* **60**, 1964 (1964).
69. Sinanoglu, O., and Pitzer, K. S., *J. Chem. Phys.* **32**, 1279 (1960).
70. McLachlan, A. D., *Mol. Phys.* **7**, 381 (1964).

VII

THE CHEMICAL FORCES OF ADSORPTION—METALS

7.1 Introduction

In physical adsorption, it is generally assumed that both adsorbate and adsorbent suffer only minor distortions as a result of the van der Waals forces between them. In chemisorption, where electrons are shared or transferred and drastic redistributions of their equilibrium positions occur, the situation is far more complex. Consequently, progress in the theory of chemisorption has been slower; adequate models have been difficult to devise. Indeed, there has been considerable pessimism concerning the state of chemisorption theory. For example, Ehrlich (*1*) says:

> The object of any theory is to unify observations by relating them to a common basis, as well as to predict and explain new effects. Both of these aims are extremely important for surface studies, in which the proliferation of observed phenomena is matched by the difficulty of defining all the important variables. In the years immediately following the war the newly enlarged understanding of electronic phenomena in solids (both in metals and non-metals) gave promise of important simplifications in the theory of chemisorption phenomena as well. It will appear . . . that these hopes were ill-founded; indeed, at the moment it is easier to document deficiencies than to point to new areas clarified by theory.

Ehrlich believes that experimental studies of the surface problems of chemisorption have been far more successful than the theoretical. He points especially to the recent experiments on surface diffusion (*2*), rates of chemisorption (*2–8*) and direct observation of surface structures (*9*).

Specific reasons for the low degree of progress in developing quantitative theories of chemisorption based on electronic concepts of surface processes are not difficult to discover. Since theoretical techniques for dealing with electronic processes in bulk solids have been developed for the most part during and after World War II, the application of these techniques to chemisorption is in reality just getting started. The complexities created by the greater forces of chemisorption compared with those of physical adsorption are, indeed, formidable. As is always true, it will be necessary to proceed from simple to more complex models. We need not become too pessimistic over the failure of simple models, for the weaknesses in the simplifying assumptions are recognized and more realistic models will be constructed in time. Part of the difficulty lies in making assumptions which follow too closely analogies with the simple chemical bond. Another troublesome factor arises from the differences between electronic states near the surface of a solid and those associated either with the solid as a whole or the isolated atoms of the solid. As we shall discover, the theoretical treatment of surface states is a very recent development, still in its infancy.

Only recently has it been possible to observe directly the disruption of metal surfaces by chemisorption. An interesting example is the chemisorption of hydrogen on a nickel crystal studied by Germer and MacRae (*10*) with the aid of low-energy electron diffraction techniques. It might be expected that hydrogen would form a "lattice gas" when chemisorbed on nickel, and probably it does at low concentrations. At higher concentrations, however, Germer and MacRae found that a reconstructive transition occurred in which a Ni (110)—2 × 1—H structure was produced. In the 2 × 1 structure, the repeat distance of nickel is twice that of the underlying nickel surface in the [10] direction and equal to that of the nickel in the [01] direction. When hydrogen is desorbed by raising the temperature, the original structure reappears. The position of the hydrogen atoms is not known. In other systems, such as oxygen on nickel, adsorbent atoms cooperate with adsorbate atoms in the formation of one or more layers of the surface phase. Unfortunately, we lack the space to do more than mention these interesting surface studies. Lander (*11*) has written an excellent review article. In some instances, our conception of an adsorbed monolayer may have to be revised. Cooperation between adsorbent atoms and adsorbed atoms as they seek a condition of minimum energy may lead to highly complex and novel configurations. The number of possible surface structures may exceed normal substrate structures by an order of magnitude (*11*). However, it is probable that such reconstructed surfaces are relatively rare and that normal surfaces with insignificant changes in lattice distances are by far the most common.

We shall consider first the simplest methods for describing chemical forces of adsorption at metal surfaces, which employ analogies with simple chemical bonds. The bond is pictured as existing between the adsorbed atom and a single atom of the metal without influence of the surrounding metal atoms. Empirical rules were developed by Pauling for bonds between single atoms and extended to adsorption bonds in the case of covalent bonding. The results have not been very successful. Pauling's *d*-character has also been invoked and, again, the results are disappointing. In the case of highly ionic bonds, some successes have been obtained in checking experimental adsorption heats against values calculated using classical electrostatic theories. Next, we shall consider a more complex method of viewing surface bonding which involves ligand-field theory. In this theory the effects of local, ordered environment in the metal on the adsorption bond are considered. The results to date are qualitative and it is difficult to see how they can be made quantitative in view of the complexities. However, the results do have heuristic value. Initial studies of a theoretical nature on the surface quantum states of metal and metal–adsorbate systems are reviewed. These studies represent the most fundamental approach to adsorption systems; but the difficulties are tremendous. At present, the studies deal entirely with idealized systems, often one-dimensional, and employ the LCAO approximation of the molecular orbital theory. Therefore, they are only of very general value. But these studies do lead to some interesting and provocative qualitative results such as general criteria for the existence of bonds of surface states outside the region of normal crystal bands. A further point that will be discussed is the variation of heats of adsorption with coverage. When the heats fall rapidly at low coverage, the most reasonable explanation appears to be associated with the inherent heterogeneity of the surface.

7.2   Analogies with Simple Chemical Bonds

COVALENT BONDING

First we shall discuss Eley's method (*12, 13*) of calculating adsorption bond energies. This method has been applied largely to the dissociative adsorption of hydrogen on metals,

$$2M + H_2 \rightarrow 2M - H. \tag{1}$$

The differential heat of adsorption (see Section 1.4) at zero coverage is given by

$$q_d = 2D_{MH} - D_{HH}, \tag{2}$$

where D_{MH} and D_{HH} are the dissociation energies of the metal–hydrogen bond and the hydrogen molecule. Equation (2) assumes that no metal–metal bonds are broken upon chemisorption, which is not always true as we have seen in Section 7.1. To determine D_{MH}, Eley extends Pauling's approximation (14) of covalent bonds between free atoms to adsorption bonds. Thus, the bond energy [in kilocalories per mole with $(\chi_M^P - \chi_H^P)^2$ in electron volts] is

$$D_{MH} = \tfrac{1}{2}(D_{MM} + D_{HH}) + 23.06\,(\chi_M^P - \chi_H^P)^2, \qquad (3)$$

where the first term on the right is the approximation for a strictly covalent bond and the second term corrects for slight polarization effects resulting from different electronegativities of metal and hydrogen, χ_M^P and χ_H^P, as defined by Pauling. The second term in Eq. (3), therefore, determines the small contribution of the ionic character of the bond. Substituting Eq. (3) for D_{MH} into Eq. (2), the differential heat of adsorption becomes

$$q_d = D_{MM} + 46.12\,(\chi_M^P - \chi_H^P)^2. \qquad (4)$$

The problem now is to determine D_{MM}, χ_M^P and χ_H^P. The metal–metal bond energy D_{MM} is estimated from the latent heat of sublimation, λ. In a face-centered cubic metal, each atom has 12 nearest neighbors and remembering that each bond involves two atoms we have

$$D_{MM} = \lambda/6. \qquad (5)$$

This equation is satisfactory for body-centered cubic lattices within the approximation of the calculation, where each atom has eight nearest neighbors and six next nearest.

Eley calculated $(\chi_M^P - \chi_H^P)$ from the approximation (15),

$$(\chi_M^P - \chi_H^P) = \mu, \qquad (6)$$

where μ is the dipole moment of the chemisorbed bond in debyes at zero coverage. This equation is limited to very small values of μ. The dipole moment at full monolayer coverage is given by

$$\mu_f = V/300(2\pi B), \qquad (7)$$

where V is the surface potential in volts and B the number of sites per square centimeter. Surface potentials at full monolayer coverage are available particularly from the work of Mignolet (16–18). If depolarization effects are assumed to be small, then μ_f may be equated to μ_0 the dipole moment at zero coverage. Otherwise, the equation given by Topping may be used,

$$\mu_0 = \mu_f[1 + (9\alpha/a^3)], \qquad (8)$$

where α is the longitudinal polarizability of the bond and a is the lattice constant of the surface, which may be taken to be 3 Å. Values of $(\chi_M^P - \chi_H^P)$ may also be obtained roughly from Pauling's table of electronegativities.

Both of the above methods of determining $(\chi_M^P - \chi_H^P)$ are crude approximations even for free diatomic molecules. In an attempt to improve the method, Stevenson [19] has made use of Mulliken's electronegativity values which are defined by

$$\chi^M = \tfrac{1}{2}(Ie + A_0 e), \qquad (9)$$

where I, A_0, and e are the ionization potential, electron affinity, and electronic charge. For free atoms, the relation between Mulliken's and Pauling's values are $\chi^M \cong 65\chi^P$. For metals, both I and A_0 are set equal to φ_M the work function of the metal, since the highest occupied and lowest unoccupied levels are at the Fermi surface. Stevenson writes

$$\chi_M^S = 0.355\varphi_M, \qquad (10)$$

with φ_M in eV and $0.355 = (23.06 \text{ kcal/mole eV})/65$, a scaling factor. Stevenson's quantity χ_M^S is then directly used in Eq. (3) in place of χ_M^P and along with Pauling's electronegativities for adsorbate. The agreement between calculated and experimental values of D_{MH} is somewhat better using Stevenson's modification. But neither the original Eley method nor its modification gives completely satisfactory agreement with experimental values of initial heats of adsorption. For hydrogen on Fe, Co, Ni, Cu, Mo, Ta, and W, the agreement using Stevenson's procedure is moderately good; values for Pd and Pt are rather low; and the value for Cr is very low, while the values for Ru, Rh, and Ir are too high. In general, the values for other gases are too low [20].

Various criticisms have been leveled at details of the Eley method. Ehrlich [1, 21] believes that the values of D_{MM} estimated from heats of sublimation, Eq. (5), should be lower. He points to the work of Oriani [22] who showed that the surface excess energy (energy to bring an atom from interior to surface) for liquid metals calculated from the heat of evaporation assuming pairwise bonding is high compared to values calculated from surface tension measurements and their temperature coefficients. Bonds in the surface layer were found to be about 13% stronger than in the bulk, and this "redistribution" of binding energy for the free surface suggests a lower value of D_{MM} for use in calculating adsorption bonds. Similar considerations for polycrystalline filaments of copper, silver, and gold, close to their melting points, whose surface tensions and their temperature coefficients were determined from creep measurements, lead to surface bonds about 25% stronger. However, this factor alone is not considered a serious limitation to the method. Ehrlich [1] is especially concerned with the fact that the heat of adsorption given by Eq. (4) is always positive, and thus predicts that adsorption will always be exothermic even for metals of low cohesive energy (D_{MM}). He states that chemisorption of hydrogen on metals is detectable for $q_d \gtrsim 3$ kcal/mole $(D_{MH} \gtrsim 53$ kcal/mole), and therefore, contrary to experiment, molecular hydrogen should be adsorbed on aluminum, silver, gold, zinc, cadmium, tin, and lead. Pauling found that bond strengths for metal hydrides obtained from the geometric mean $(D_{MM} D_{HH})^{1/2}$ give values in much better agreement than those obtained from the arithmetic mean, $(D_{MM} + D_{HH})/2$. Ehrlich investigated the use of the geometric mean in the chemisorption of hydrogen on metals. He found the agreement poorer for the transition metals, though chemisorption of hydrogen on zinc, cadmium, and mercury were correctly predicted as endothermic with values of 23.06 $(\chi_M - \chi_H)^2 \gtrsim$ 2–3 kcal/mole.

Attempts to make minor repairs on the Eley equations do not appear worthwhile. The foundations are shaky. As we mentioned in Section 7.1, metal–metal bonds are broken in some chemisorptions of hydrogen on transition metals, which adds an endothermic term. The hydrogen atom may not be bonded to a single atom of the solid. There is some evidence *(23–25)* indicating that the atom occupies the central point of a lattice square of surface metal atoms with bonding to at least two of them, which, as Eley *(13)* has pointed out, introduces an additional term for pivotal resonance of bonds. Perhaps the most fundamental criticism is that representing covalent bonds by either the arithmetic or the geometric mean is a gross approximation even for free atoms.

Trapnell *(26)* finds that heats of adsorption on different metals for the gases H_2, O_2, N_2, CO, CO_2, C_2H_4, and NH_3 follow a common pattern

$$\text{Ti, Ta} > \text{Nb} > \text{W, Cr} > \text{Mo} > \text{Fe} > \text{Mn} > \text{Ni, Co} > \text{Rh} > \text{Pt, Pd} > \text{Cu, Au.} \quad (11)$$

The only definite reversal is weak chemisorption of hydrogen on Mn. Some single property of the metals apparently determines activity for chemisorption. From Eley's theory, this factor could be either the heat of sublimation or the difference in electronegativities. Neither gives a good correlation. For example, the order of heats of sublimation are

$$\text{W} > \text{Nb, Ta} > \text{Mo} > \text{Rh} > \text{Pt} > \text{Ti} > \text{Co, Ni} > \text{Fe, Pd} > \text{Cr} > \text{Mn} > \text{Au} > \text{Cu.} \quad (12)$$

The continued search for a single property of metals responsible for chemisorption has also led to the use of Pauling's valence-bond theory. In this theory, cohesion in metals is assumed to stem from localized *dsp* hybrid bonds in contrast to the electron band theory which treats the metal as an assembly of positive nuclei through which valency electrons move more or less freely. In a metal lattice, an atom may have as many as twelve nearest neighbors (face-centered cubic) and there are not enough electrons to form localized bonds between all nearest-neighbor pairs simultaneously. Therefore, Pauling assumes that the electrons resonate between pairs as explained below. He postulates that in transition metals, both *d* and *s* electrons are potentially available for covalent bonding. For Cr, Ni, and Co, which Pauling studied in detail, he assumes a covalence of six in the metallic state. The evidence behind this assumption is largely empirical, briefly as follows. In going from left to right across the periodic table, there is a gradual increase in electrons in metals available for bonding in the sequences K–Cr, Rb–Mo, and Cs–W. There is also a gradual decrease in atomic radii and, therefore, a decrease in bond length. The decrease in bond length is related to an increase in bond strength and to an increase in the number of bonding electrons per atom. When Cr, Mo, and W are reached, the atomic radii continue to be roughly constant throughout the transition series. Pauling interprets this to mean that the bond lengths are constant as are the bond strengths and the number of bonding electrons. Since Cr has only six available electrons, five $3d$ and one $4s$, he assumes that six is the common valence for transition metals with more than six electrons in the d and s orbitals.

Taking cobalt metal as an example, the problem is to fit nine electrons into electronic orbitals, six of them being bonding electrons. It is obvious that some of the electrons must be lifted to a higher state—the $4p$ state; otherwise, because of the rule concerning the pairing of electrons, there would be less than six unpaired electrons using only $3d$ and $4s$ orbitals. Thus there are nine available orbitals: five d, one s, and three p. In order to explain some of the facts, Pauling sets up the following types of electrons and orbitals, and semiempirical rules governing their disposition:

(a) ↑ or ↓ unpaired, nonbonding electrons, responsible for the magnetic moment, restricted to the d-orbital;

(b) ↑↓ paired nonbonding electrons, restricted to the d-orbital;

(c) · bonding electrons (unpaired before bonding occurs), six available, which fill orbitals in order of energy, d, s, p;

(d) ∘ metallic orbitals, available locations for bonding electrons.

With these rules, we find that cobalt has six bonding electrons, three remaining electrons of which 1.7 (magnetic moment in Bohr magnetons) are unpaired, giving 2.35 ↑ electrons and 0.65 ↓ electrons for a net number of 2.35 − 0.65 = 1.7 unpaired, nonbonding electrons. The occupied orbitals are thus $6 + 2.35 = 8.35$, leaving $9 − 8.35 = 0.65$ metallic orbitals. The fractional values may be considered as the average of two states, for example, cobalt may be depicted as in Fig. 7.1. When the sum, six bonding electrons plus the number of unpaired nonbonding electrons, exceeds the total number of electrons, a

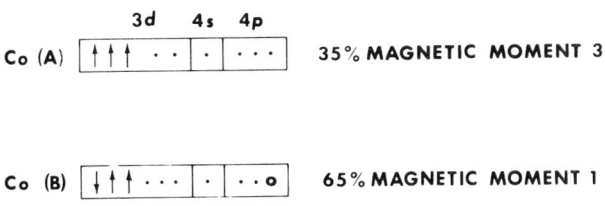

FIG. 7.1. Valence-bond representation of two forms of cobalt metal atoms.

slight change in the rules is necessary. In the case of iron, for example, which has a total of eight electrons and a magnetic moment of 2.22, the number of bonding electrons is reduced to 5.78. The number of ↑ electrons is 2.22, the number of ↓ electrons zero, and the number of metallic orbitals 1.

Pauling assumes that the dsp bonds between pairs of atoms in the metal are continually resonating,

$$\begin{matrix} M:M \\ M:M \end{matrix} \rightleftharpoons \begin{matrix} M\ M \\ \overset{..}{M}\ \overset{..}{M} \end{matrix}$$

and the phenomenon is called "synchronized" resonance. However, this picture alone is not sufficient to explain even qualitatively the experimental facts. For example, it does not explain the electrical conductivity of a metal. The model has no electrons available for conduction; a switch in one pair of electrons is always balanced by a compensating switch of another pair. A further type of resonance is postulated

$$\begin{matrix} M:M \\ M:M \end{matrix} \rightleftharpoons \begin{matrix} M:M^- \\ M^+\overset{..}{M} \end{matrix}$$

and is called pivotal resonance as the bond pivots about the lower right atom. A shift of one bond does not involve the simultaneous shift of any others. The metallic orbitals are used to accumulate additional electrons or atoms which receive an extra charge. Such resonance can qualitatively explain conduction.

Pauling defines a new quantity, the d-character, as the ratio of d bonding electrons to total bonding electrons plus metallic orbitals. For example, the d-character of 35% Co(A) plus 65% Co(B) is $\frac{2}{6} \times 35 + \frac{3}{7} \times 65 = 39.4\%$. High d-character indicates a high participation of electrons in cohesion and consequently could indicate a low availability for formation of covalent bonds at the surface. The order of heats of chemisorption should therefore be the inverse order of d-characters which is (26)

$$Ti > Ta, Cr, Nb > Co > Fe > Ni, Mn > Mo > W > Pt > Pd > Rh > Cu, Au. \quad (13)$$

Again, the order does not agree well with the true order, Eq. (11).

Certain factors previously mentioned, such as breaking of adsorbent bonds upon chemisorption, sharing of the adsorbate by more than one adsorbent atom, and surface free valences differing from bulk orbitals, may obscure the actual role of d orbitals in chemisorption. It is not possible to rule out theories unequivocally that claim chemisorption is the result of covalent bond formation with empty d orbitals or d-band vacancies, or with dsp hybrid surface orbitals. In fact, the greater activity of d-metals in general for chemisorption, appears to indicate that d-orbitals are involved. But the problem has taken on an added complexity recently with the finding (23–25) that preformed hydrogen atoms can be chemisorbed on Group IB metals while molecular hydrogen cannot. This would seem to indicate that chemisorption, in some cases, might be limited by a prohibitively high activation energy and not necessarily by the inability of the metal to chemisorb. There may be little difference between the potential energy curves of chemisorption of, say, hydrogen atoms on *sp* (nontransition) and *dsp* (transition) metals. It may be as Dowden (*27*) points out that there is an initial state of chemisorption called Type C chemisorption intermediate between physical adsorption and strong chemisorption and that there is a difference between the potential energy curves for this weak chemisorption. In the case of transition metals, the Type C potential curve, may intersect the curve for strong atomic chemisorption in such a way that very little activation energy is required for the hydrogen to transfer to the atom curve. Evidence for weak chemisorption of this kind, accompanying strong chemisorption, has been found. In the case of a non-transition metal, the crossing point may be located in a position that requires a large activation energy for the transfer.

IONIC BONDING

Rough estimates of the energy change associated with the ionic mechanism for adsorption of A on metal M

$$M + A \rightarrow M^- A^+$$

may be made by breaking down the energy as follows: (i) remove an electron from the highest occupied level of an isolated atom of A to infinity $(-Ie)$; (ii) transfer the electron to the Fermi surface of the metal M, (φe); (iii) bring A^+ to its equilibrium distance z^* from the surface of the metal $(e^2/4z^*)$. The heat of adsorption per gram atom at zero coverage is then considered to be given by

$$q_d = [-Ie + \varphi e + (e^2/4z^*)]N, \tag{14}$$

where I and φ are the ionization potential of A and work function of M, respectively, and N is Avogadro's number. The last term on the right is the classical image energy of attraction of the ion A^+. A similar expression for the negative ion mechanism

$$M + A \rightarrow M^+ A^-$$

can be written

$$q_d = [-\varphi e + A_0 e + (e^2/4z^*)]N, \tag{15}$$

where A_0 is the electron affinity of adsorbate A. If the adsorption is dissociative, as in the case of hydrogen and other diatomic molecules, then $-\frac{1}{2}D_{AA}$, the heat of dissociation of A_2 per gram atom of A must be added to the right-hand side of Eqs. (14) and (15). The metal is considered to be a semi-infinite plane and z^* is assumed to be the radius of the adsorbed ion. In addition to the question of the proper value for z^*, other criticisms of the accuracy of these equations can be made. For example, (i) the classical image law is not adequate at distances comparable with lattice spacing and (ii) short-range quantum mechanical forces of attraction and repulsion are ignored.

Values of $N\varphi e$ lie between about 90 and 140 kcal/gm atom for metals other than the alkali and alkaline earth metals; values of NIe for oxygen, nitrogen, and hydrogen atoms are 312.0, 333.5, and 312.0 kcal/gm atom, and the classical image energy $Ne^2/4z^*$, is in the neighborhood of 50 kcal/gm atom. Using values in this range for $N\varphi e$, NIe, and the image energy in conjunction with the dissociation energies of oxygen, nitrogen, and hydrogen (117, 225, and 103 kcal/mole, respectively), we find extremely high endothermic values. For example, we find for hydrogen that the heat of adsorption for

$$W + \tfrac{1}{2}H_2 \rightarrow W^-H^+$$

is approximately -210 kcal/gm atom. Such processes are highly improbable. Similar conclusions can be reached with respect to adsorption in which the adsorbate ion is negative. We may conclude that a completely ionic bond will not form.

On the other hand, the adsorption of alkali metals on tungsten appears to proceed with the formation of positive adsorbed ions as may be seen in Table 7.1 from Trapnell and Hayward (26).

TABLE 7.1

CALCULATED AND EXPERIMENTAL INITIAL HEATS OF CHEMISORPTION[a]

System	$N\varphi e$	NIe	$Ne^2/4z^*$	q_d calculated	q_d experimental	Reference
Na on W	104.0	118.0	44.5 (1.83 Å)[b]	30.5	32.0	(28)
K on W	104.0	99.6	35.9 (2.27 Å)	40.3	—	
Cs on W	104.0	89.4	31.1 (2.62 Å)	45.7	64.0	(29)

[a] Values in kilocalories per gram atom.
[b] Values in parentheses, z^*.

Bennett and Falikov (61) have studied the adsorption of alkali metal atoms on tungsten in more theoretical detail, which we will not describe here. They treat the metal–adsorbate system in terms of a simple model allowing calculation of the effective charge on the adsorbed atom. The Slater approximation is used for the adsorbed alkali metal atom and a free-electron model is used for the tungsten. Numerical results for K, Rb, and Cs on W are only in fair agreement with experiment. The important point in this and other similar studies (62–65) now in their infancy is that the fundamental mechanisms involved can be understood in some detail by such treatments.

A QUANTUM MECHANICAL APPROACH

Higuchi *et al.* (*30*) using a simple quantum mechanical approach based on a proposal of Wall (*31*), have calculated the percentage ionic character of covalent adsorption bonds and heats of adsorption. They approximate the surface complex by a diatomic molecule consisting of the adsorbed atom A and a surface atom M. The eigenfunction of the bond is given by

$$\psi = C_i \psi_i + C_c \psi_c, \tag{16}$$

where ψ_i and ψ_c are eigenfunctions of the ideal ionic and covalent bonds, respectively, and C_i and C_c are constants. The wave equation $H\psi = E\psi$ is put into the form $(\int \psi \, H\psi \, d\tau)/\int \psi^2 \, d\tau = E$, and E, the energy of the adsorption bond, is minimized by the usual variational procedure with respect to C_i and C_c. The result is

$$1/C_i^2 = 1 + (E - H_{ii})/(E - H_{cc}), \tag{17}$$

where $H_{ii} = \int \psi_i H\psi_i \, d\tau$ and $H_{cc} = \int \psi_c H\psi_c \, d\tau$, and $C_i^2 = 1 - C_c^2$ is the fractional ionic character of the bond. It is assumed, in deriving Eq. (17), that ψ_i and ψ_c are separately normalized. When $C_i^2 = 1$, it is seen from Eq. (17) that $E = H_{ii}$, and the bond is completely ionic. Similarly, for a completely covalent bond, $E = H_{cc}$. To a good approximation, the authors write

$$H_{ii} = A_0 e - Ie + \tfrac{8}{9}(e^2/r^*), \tag{18}$$

where A_0 is the electron affinity of atom M and I the ionization potential of atom A, if the electron transfer is from A to M. When the direction of transfer is reversed, then A_0 is the electron affinity of atom A and I the ionization potential of atom M. The last term on the right in Eq. (18) is obtained by assuming that the energy u_s between a positive ion and a negative ion is

$$u_s = (-e^2/r) + (B/r^9). \tag{19}$$

When the energy is minimum, $(\partial u_s/\partial r) = 0$ at r^*, and

$$u_s(r^*) = \tfrac{8}{9}(e^2/r^*). \tag{20}$$

The equilibrium distance is taken as the sum of the metallic radius of M and the *covalent* radius of A. [Compare with the definition of z^* in Eqs. (14) and (15).] Equation (20) corresponds to the mirror-image term of Eqs. (14) and (15) for ionic adsorption where the ion is assumed to interact with a semi-infinite plane metal surface rather than with a single metal atom as in the present case of covalent adsorption with ionic character. Following Eley and Pauling,

$$H_{cc} = \tfrac{1}{2}(D_{MM} + D_{AA}), \tag{21}$$

the energy for an ideal covalent bond, which corresponds to Eq. (3) when the bond is completely covalent. The fractional ionic character of the bond C_i^2 is calculated from the expression

$$\mu = C_i^2 er^*, \tag{22}$$

where μ is the bond moment. Using this value of C_i^2 and values of H_{ii} and H_{cc} from Eqs. (18) and (21), the bond energy E may be calculated from Eq. (17). The bond moment μ is obtained from measurements of surface contact potentials as in Eqs. (7) and (8). (See Section 7.3 for further details.) The heat of adsorption per gram atom for monatomic gases at approximately zero coverage is

$$q_d = NE, \tag{23}$$

and per gram mole for homonuclear diatomic gases A_2 is

$$q_d = N(2E - D_{AA}). \tag{24}$$

The results of Higuchi *et al.* show that bonds such as cesium on tungsten and sodium on tungsten are purely ionic, but bonds of barium on tungsten and strontium on tungsten are ionic with a small amount of covalent character ($C_i^2 = 0.97$ for Ba and 0.67 for Sr). For gases on metals, the bonds are covalent with a small amount of ionic character ($C_i^2 = 0.02$ to 0.09). As would be expected in the latter strongly covalent systems, they find bond energies differing little from those obtained by Eley (*12*), since the present method offers an alternative method of calculating ($\chi_M^P - \chi_A^P$) which makes only a small contribution to q_d. The method is more versatile, for all types of adsorption, covalent → ionic, are treated. However, it suffers from the same general weaknesses discussed under the Eley method. Manes and Molinari (*32*) have presented evidence that the ionic term is usually underestimated; i.e., the value of C_i^2 is usually too small. The authors propose an empirical method for improving the value of C_i^2 and of introducing the perturbation of the metal–metal bonds of the adsorbent which involves adding a term proportional to D_{MM}. Although greater accuracy of calculated heats is obtained the information gained in understanding the nature of the chemisorption bond is limited because the method is wholly empirical.

7.3   Variation in Heats of Adsorption with Coverage

Metals generally show a pronounced decrease in heat of adsorption with coverage. The decrease has been attributed to various factors including, (i) adsorbate–adsorbate interactions, (ii) change in work function of the metal by chemisorption, (iii) alteration of bonding with coverage, and (iv) inherent nonuniformity of the metal surface.

Adsorbate–adsorbate interactions may be the result of mutual repulsion of parallel oriented dipoles of adsorbate, of short-range repulsion forces, or of attraction between adsorbate atoms caused by van der Waals forces. However, these interactions can account for only a small part of the total decrease. For example, the bond moment of a hydrogen atom on tungsten is $\mu = -0.5$ D and the distance of separation of neighboring dipoles is $r^* = 2.73 - 3.8$ Å. Summing the interaction energy ($\mu^2/2r^3$) over the surface at full coverage gives 2–3 kcal/mole compared with an observed decrease of 25–30 kcal/mole (*33*). In a similar example for nickel, the decrease due to dipole repulsion of hydrogen is at most 2 kcal/mole, while the experimental decrease is about ten times as large (*34*). Only in the case of alkali–metal atoms is the dipole moment large enough for repulsive interactions to form an appreciable fraction of the observed decrease. For example, cesium and sodium on tungsten ($\mu = 3.6$ and 5.4 D, respectively) give calculated decreases of 12 and 10 kcal/mole compared with observed decreases of 24.5 and 15 kcal/mole, respectively. Short-range repulsive forces are generally thought to be too small to give appreciable heat decreases, because chemisorbed particles are not sufficiently near for electron clouds to overlap. If, as Grimley (*35*) suggests, surface orbitals extend over several atomic diameters, then such repulsive forces may be important. Short-range repulsive forces are in general difficult to evaluate quantitatively. Van der Waals forces between physically adsorbed molecules are observed at higher coverages, and lead to increases in the heat of adsorption roughly equal to the heats of condensation. No estimates of such forces between chemisorbed particles have been made, but they are probably of the same magnitude as those found in physical adsorption. Many experimental heats of adsorption show a drastic decrease at low coverages, whereas changes due to adsorbate–adsorbate interactions would be expected to show largest effects at higher coverages ($\theta \gtrsim 0.5$).

Several workers have suggested that the fall in heat of adsorption can be attributed to $\Delta\varphi$ the decrease in the work function of the metal. Boudart (*36*) suggested that the decrease Δq_d is equal to $\frac{1}{2}Ne\Delta\varphi$; Mignolet (*37*) and de Boer and co-workers (*38*) give $\Delta q_d = Ne\Delta\varphi$. The problem has been discussed in some detail by Higuchi *et al.* (*30*). They start with the Helmholtz equation

$$V = 4\pi B\theta ed/\epsilon, \tag{25}$$

where V, the contact potential, is the potential difference between the adsorbed layer and the metal surface, and is therefore equal to $\Delta\varphi$ the change in work function; B is the number of adsorption sites per unit area, θ the fraction of sites covered, ϵ the dielectric constant of the medium in the electric double layer formed by the adsorbed film, and d the distance shown in Fig. 7.2a for the adsorption of an ion with its mirror image. The Helmholtz equation holds also for the electrical double layer formed by a covalent adsorption

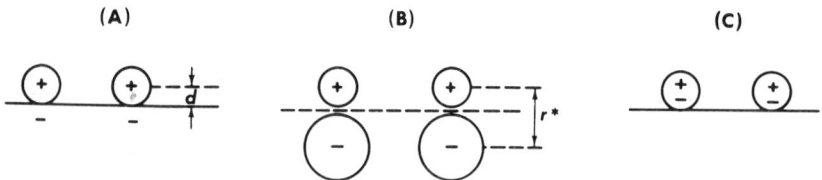

FIG. 7.2. (a) Adsorption of an ion; the negative charges are the electrical mirror images of the positive charges of the adions. (b) The ionic part of a covalent adsorption with positively charged adatoms. (c) Van der Waals adsorption with adatoms polarized by a surface field (*30*). [After Higuchi *et al.*, *J. Am. Chem. Soc.* **79**, 1330 (1957). By permission of The American Chemical Society.]

with partial ionic character. To put Eq. (25) into appropriate form for the covalent case, we use, in place of d, distance r^* which is the sum of the covalent radius of the adatom and the metallic radius of the metal atom, as in Fig. 7.2b; and we set $\mu = er^*$ and $\epsilon = 1$. We obtain

$$V = 2\pi B\mu\theta, \tag{26}$$

which is equivalent to Eq. (7) of Section 7.2.

Now the contact potential V is the result of an electron transfer accompanying covalent adsorption. In desorption, the electron is transferred back to the original position through the potential difference $V = \Delta\varphi$. If the adatom is positive, i.e., an electron is transferred to the metal, the potential as it increases with coverage favors more and more the transfer of the electron back to the adatom. If the adatom is negative, the potential as it increases with coverage favors more and more the transfer of the electron back to the metal. Thus, in both cases the heat of desorption decreases with coverage and is given approximately by

$$\Delta q_d = Ne\Delta\varphi \tag{27}$$

in agreement with deBoer and Mignolet. In the special case where homonuclear diatomic molecules chemisorb as atoms, but desorb as molecules, Eq. (27) becomes

$$\Delta q_d = 2Ne\Delta\varphi. \tag{28}$$

Higuchi *et al.* show good agreement between observed decreases and calculated decreases for hydrogen on iron and on nickel, using Eq. (28). There are not enough data available to determine whether it is generally true that the decrease in heat is caused chiefly by the change of φ.

Dowden (*27*) suggests that chemisorption bonds are altered as coverage increases. He reasons that some *dsp* hybrid bonds at the surface of a *d*-metal must exist essentially free, otherwise the Eley–Stevenson equation for heats of chemisorption discussed in Section 7.2 should contain an endothermal term for initial breaking of metal–metal bonds. He further postulates that these free *dsp* bonds provide a reservoir of metallic orbitals which are used only to a minor extent in pivotal resonance of metal–metal bonds. (See the discussion of the valence-bond theory of Pauling in Section 7.2.) These orbitals are largely available for chemisorption. As chemisorption proceeds, the metallic orbitals of the free *dsp* bonds will be used up. Thus, with increasing coverage, heats of chemisorption will fall, because there will be increasing interference with pivotal resonance in the metal until a point is reached at which an endothermal term for breaking of metal–metal bonds must be included in any calculations of the heat of adsorption.

Although no quantitative verifications are possible, Dowden's theory cannot be categorically denied. Neither can it be denied that decreasing heats of chemisorption may be the result of a composite of the mechanisms described above, some of which alone are incapable of producing the observed decrease. Finally, there is considerable experimental evidence that an inherent nonuniformity of surfaces, associated with point imperfections, surface dislocations, different crystal faces, impurities, etc. is a significant cause of falling heats. Indeed, it appears to be the most reasonable cause when the heat falls rapidly at low coverages (*39*).

7.4   Ligand-Field Treatment of Surface Bonding

In the valence-bond theory of metals, all bonding is considered to be localized. Properties such as cohesion and conduction are explained by means of synchronized resonance and pivotal resonance. Other theories involve collective electrons which are assumed to roam more or less freely through the volume of the metal. The simplest theory of collective electrons in metals is the *Free Electron Theory*. In this theory, a metal is pictured as a box containing the valence electrons in unhindered motion, and it is assumed that there are no repulsive forces between electrons. The problem is solved essentially by the familiar particle-in-a-box method of quantum mechanics. If there are N free (valence) electrons, there will be $N/2$ energy states filled according to the Pauli principle with two electrons per state. At absolute zero, electrons will fill the lowest energy state first, and then proceed to some higher level until all electrons are placed. The highest filled energy level is called the Fermi level. The distribution of energy states $N(E)$, where $N(E)\, dE$ is the number of energy states per unit volume with energies between E and $E + dE$, is proportional to $E^{1/2}$. Figure 7.3a gives a plot of $N(E)$ versus E, where E_{max} is the Fermi level.

Another collective electron approach is the band theory. It was developed by Bloch (*40*) who introduced molecular orbital theory for this purpose—a theory which was later applied to free molecules. In molecular orbital theory,

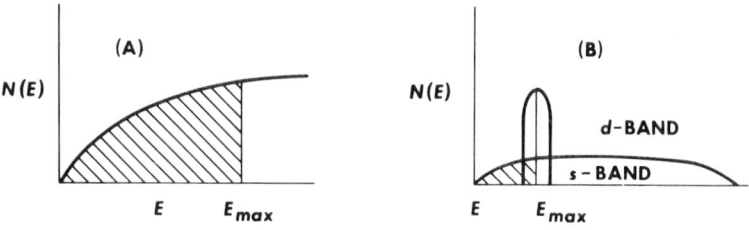

FIG. 7.3. (a) Distribution of energy states, free-electron model. (b) Distribution of energy states for nickel, band model.

each electron is presumed to move in a potential field that extends over all atoms. By suitable averaging, one-electron functions may be developed, and solutions of the wave equation are described approximately by linear combinations of atomic orbitals. The wave function for a single atom "A" in a given state is given by one atomic orbital ψ_A. If we add a second atom " B," the approximate wave function becomes the two diatomic orbitals $\psi_A \pm \psi_B$ each with its distinct energy one above and one below the original energy. Each successive addition of an atom and its corresponding atomic orbital to the wave function adds another energy value and alters slightly the energies of the previous set. The orbitals for an infinite array, as a metal is considered to be, differ in one very important respect from those of a free molecule. In the limit of an infinite array of atoms, the original atomic orbital gives rise to a band of energies. There is one band for each of the allowed orbitals of the original atom. The levels within a band are so close together that they may be called a continuum. When the atoms are at infinite separation from one another, the energies are just the energies of the separate atomic orbitals. As the lattice constant diminishes, interaction occurs and the band width increases. When the lattice constant is sufficiently small, some atomic orbitals will show strong overlap with similar orbitals on neighboring atoms and the band may become so broad that it merges into other bands. Considerable band merging occurs in metals. There is some overlap of nd orbitals in d-metals and even more for $(n + 1)s$ orbitals, but very little overlap for deeper levels. Thus the d-orbitals usually produce a narrow band, the s-orbital a broad band, and deeper levels retain the characteristics of isolated atomic orbitals. Figure 7.3b shows schematically the band structure of nickel. Explicit calculations for nickel indicate that there are about 9.45 electrons in the d-band giving 0.55 holes, and 0.55 electrons in the s-band. The Fermi level, which is shown in Fig. 7.3b, is thus close to the upper edge of the d-band.

The band model ignores crystal structure. Recently electron bands have been described in terms of the overlap of molecular orbitals which are directional because of the crystal field of the metal (41, 42). This approach is an extension of the ligand-field theory employed in the study of complex molecules *(43–45)*. According to the ligand-field theory, the five-fold degenerate *d*-level of an isolated atom splits in cubic coordination, which we use as an example, into a three-fold and a two-fold degenerate level. The three-fold degenerate level (t_{2g} electrons) contains the d_{xy}, d_{zy}, and d_{zx} orbitals, and the two-fold degenerate level (e_g electrons) contains the $d_{x^2-y^2}$ and d_{z^2} orbitals. In a metal, the five-fold degenerate *d*-level not only splits into a threefold and a twofold degenerate level, but each level tends to broaden into a sub-band, the t_{2g} and the e_g, of the *d*-band. Following Goodenough *(42)*, we consider a face-centered cubic structure with a Cartesian frame of reference placed as in Fig. 7.4. The orientation of the t_{2g} electrons bonds (d_{xy}, d_{zy}, and d_{zx} orbitals) and the e_g electron bonds (d_{z^2} and $d_{x^2-y^2}$ orbitals) are also shown.

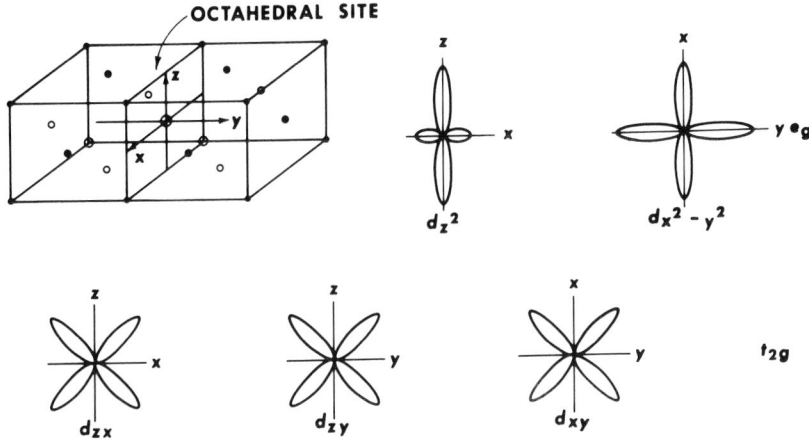

FIG. 7.4. Cartesian frame of reference for face-centered cubic metal and the e_g and t_{2g} orbitals.

It may be seen from Fig. 7.4, that there are twelve nearest-neighbor atoms and that the central atom may be bonded to each of them as a result of each of the twelve lobes of its t_{2g} orbitals (four in each of the three planes) overlapping with the lobe of a nearest-neighbor. There are six next-nearest-neighbor atoms and each of the six lobes of the e_g orbitals of the central atom are directed toward one of them along the coordinate axes, but not necessarily bonded to them because of the greater distance of separation. Goodenough gives the density of states curve, $N(E)$ versus E, for nickel, including the crystal field modification of the bands. In Fig. 7.5, we see that the e_g sub-band is narrow, indicating that the orbitals are essentially localized and that they do not overlap neighboring orbitals significantly. The localized e_g electrons occupy two levels (narrow bands) resulting from intra-atomic exchange splitting (Hund's rule). The t_{2g} sub-band, on the contrary, is broadened appreciably by

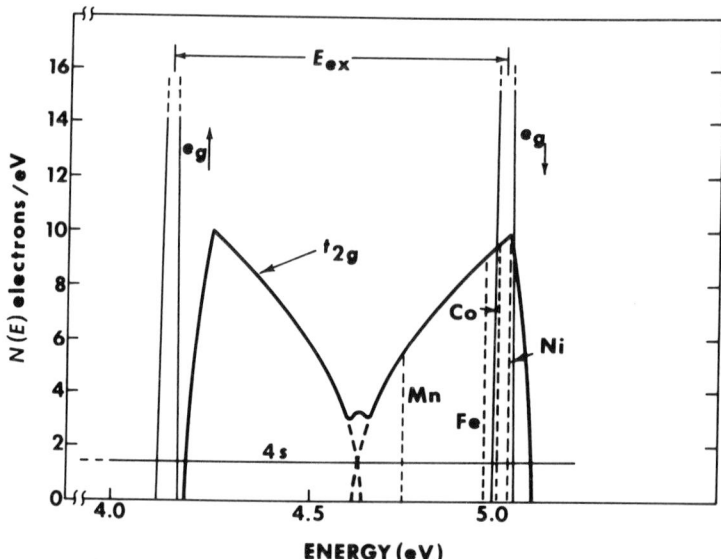

Fig. 7.5. Schematic density of states curve for face-centered cubic nickel. Fermi levels are given for Ni., Mn, Fe, and Co. [After Goodenough, "Magnetism and the Chemical Bond," Wiley (Interscience), New York, 1963.]

orbital overlapping, though it is still narrow compared to the s-band (Fig. 7.3b). From polarized neutron experiments, Goodenough estimates that the 0.55 holes of the d-band are split 0.41 t_{2g} holes and 0.14 e_g holes.

Bond (*46*) applies the theory qualitatively to chemisorption. From the information given in Fig. 7.4, we can describe the emergence of orbitals on the various faces of a face-centered cubic metal. For example, two t_{2g} orbitals emerge at 45° (d_{yz}, d_{xz}) on the (100) face; one t_{2g} orbital (d_{xy}) is parallel to the 100 face and does not emerge (0°); one e_g orbital (d_{z^2}) emerges normal to the 100 surface (90°), and one e_g orbital ($d_{x^2-y^2}$) is parallel and does not emerge. In a similar manner, Bond describes the emergence on the (110) and (111) faces of a face-centered cubic metal. The information is summarized in Table 7.2. The (111) plane is the only one which has no vertically emerging orbitals.

A possible site for adsorption of a hydrogen atom on a (100) surface is the vertically emerging e_g orbital which could overlap with the $1s$ orbital of hydrogen. An alternative site is the position marked octahedral site in Fig. 7.4, where a $1s$ orbital could be overlapped by five e_g orbitals (*47*) giving a strongly bound state.

Adsorption of mono-olefins should take place most readily on the (100) face of nickel using the partially vacant e_g (90°) and t_{2g} (45°) orbitals to form a π-adsorbed complex as shown in Fig. 7.6. On the (110) face, a coordinate bond could form between the olefin and an e_g orbital with the bond axis at 45° to the

TABLE 7.2

ANGLE OF EMERGENCE OF ORBITALS ON THE CRYSTAL FACES OF A
FACE-CENTERED CUBIC METAL

Sub-band	Orbital	Face		
		(100)	(110)	(111)
t_{2g}	d_{xy}	$0°$	$30°$ (2)	$30°$ (1)
t_{2g}	d_{xz}	$45°$ (2)[a]	$30°$ (2)	$30°$ (1)
t_{2g}	d_{yz}	$45°$ (2)	$90°$ (1)	$30°$ (1)
e_g	$d_{x^2-y^2}$	$0°$	$45°$ (1)	$36°16'$ (2)
e_g	d_{z^2}	$90°$ (1)	$45°$ (1)	$36°16'$ (1)

[a] Numbers in parentheses are numbers of lobes that emerge.

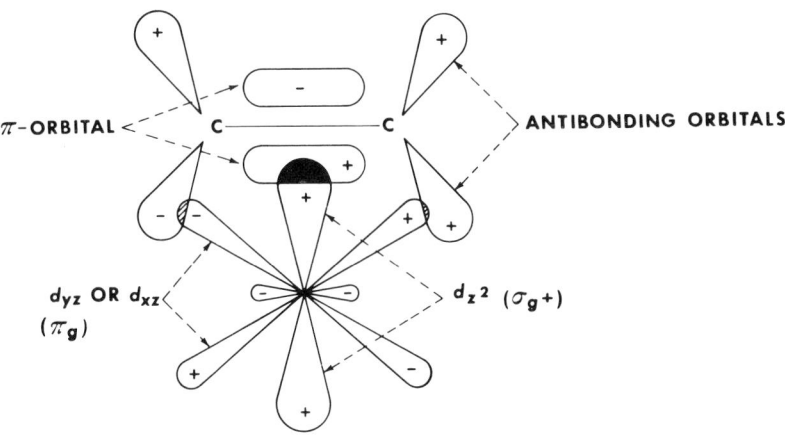

FIG. 7.6. π-adsorbed ethylene on the (100) face of nickel.

surface. If the t_{2g} and e_g orbitals can interchange functions, a bond similar to that on the (100) face is conceivable. Similarly on the (111) face there are linear sequences e_g, t_{2g}, e_g which could serve for π-adsorption of ethylene, if the functions of e_g and t_{2g} orbitals can interchange. There are no linear sequences t_{2g}, e_g, t_{2g} on the (111) face. In any event, however, the e_g orbitals make an angle of only about 36° with the surface and this could lead to considerable repulsion between the adsorbed molecule and neighboring surface atoms. Two possibilities exist for σ-diadsorbed olefins: (i) overlap with adjacent vertical e_g orbitals on the (100) face, and (ii) overlap with adjacent vertical t_{2g} orbitals or with 45°-emergent e_g orbitals on the (110) face. The (111) face again appears to be the least suited for chemisorption of olefins. Bond gives further examples of the qualitative application of the ligand-field approach to the chemisorption of carbon monoxide, diolefins, and alkynes with reasonable results.

7.5   Surface States

In all the theories discussed so far, we have assumed, tacitly at least, that the forces and bonds in a solid surface and those that emerge from the surface and are available for bonding foreign atoms are not essentially different from those operating within the metal. Of course, it has been known in theory for a long time that states peculiar to a free surface can exist (48), and it has been suspected that these free-surface states could have a serious effect on our theories of chemisorption. Nevertheless, the complexity of the problem has deterred progress. Many qualitative speculations have been made. For example, Oriani (22) showed that pairwise bonding of surface atoms to their nearest and next-nearest neighbors is stronger and the number of ligands smaller than in the bulk. And, on the basis of this evidence, Dowden postulates that nearest-neighbor bonds in α iron could rearrange to give surface states as follows:

$$d^3sp^2 \to d^2p \text{ (trigonal pyramid)} + dsp \text{ (bent planar)}$$

and for next-nearest neighbors:

$$d^2sp^2 \to d^2sp^2 \text{ (distorted tetragonal pyramid).}$$

In order to get a clear, quantitative understanding of the surface properties of solids, two important problems must be solved. First, there is the problem of a free surface, and second, the problem of a surface contaminated by adsorbed atoms or molecules. The problem of a free surface goes back to a paper by Tamm (48), but the second problem is more recent (49, 50). Only a small amount of work has been done on real surfaces in either problem. Most of the work deals with simple models, often one dimensional, and employs the LCAO approximation of the molecular-orbital theory. No numerical values of energies have been calculated as yet, but the results are interesting and provocative. The energy patterns of the one-electron wave functions constructed in this manner are evaluated by the variational procedure using the tight-binding approximation which includes resonance integrals between nearest neighbors and the Coulomb integrals, but ignores overlap integrals. Electron-electron interactions are also ignored. In the usual crystal problem, an infinite crystal is assumed, so that surface effects are not encountered; in the present case, we must deal with a finite crystal. The main contributions to the theory of free surfaces are from Tamm (48), Shockley (51), Goodwin (52), Artmann (53), Koutecky (54–56) and Grimley (57, 58). The simplest model is a one-dimensional chain of N atoms, numbered 0, 1, ..., N, the two ends of which represent the free "surfaces." This model possesses all the essential features of the problem. Following Grimley (*58*), we shall outline the solution to illustrate the principles involved. Associated with each atom m is an atomic orbital $\varphi_m(r)$, and the LCAO wave function of an electron in the chain is

$$\psi(r) = \sum_m \varphi_m(r)c_m. \tag{29}$$

Using the effective one-electron Hamiltonian operator for the chain, $\psi(r)$ satisfies the equation

$$\mathcal{H}\psi(r) = E\psi(r). \tag{30}$$

If we substitute Eq. (29) into Eq. (30) and minimize the energy by the variational procedure, we obtain the usual system of linear equations for the wave function coefficients c_m and the energies E,

$$(E - H_{mm})c_m = \sum_{n(\neq m)} H_{mn}c_n, \tag{31}$$

where

$$H_{mn} = \int \varphi_m^*(r)\mathcal{H}\varphi_n(r)\,d\mathbf{r} \tag{32}$$

is the resonance integral between atoms m and n, and H_{mm} is the Coulomb integral. Using the tight-binding approximation (which is not particularly satisfactory for metals), we include the resonance integral only between nearest neighbors, $H_{m,\,m\pm1}$, and the overlap integrals $\int \varphi_m \varphi_n\,d\mathbf{r}$ are omitted. We now set

$$H_{m,\,m\pm1} = \beta; \qquad H_{mm} = \alpha, \qquad m \neq 0; \qquad H_{00} = \alpha', \tag{33}$$

in which we recognize the existence of a free surface by making the Coulomb integral α' on the end atom, $m = 0$, and α otherwise. Equation (31) now becomes

$$(E - \alpha)c_m = \beta(c_{m+1} + c_{m-1}), \qquad m \neq 0, \tag{34}$$

and the boundary condition

$$(E - \alpha')c_0 = \beta c_1, \qquad m = 0. \tag{35}$$

We need a boundary condition for the other end of the chain, $m = N$. If N is very large, this boundary condition cannot affect conditions near $m = 0$ significantly. Thus, we set $c_N = 0$. Now let

$$c_m = \sin(N - m)\theta, \qquad m = 0, 1, \ldots, N. \tag{36}$$

This satisfies the boundary condition at $m = N$. By substitution, it satisfies Eq. (34) if

$$E = \alpha + 2\beta \cos \theta \tag{37}$$

and Eq. (35), the boundary condition at $m = 0$, if θ is one of the N roots of

$$z + \cos \theta + \sin \theta \cot N\theta = 0, \tag{38}$$

where

$$z = (\alpha - \alpha')/\beta. \tag{39}$$

Thus, Eq. (36) is a solution to Eq. (34) provided E and θ are expressed by Eqs. (37) and (38), respectively. An analysis of Eq. (38) shows that it has at least $N - 1$ real roots. The wave functions corresponding to Eq. (36) are periodic, and writing $E' = (E - \alpha)/2\beta$, we find from Eq. (37) that the energy levels lie in the range

$$-1 < E' < 1. \tag{40}$$

This equation defines a band of nonlocalized states of width $4|\beta|$. All of these levels arise from a single state $\varphi(r)$ of the isolated atoms. If $|z| < 1 + N^{-1} \cong 1$, the remaining root will also be real and lie in the band. In this case, the chain of atoms has only nonlocalized states. If, however, $|z| > 1 + N^{-1} \cong 1$, then the remaining root is of the form $\theta = i\xi$ or $\theta = \pi + i\xi$, where ξ is real and positive. These states are localized, associated with the free end $m = 0$, and their energies fall outside the band of levels given in Eq. (40). For the state with $\theta = i\xi$, E' is positive and is called a $\mathscr{P}$ state; for the state with $\theta = \pi + i\xi$, E' is negative and is called an $\mathscr{N}$ state. We find from Eq. (38) that a $\mathscr{P}$ state exists if

$$-z = \cosh \xi + \sinh \xi \coth N\xi, \tag{41}$$

where we have used relations such as

$$(e^{i\theta} + e^{-i\theta})/2 = \cos \theta = (e^{\xi} + e^{-\xi})/2 = \cosh \xi.$$

If N is very large, we have

$$-z = e^{\xi}, \qquad (\xi > 0), \tag{42}$$

and, therefore, a $\mathscr{P}$ state exists if $z < -1$ with wave function and energy given by

$$c_m = c_0 e^{-m\xi}, \tag{43}$$

$$E = \alpha + 2\beta \cosh \xi. \tag{44}$$

Similarly for an $\mathscr{N}$ state, $z = e^{\xi}$ when N is very large, and this state exists if $z > 1$ with wave function and energy given by

$$c_m = c_0(-1)^m e^{-m\xi}, \tag{45}$$

$$E = \alpha - 2\beta \cosh \xi. \tag{46}$$

For a specific case, we assume $\beta < 0$ which applies when the orbitals $\varphi(r)$ have the symmetry of atomic s-states. Then in summary, we find when $z = 0$ that the chain has N nonlocalized states with a band width of $4|\beta|$. As z decreases (chain end electron attracting), all of these states decrease in energy. At $z = -1$ a type $\mathscr{P}$ end-state separates below the band and continues to drop with further decrease in z. Its wave function concentrates more and more on the end atom. As $z \to -\infty$, the $\mathscr{P}$ state becomes the completely discrete atomic orbital $\varphi(r, 0)$, and the wave functions for the nonlocalized states are zero on the end atom. A corresponding situation occurs with the $\mathscr{N}$ state as z increases above zero (chain end electron repelling), except that the level separates above the band instead of below.

The two-dimensional case is a straightforward extension of the method just outlined. There is one difference—the discreet end state of the one-dimensional case for $|z| > 1$ now becomes a band of surface states of width $8|\beta|$ and contains N^2 levels. At large $|z|$, the band of surface states is nearly separate from the crystal bands, and the normal crystal bands are vanishingly small on the surface atoms. Such bands of surface states are called Tamm states, and they stem from the difference in energy parameters between the surface and the interior, as we have seen. In a more realistic picture of the electronic structure of a crystal, more than one atomic state would have been used, and we would have obtained more than a single crystal band. Metals, for example, have a complex structure of overlapping bands. With more than one atomic orbital on each crystal atom producing an overlapping band system, then bands of surface states can be formed even without a difference in the energy parameters between the surface and interior regions. The surface bands are called Shockley states. They are difficult to treat even in the simple LCAO approximation. For a critical discussion of the various methods of solution of the LCAO wave function, the reader is referred to a publication of Levine and Mark (*66*).

When a single foreign atom is adsorbed on a crystal surface, the wave function can be written as in Eq. (29), but the summation over atomic orbitals contains an additional term for the foreign atom. If we let the position of the foreign atom be designated by $m = \lambda$, there will be an additional atomic orbital $\varphi_\lambda(r)$. The Coulomb integral and resonance integrals will be different for the foreign atom and we represent them by $\alpha'' = H_{\lambda\lambda}$ and $\beta' = H_{0\lambda} = H(_{\lambda, 0})$, respectively. Along with Eq. (30), we now have two boundary conditions

$$(E - \alpha')c_0 = \beta c_1 + \beta' c_\lambda, \tag{47}$$

$$(E - \alpha'')c_\lambda = \beta' c_0. \tag{48}$$

With the new boundary conditions in three dimensions, simple solutions are not possible. However, methods developed by Baldock (*59*) and by Koster and Slater (*60*) have been found applicable. We summarize the results.

When $|z| < 1$, so that the crystal has no free surface states, it turns out that the system crystal plus foreign atom may have localized states whose wave functions fall to zero in the interior of the crystal. A maximum of two localized states can be formed—two $\mathscr{P}$ states, two $\mathscr{N}$ states, or one $\mathscr{P}$ and one $\mathscr{N}$ state. There may be only one localized state or even none. In the latter case, all states are nonlocalized and their wave functions extend throughout the crystal with energies lying in the normal crystal band. With $\beta < 0$, $\mathscr{P}$ states will have energies lying below the normal band, and $\mathscr{N}$ will be above it. When $|z| > 1$, the crystal has a band of free surface states which have a significant effect on the locations of any localized states formed with the foreign atom. Any such localized states must be not only outside the surface band, but outside the normal crystal band as well.

Actually, the maximum number of localized states depends on the assumption that only one orbital on the foreign atom, and one band of crystal orbitals are in interaction and perturbation by the foreign atom does not extend beyond the first crystal atom. Extending the perturbation modifies the Coulomb integrals (α, α', α'') beyond the first crystal atom and increases the maximum number of localized states.

When two foreign atoms are adsorbed on the crystal surface, one of several possible situations may prevail. If the adsorbed atoms are infinitely far apart, any localized levels which are formed will be doubly degenerate. At finite separations, each localized level will split into an even and an odd level, and the magnitude of the splitting increases with decreasing distance of separation. If the crystal alone has no surface states, the even localized states involving the foreign atoms are located farther from the normal crystal band and the odd states are located closer to it than the original doubly degenerate localized level. There may be a critical value of separation of the adsorbed atoms below which the odd state merges into the normal crystal band. In this case, the odd state ceases to be a localized level. Thus it is possible to have a single localized state when one atom is adsorbed and not to produce a second one when two molecules are adsorbed. It is important to note that the interaction between two adsorbed atoms takes place *through* the crystal and not directly. We have a possible mechanism here for explaining falling heats of adsorption with coverage.

We have mentioned that most of the work on the molecular orbital theory of chemisorption is concerned with hypothetical, perfect crystals. Grimley (35) has given a few speculations on the real system hydrogen atom plus a metal. Methods similar to those outlined above were used with particular attention to the characteristics of orbitals in a metal. He shows that the adsorption of hydrogen as H^- ions cannot be justified for metals. But localized levels between a hydrogen atom and a metal atom located directly below it in the surface appear to exist. If such a level lies below the Fermi level of the metal, it could be doubly occupied with the formation of a covalent bond between the hydrogen atom and the metal. For similar work, the reader is referred to Blyholder and Coulson (*67*), Janson (*68*), Dunken and Optiz (*69*), and Ruiperez (*70*).

REFERENCES

1. Ehrlich, G., *Proc. 3rd Intern. Cong. Catalysis* **1**, 113 (1964).
2. Erhlich, G., *Brit. J. Appl. Phys.* **15**, 349 (1964).
3. Ehrlich, G., *J. Chem. Phys.* **34**, 29 (1961).
4. Nasini, A. G., and Ricca, F., *Trans. N. Y. Acad. Sci.* **101**, 791 (1963).
5. Ehrlich, G., *J. Chem Phys.* **34**, 39 (1961).
6. Redhead, P. A., *Trans. Faraday Soc.* **57**, 641 (1961).
7. Hickmott, T. W., *J. Chem. Phys.* **32**, 810 (1960).
8. Pasternak, R. A., and Wiesendanger, H. U. D., *J. Chem. Phys.* **34**, 2062 (1961).
9. Gomer, R., "Field Emission and Field Ionization," Harvard Univ. Press, Cambridge, Massachusetts, 1961.
10. Germer, L. H., and MacRae, A. U., *J. Chem. Phys.* **37**, 1382 (1962).
11. Lander, J. J., *Surface Science* **1**, 125 (1964).
12. Eley, D. D., *Discussions Faraday Soc.* **8**, 34 (1950).
13. Eley, D. D., "Catalysis and the Chemical Bond," Univ. of Notre Dame Press, Indiana (1954).
14. Pauling, L., "The Nature of the Chemical Bond," Cornell Univ. Press, Ithaca, New York, 1939.
15. Malone, M., *J. Chem. Phys.* **1**, 197 (1933).
16. Mignolet, J. C. P., *Discussions Faraday Soc.* No. 8, 105 (1950).
17. Mignolet, J. C. P., *Discussions Faraday Soc.* No. 8, 326 (1950).
18. Mignolet, J. C. P., *Rec. Trav. Chim.* **74**, 701 (1955).
19. Stevenson, D. P., *J. Chem. Phys.* **23**, 203 (1955).
20. Bond, G. C., "Catalysis by Metals," Academic Press, New York and London, 1962.
21. Ehrlich, G., *J. Chem. Phys.* **31**, 1111 (1959).
22. Oriani, R. A., *J. Chem. Phys.* **18**, 575 (1950).
23. Culver, R. V., Pritchard, J., and Tompkins, F. C., *Z. Elektrochem.* **63**, 741 (1959).
24. Pritchard, J., and Tompkins, F. C., *Trans. Faraday Soc.* **56**, 540 (1960).
25. Culver, R. V., Pritchard, J., and Tompkins, F. C., *Proc. Intern. Conf. Surface Activity* **2**, Butterworth, London (1957).
26. Trapnell, B. M. W., and Hayward, D. O., "Chemisorption" (W. E. Garner, ed.), p. 210. Butterworth, London and Washington, D.C., 1964.
27. Dowden, D. A., "Chemisorption" (W. E. Garner, ed.), p. 9. Butterworth, London and Washington, D.C., 1958.
28. Bosworth, R. C. L., *Proc. Roy. Soc.* (*London*) *Ser. A* **162**, 32 (1937).
29. Taylor, J. B., and Langmuir, I., *Phys. Rev.* **44**, 423 (1933).
30. Higuchi, I., Ree, T., and Eyring, H., *J. Am. Chem. Soc.* **79**, 1330 (1957).
31. Wall, F. T., *J. Am. Chem. Soc.* **61**, 1051 (1939).
32. Manes, L., and Molinari E., *Z. Physik. Chem.* **39**, 104 (1963).
33. Trapnell, B. M. W., *Proc. Roy. Soc.* (*London*) *Ser. A* **206**, 39 (1951).
34. Gundry, P. M., and Tompkins, F. C., *Trans. Faraday Soc.* **52**, 1609 (1956); **53**, 218 (1957).

35. Grimley, T. B., "Chemisorption" (W. E. Garner, ed.), Butterworth, London and Washington, D.C. 1957.
36. Boudart, M., *J. Am. Chem. Soc.* **74**, 3556 (1952).
37. Mignolet, J. C. P., *Bull. Soc. Chim. Belges* **64**, 126 (1955).
38. DeBoer, J. H., and Veenemans, C. F., *Physica* **1**, 960 (1934).
39. Halsey, G., and Taylor, H. S., *J. Chem. Phys.* **15**, 624 (1947).
40. Bloch, F., *Z. Physik.* **52**, 555 (1928).
41. Trost, W. R., *Can. J. Chem.* **37**, 460 (1959).
42. Goodenough, J. B., "Magnetism and the Chemical Bond," Wiley (Interscience), New York, 1963.
43. Orgel, L. E., "An Introduction to Transition-Metal Chemistry, Ligand-Field Theory," Wiley, New York, 1960.
44. Ballhausen, C. J., "Introduction to Ligand-Field Theory," McGraw-Hill, New York, 1962.
45. Griffith, J. S., "The Theory of Transition Metal Ions," Cambridge Univ. Press, Cambridge, Massachusetts, 1961.
46. Bond, G. C., *Discussions Faraday Soc.* **41**, 200 (1966).
47. Dowden, D. A., *in* "Coloquio sobre Quimica Fisica de Processos en Superficies Solidas" p. 177. Liberia Cientifica Medinaceli, Madrid, 1965.
48. Tamm, I., *Phys. Z. Sowjet,* **1**, 733 (1932).
49. Koutecky, J., *Trans. Faraday Soc.* **54**, 1038 (1958).
50. Grimley, T. B., *Proc. Phys. Soc. (London) Ser. B* **72**, 103 (1958).
51. Shockley, W., *Phys. Rev.* **56**, 317, (1939).
52. Goodwin, E. T., *Proc. Cambridge Phil. Soc.* **35**, 221 (1939).
53. Artmann, K., *Z. Physik* **131**, 224 (1952).
54. Koutecky, J., *Phys. Rev.* **108**, 13 (1957).
55. Koutecky, J., *Advan. Chem. Phys.* **9**, 85 (1965).
56. Koutecky, J., *J. Phys. Chem. Solids* **14**, 233 (1960).
57. Grimley, T. B., *J. Phys. Chem. Solids* **14**, 227 (1960).
58. Grimley, T. B., *Advan. Catalysis* **12**, 1 (1960).
59. Baldock, G. R., *Proc. Cambridge Phil. Soc.* **48**, 457 (1952).
60. Koster, G. F., and Slater, J. C., *Phys. Rev.* **95**, 1167 (1954).
61. Bennett, A. J., and Falikov, L. M., *Phys. Rev.* **151**, 512 (1966).
62. Gyftopoulos, E. P., and Levine, J. D., *Surface Sci.* **1**, 171, 225, 349 (1964).
63. Toya, T., *J. Res. Inst. Catalysis, Hokkaido Univ.* **VIII**, 209 (1961).
64. Gomer, R., and Swanson, L. W., *J. Chem. Phys.* **38**, 1613 (1963).
65. Gadzuk, J. W., *Surface Sci.* **6**, 133, 159 (1967).
66. Levine, J. D., and Mark, P., *Phys. Rev.* **182**, 926 (1969).
67. Blyholder, G., and Coulson, C. A., *Trans. Faraday Soc.* **63**, 1782 (1967).
68. Jansen, L., Mol Processes Solid Surfaces, Battelle Inst. Mater. Sci. Colloq. 3rd (1968), 49. Publ. (1969).
69. Dunken, H. H., and Optiz, Ch., *Z. Phys. Chem.* **60**, 25 (1968).
70. Ruiperez, G., *Ann. Fiz.* **64**, 9 (1968).

CHAPTER

VIII

THE CHEMICAL FORCES
OF ADSORPTION—SEMICONDUCTORS

8.1 Introduction

In this chapter, we shall discuss the boundary-layer theory and the theories of the Russian school, especially Wolkenstein's concepts, all related to chemisorption on semiconductors. The boundary-layer theory follows closely the theory of the contact between a metal and a semiconductor. It applies strictly to ionic adsorption on *n*- or *p*-type semiconductors, involving a transfer of electrons from conduction levels of the adsorbent to the adsorbate or vice versa. As a result of the charged adsorbed particles, a space charge develops in the semiconductor and a boundary layer exists at the surface of the semiconductor across which there is an increasing or decreasing electron concentration in going from the inner edge of the layer toward the surface. Adsorption of this type causes changes in conductivity in the solid and applies only in cases where a transfer of charge occurs. There are examples of systems in which adsorption occurs with no change in conductivity at all, thus, presumably, with no transfer of charge from conduction levels. Such cases fall outside the realm of the boundary-layer theory. The Russian work, particularly that of Wolkenstein, considers adsorption with and without charge transfer. He assumes that the various types of adsorption may occur simultaneously and calculates, using Fermi–Dirac statistics, the equilibrium existing between them. The discussion of these theories is preceded by a survey of the electron theory of semiconductors and insulators upon which the adsorption theories are based.

8.2   Electron Theory of Semiconductors and Insulators

We have seen that the theories of chemisorption on metals are based on qualitative applications of the valence-bond theory and the band theory as well as on semiempirical theories which assume a localized bond between an adsorbate molecule and an atom of the metal. The theory of surface states has not yet developed to the point where numerical calculations of energies can be made for either metals or nonmetals. Theories of chemisorption on semiconductors are based rather firmly on the band theory and we wish now to consider this theory in a little more detail. As we saw in Sections 7.4 and 7.5, the band theory of solids is an extension of the molecular orbital theory for molecules. In Section 7.5, a wave function for an orbital extending over a *finite* linear array of atoms was formed from a linear combination of atomic orbitals. In this section, we shall ignore free surface states and their peculiar boundary conditions, and consider only those states existing within the solid. It is then possible to obtain numerical values for energies using the technique of Bloch molecular orbitals. Although it has already been shown formally in Section 7.4 that energy band structures exist in the solid, we wish to sketch briefly how Bloch orbitals are derived, since they provide a useful extension into the realm of practical calculations when free surfaces are ignored.

BLOCH ORBITALS[1]

Consider the one-dimensional, linear array of N atoms shown in Fig. 8.1. Difficulties introduced by the existence of boundaries may be avoided by imagining this finite array to be part of an infinite array and imposing periodic boundary conditions on the wave functions,

$$\psi(x + Na) = \psi(x). \tag{1}$$

The Hamiltonian for a particle of mass m moving in this crystal is

$$H(x) = - (\hbar^2/2m)(d^2/dx^2) + V(x), \tag{2}$$

where the potential $V(x)$ is a periodic function of x by the nature of the problem,

$$V(x + a) = V(x). \tag{3}$$

[1] See Bloch (*1*).

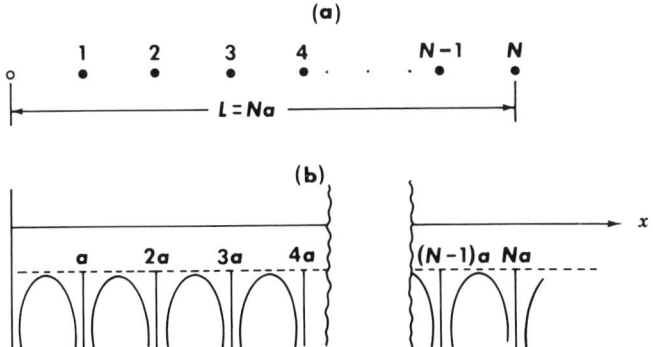

FIG. 8.1. A one-dimensional crystal: (a) arrangement of atoms, (b) potential energy. [After Leighton, "Principles of Modern Physics" McGraw-Hill, 1959. Used with permission of McGraw-Hill Book Company.]

Thus, the Hamiltonian operator is also periodic in x,

$$H(x + a) = H(x), \tag{4}$$

that is, $H(x)$ is invariant under any translation of the lattice which is an integral multiple of a. Now, we introduce an operator $T(x)$ such that

$$T(x)f(x) = f(x + a). \tag{5}$$

The action of T upon functions including the energy eigenfunctions $\psi(x)$ is to perform the transformation under which the Hamiltonian operator is invariant. These mathematical properties assure that any nondegenerate energy eigenfunction must also be an eigenfunction of T. Thus, if T_0 is an eigenvalue, we may write

$$T\psi(x) = T_0\psi(x) = \psi(x + a). \tag{6}$$

Applying the operator T again, we obtain

$$T^2\psi(x) = TT_0\psi(x) = T_0 T\psi(x) = T_0^2\psi(x) = \psi(x + 2a). \tag{7}$$

Applying the operator N times gives

$$\psi(x + Na) = T_0^N\psi(x). \tag{8}$$

But the boundary condition, Eq. (1), requires that

$$T_0^N = 1,$$

or

$$T_0 = 1^{1/N} = \exp(2\pi i l/N), \qquad l = 0, 1, \ldots, N - 1. \tag{9}$$

There are N distinct eigenvalues of T and they are the Nth roots of unity. The eigenfunctions of T must be those functions which change only by the complex phase factor, Eq. (9), when x is changed to $x + a$. They must be of the form

$$\psi(x) = g_l(x) \exp(2\pi i l x/L) = g_k(x) \exp(ikx), \tag{10}$$

where $L = Na$, $k = 2\pi l/Na$, and $g_k(x)$ is periodic in x with period a,

$$g_k(x) = g_k(x + a). \tag{11}$$

Thus in place of a linear combination of atomic orbitals, we have in this case a much simpler expression for the molecular orbital of an electron associated with N atoms of a one-dimensional crystal. Equation (10) is called a Bloch molecular orbital. Wigner and Seitz (2, 3) developed a method of solving the Schrödinger equation for a Bloch-type function. For a free particle, g becomes a constant and the energy is

$$E = \hbar^2 k^2/2m, \tag{12}$$

that is, the energy is proportional to k^2. A plot of E versus k gives a parabola. The mathematics for the case of an electron which is appreciably influenced by the periodic field is complicated. We omit the details of the evaluation of $g_k(x)$ and the solution. But it is found that regardless of the shape of the potential curve, a small periodic perturbation such as $g_k(x)$ causes the energy levels to take sudden jumps at multiples of $k = \pm \pi/a$. Thus the presence of energy bands and forbidden zones are clearly established. The three-dimensional problem is not essentially different from the one-dimensional problem, although it is considerably more complex.

CONDUCTORS, SEMICONDUCTORS, AND INSULATORS[2]

The band theory provides a good model for describing differences between conductors and insulators (Fig. 8.2). In a conductor, the top (valence) band is only partially filled with electrons. An EMF applied to the conductor gives

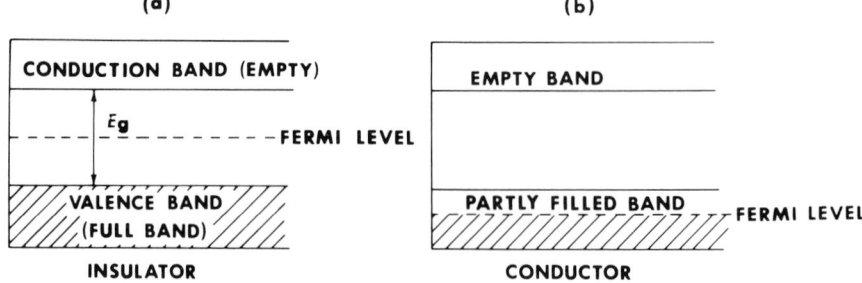

FIG. 8.2. Insulator and conductor. Filled band in (a) makes it an insulator. Partly filled band in (b) makes it a conductor.

[2] See Mott and Jones (4) and Wilson (5).

electrons flowing in the direction of the field more energy, raising them to some of the previously empty levels of the band. In this way, a current flows. In an insulator, the valence band and all deeper bands are full, and the conduction band (next band above the valence band) is completely empty. If the energy gap between the two bands is large compared to kT, then it is difficult to raise electrons to the empty band and there is essentially no conduction. In both cases, delocalized molecular orbitals exist. In the insulator, however, a flow of electrons in one direction must be balanced by a flow in the opposite direction. There can be no net flow, for the band is completely full. Divalent metals, such as magnesium, might be considered to have a completely filled valence band, but conduction occurs in such cases because of overlapping with a higher empty band.

   If the energy gap between a filled valence band and the conduction band is small enough so that a reasonable number of electrons can be excited across it at various temperatures, we have an intrinsic semiconductor. Conduction occurs by electrons in the conduction band and by the " positive holes " left in the valence band. More important in chemisorption and catalysis are extrinsic semiconductors, which rely on energy levels created by imperfections or impurities in the lattice. There are two types of extrinsic semiconductors: n-type and p-type. In n-type (normal or excess) semiconductors (Fig. 8.3a),

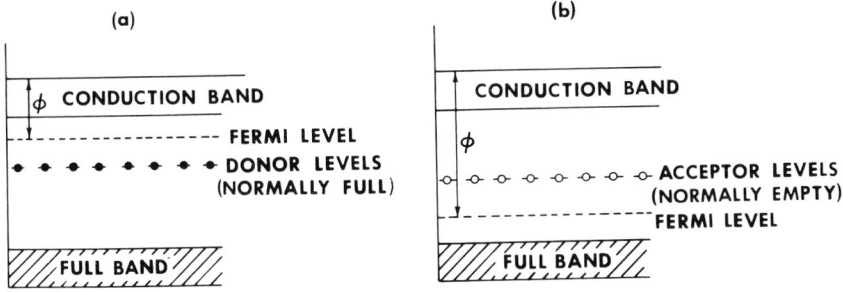

FIG. 8.3. Types of semiconductors. (a) n-type, (b) p-type.

the imperfection or impurity produces donor levels just below the conduction band into which electrons associated with the impurity atom may be excited. In p-type (abnormal or deficit) semiconductors (Fig. 8.3b), acceptor levels lie just above the valence band. Electrons are excited from the valence band into the acceptor impurity levels and conduction occurs as a result of the positive holes left in the valence band. The impurity levels are small in number and widely separated in the crystal so that they are essentially discrete. Therefore, conduction does not occur in the impurity levels.

   There are many types of imperfections or defects in crystals which may have a pronounced effect on conduction. There are reversible imperfections consisting of Frenkel defects in which interstitial atoms occur, or Schottky defects in which vacant lattice sites occur. There are irreversible imperfections such as Smekal cracks and dislocations. But we are most interested in the irreversible imperfections which arise from chemical origins rather than physical. The former may stem from nonstoichiometric compositions or from the presence of actual foreign-ion impurities. Nonstoichiometric compositions with anion vacancies or interstitial cations show an excess of cations and are *n*-type; and those with cation vacancies or interstitial anions show a deficit of cations and are *p*-type. Because of their large size, interstitial anions are seldom observed. Similarly, the presence of foreign ions can cause either *n*-type or *p*-type conductivity depending on the system. In Fig. 8.4, we illustrate some of the various types of chemical imperfections. Overall electrical neutrality is maintained in all types.

In heteropolar semiconductors, such as oxides, with which we shall be concerned mostly, the metal ions form the deeper energy states, and the wave functions of neighboring metal ions do not overlap. Thus, broadening of the energy states of the metal ions into bands is insignificant. However, oxygen ion bands are formed. The hybrid *sp* band formed from the filled 2*s* and 2*p* levels of isolated ions (O^{2-}) should be filled and separated from the hybrid

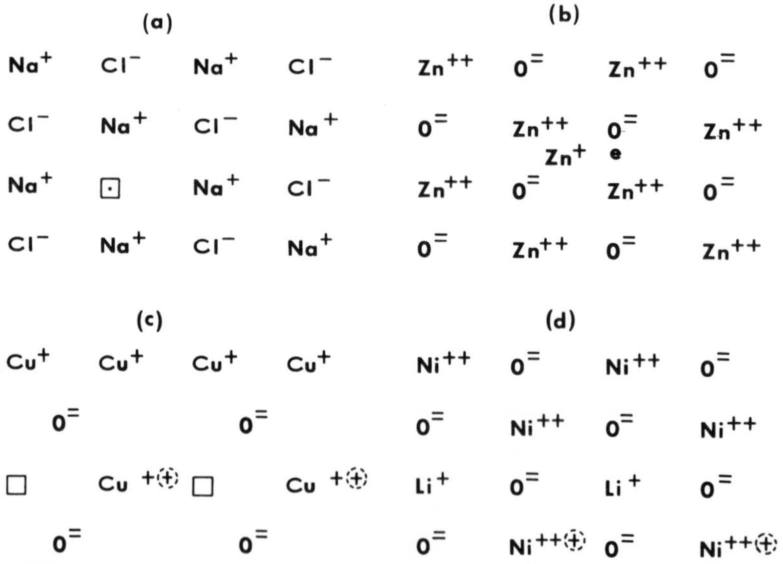

Fig. 8.4. Semiconductors with chemical imperfections. (a) NaCl *n*-type with electron trapped in anion vacancy, ⊡, occupying donor impurity level; (b) ZnO *n*-type with interstitial Zn^+ and quasi-free electron *e* occupying donor impurity level; (c) Cu_2O *p*-type with cation vacancies and quasi-free positive holes ⊕ at acceptor impurity level; (d) NiO *p*-type with foreign-ion and quasi-free positive holes ⊕ at acceptor impurity level.

sp band formed from the empty 3*s* and 3*p* states. Therefore, in the stoichiometric oxide, no appreciable conduction should be observed. Now, in the nonstoichiometric oxide, zinc oxide for example, the quasi free electrons from interstitial zinc atoms cannot enter the filled *sp* band of oxygen, but they can enter the higher, empty *sp* band giving conduction.

In Chapter VII, the Fermi level of a metal E_{max} was defined as the level below which all energy states are full of electrons and above which all states are completely empty at absolute zero. It will be more convenient to have a definition applicable at any temperature for metals as well as nonmetals. In the free electron theory of metals using Fermi-Dirac statistics, it is shown that the probability f of an electron being in a given energy state E is

$$f = \{[\exp(E - E_{max})/kT] + 1\}^{-1}. \tag{13}$$

This equation confirms the definition of the Fermi level given above, for at absolute zero when $E > E_{max}$, $f = 0$ and when $E < E_{max}$, $f = 1$. At any temperature, when $E = E_{max}$, $f = \frac{1}{2}$. At temperatures above absolute zero, the boundary between completely filled states and completely empty states is diffuse and becomes more diffuse the higher the temperature as electrons from levels just below E_{max} are excited into levels just above E_{max} only partially filling them. Thus, a new definition, more general in scope, for the Fermi level is that energy at which $f = \frac{1}{2}$. This definition is true for any temperature and we designate the Fermi level as E_F where $E_F \rightarrow E_{max}$ as $T \rightarrow 0$.

In the case of an intrinsic semiconductor, there will, normally, be very few electrons in the conduction band because only those at the top of the valence or filled band can receive enough thermal energy to bridge the energy gap. Since only the lower levels of the conduction band will be occupied, the electrons in them may be considered to behave as free electrons. Therefore, we may use Fermi–Dirac statistics. From the energy density of states (see Section 7.4) and the probability f that a state will be filled, we can calculate n_e, the number of electrons in the conduction band. Similarly, we can calculate n_h, the number of holes in the valence band. (We use the fact that $1 - f$ is the probability that a state will be empty, i.e., have positive holes.) Since n_e must equal n_h, we equate the two expressions and solve for E_F, the Fermi level. It turns out that the Fermi level lies in the energy gap, midway between the top of the valence band and the bottom of the conduction band (see Fig. 8.2a). This is the result of the free-electron approximation. More exact considerations show that the Fermi level will indeed be between the two bands, but not necessarily midway. In a similar manner (see Fig. 8.3a, b), the Fermi level for defect semiconductors will be approximately midway between the discrete impurity level (donor or acceptor) and the bottom of the conduction band (*n*-type) or the top of the valence band (*p*-type).

The conductivity of semiconductors will also be of interest to us in relation to chemisorption. In general, conductivity σ is expressed

$$\sigma = nev, \tag{14}$$

where n is the number of conducting electrons per unit volume or holes in the normally filled band, v is their mobility, and e their charge. For a given substance, in which v is constant, the conductivity should be proportional to the number of conducting electrons. In an intrinsic semiconductor, using Fermi–Dirac statistics, the number of conducting electrons is

$$n = (2(2\pi mkT)/h^2)^{3/2} \exp(-E_g/2kT), \tag{15}$$

where E_g is the width of the energy gap from the top of the filled band to the bottom of the conduction band. Thus we may write for the conductivity

$$\sigma = A \exp(-E_g/2kT). \tag{16}$$

Experiments confirm this equation in many cases, a plot of $\ln \sigma$ versus $1/T$ giving a straight line of slope $E_g/2k$ and intercept $\ln A$. Similarly, for defect semiconductors, straight-line plots are frequently obtained. However, the situation may be more complex. In general, conductivity of defect semiconductors is expressed by two straight lines on a $\ln \sigma$-versus-$1/T$ plot, the general form being

$$\sigma = A_1 \exp(-E_1/kT) + A_2 \exp(-E_2/kT), \tag{17}$$

where one member on the right-hand side is presumed to represent the intrinsic conductivity due to electrons excited from the filled band to the conduction band, and the other defect conductivity due to excitation of electrons from impurity levels to the conduction band. The activation energy for defect conductivity is considerably smaller than that for intrinsic conductivity.

8.3   The Boundary-Layer Theory of Chemisorption

Early work on this theory was done by Aigrain and Dugas (*6*), Hauffe and Engell (*7*, *8*), and Weisz (*9*, *10*). They followed closely the earlier theory of the contact between a metal and a semiconductor, a good discussion of which is given by Mott and Gurney (*11*). We first consider the case of an n-type semiconductor and a neutral atom C which is adsorbed as a negative ion C^-. An example of such behavior is the adsorption of oxygen as either O^- or O^{2-} on zinc oxide.

$$\tfrac{1}{2}O_2 + 2e \to O^{2-}. \tag{18}$$

Before adsorption, as shown in Fig. 8.5a, the electron-accepting level of C is situated below the bottom of the conduction band of the semiconductor.

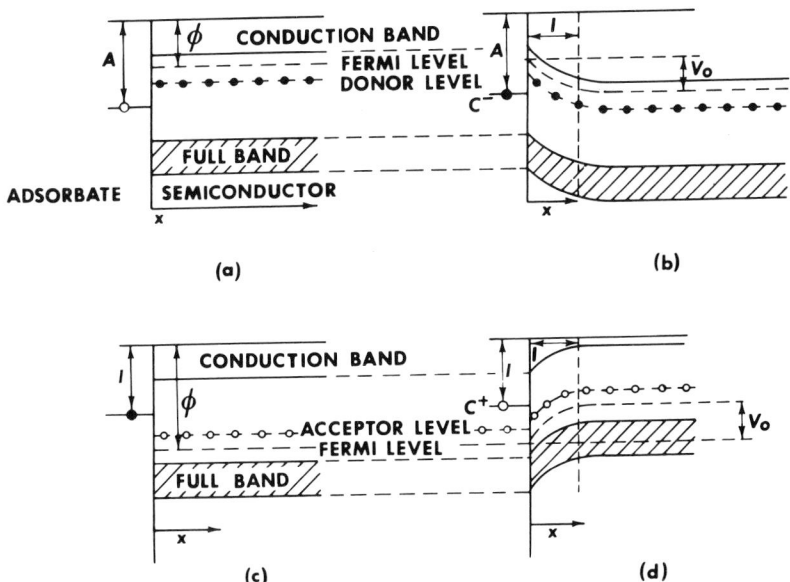

FIG. 8.5. Adsorption of C as C⁻ on *n*-type semiconductor: (a) before adsorption, (b) after adsorption. Adsorption of C as C⁺ on *p*-type semiconductor: (c) before adsorption, (d) after adsorption.

Then, when an atom of C is brought to the surface of the solid, an electron near the surface is transferred from the conduction band to the electron-accepting level of C giving C^-. As the process of adsorption continues, a positive space charge develops in the semiconductor due to the loss of electrons to C, and a negative charge builds up on its surface. Therefore, a field will exist in the surface layers of the semiconductor which tends to force electrons away from the surface into the interior. The force will die off with distance so that at some short distance into the metal, the distribution of electrons in the conduction band and discrete donor levels will be normal. Thus a boundary layer exists across which there is a decreasing electron concentration in going from the inner edge of the layer toward the surface of the semiconductor. Electron transfer to C becomes progressively more difficult with increasing adsorption as the space charge builds up. Furthermore, there are normally few electrons in the conduction band; in fact, the total supply of electrons from impurity levels is small. If these are used up, electrons for additional adsorption must come from the much deeper levels of the valence band, which may require a prohibitively large activation energy. Adsorption may stop far short of a monolayer as the supply of electrons decreases and the space charge builds up. Hauffe has called this type of adsorption "depletive chemisorption." Adsorption of this type causes an increase in the work function φ of the semiconductor by the amount of additional energy required to move an electron across the potential V of the boundary layer. In effect, the Fermi level is pushed down as adsorption proceeds. Equilibrium adsorption is attained when the potential energy of the electrons (represented by the Fermi level E_F) becomes equal to the potential energy of the electron-accepting level of adsorbate C (the electron affinity A of C), as shown in Fig. 8.5b. It is obvious that conductivity should decrease with increasing adsorption in this case. Both limited adsorption and drop in conductivity have been observed (12).

We now consider the case of a p-type semiconductor and a neutral atom C which again is adsorbed as a negative ion. An example of this type of adsorption is oxygen adsorbed as O^- on nickel oxide. Electrons must come, in this case, from the filled valence band. From levels of this type near the surface, there are ample electrons to form a monolayer of C^-. As electrons are removed from the valence band, the remaining holes will cause increased conductivity. Both high adsorption and increasing conductivity have been observed ($13, 14$). Adsorption of this type is called cumulative adsorption. The band picture is similar to that in Fig. 8.5a, b except that the large supply of electrons near the surface means that the potential barrier to further adsorption is lower and the width of the barrier less.

A corresponding picture can be developed for the adsorption of C as C^+ on n-type and p-type semiconductors, i.e., for liberation of electrons instead of the capture of electrons. An example for an n-type semiconductor is the chemisorption of hydrogen on the oxygen anions of zinc oxide

$$\tfrac{1}{2}H_2 + O^{2-} \rightarrow OH^- + e. \tag{19}$$

Electrons are removed from C (ionization potential I) and enter the nearly empty conduction band (electron affinity φ) thus increasing conductivity. There is no barrier to putting large numbers of electrons into the conduction band so that monolayer coverage should be attained in accordance with cumulative-type chemisorption. On the other hand, a p-type semiconductor tends to give depletive chemisorption because electrons from C enter the nearly full valence band. As the holes in the valence band levels near the surface of the crystal are filled, further adsorption can take place only by crossing the increasing potential barrier to valence band levels deeper within the crystal. In this case, the Fermi level is pushed up and the valence and conduction bands curve downward on approaching the surface across the potential barrier as seen in Figs. 8.5c, d.

In summary, we give, in Table 8.1, the types of adsorption systems according to the boundary-layer theory of chemisorption.

We turn now to the quantitative development of the theory using the first two types in Table 8.1 as models. We assume that adsorption equilibrium has been established. Let x be the distance of any point in the boundary layer

TABLE 8.1

TYPES OF CHEMISORPTION ACCORDING TO THE BOUNDARY-LAYER THEORY

Nature of adsorbed species	Type of Semiconductor	Type of chemisorption
C^-	n-type	depletive
C^-	p-type	cumulative
C^+	n-type	cumulative
C^+	p-type	depletive

from the surface of the semiconductor, and $x = l$ be the width of the boundary layer. Also let $V(x)$ be the potential difference between the interior of the solid and point x. Then $V(0) = V_0$ is the potential difference at the surface and $V(\gtrless l) = V_l = 0$. Furthermore, let $n_e(x)$ and $n_h(x)$ be the density of free electrons in an n-type semiconductor and electron holes in a p-type semiconductor, respectively at point x. Then, at $x \gtrless l$, n_e and n_h will have constant values, $n_e(l)$ and $n_h(l)$, characteristic of the semiconductor. If each defect or impurity is assumed to give one free electron in the conduction band or hole in the valence band, then the maximum value of the densities $n_e(l)$ and $n_h(l)$ is n_d, the density of defects in either an n-type or p-type semiconductor. Now, if the density of electrons or holes is low, Boltzmann statistics may be used and we find

$$n_e(x) = n_e(l) \exp(-eV(x)/kT), \qquad (20)$$

$$n_h(x) = n_h(l) \exp(eV(x)/kT). \qquad (21)$$

In general, $n_e(x)$ and $n_h(x)$ for $0 < x < l$ will behave as shown in Figs. 8.6a and c, respectively. In order to simplify the problem, Hauffe and Engell (7, 8) make the assumption that $n_e(x)$ and $n_h(x)$ are constant for $0 < x < l$. The approximation is shown in Figs. 8.6b and d. It is obvious from Eqs. (20) and (21) that the approximation will apply best for $eV_0 \gg kT$. With this inequality, $n_e(0)$ is very small compared to $n_e(l)$ and $n_h(0)$ is very large compared to $n_h(l)$. The space charge ρ (dearth of free electrons in the conduction band or excess of holes in the valence band compared to the interior of the crystal) in the boundary layer will also be constant with these assumptions and equal to

$$\rho = e[n_e(l) - n_e(0)] \cong en_e(l), \qquad 0 < x < l, \qquad (22)$$

$$\rho = e[n_h(0) - n_h(l)] \cong en_h(0), \qquad 0 < x < l,$$
$$\rho = 0, \qquad x \gtrless l. \qquad (23)$$

If we assume that every adsorbed atom captures one electron and that the concentration of un-ionized defects is the same in the interior and in the

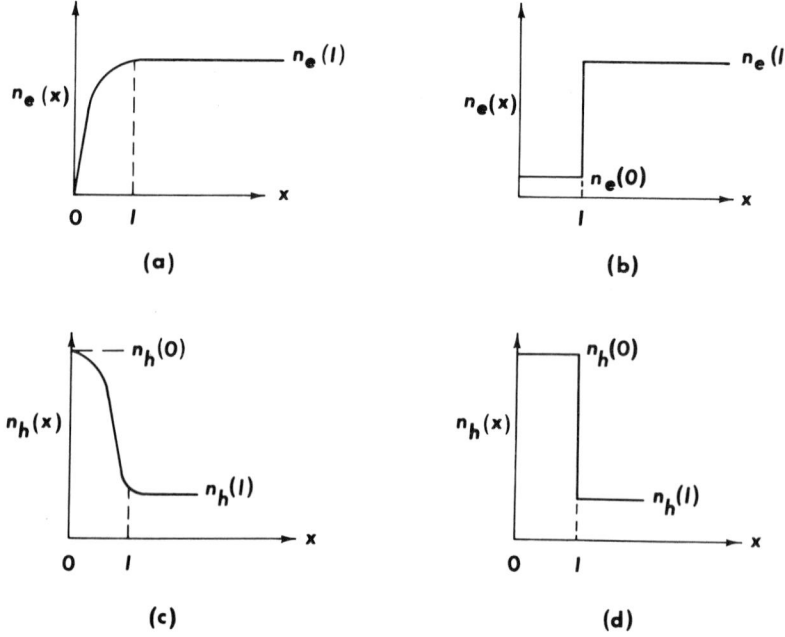

FIG. 8.6. Approximation for distribution of free electrons and holes in the boundary layer and the interior of n-type and p-type semiconductor: (a) actual distribution of free electrons (schematic), (b) approximation for free electrons, (c) actual distribution of holes (schematic), (d) approximation for holes.

boundary layer, then the number of surface charges per unit area, i.e., the number of chemisorbed atoms is

$$N_s = [n_e(l) - n_e(0)]l \cong l n_e(l) \qquad (24)$$

and similarly for a p-type semiconductor

$$N_s = [n_h(0) - n_h(l)]l \cong l n_h(0). \qquad (25)$$

Now from the Poisson equation, which holds, of course, regardless of whether equilibrium has been established,

$$d^2V(x)/dx^2 = 4\pi\rho/\epsilon, \qquad (26)$$

where ϵ is the dielectric constant. Integrating with the boundary conditions $x = l$, $dV/dx = 0$ and $V = 0$ and setting $x = 0$ in the result, we obtain,

$$V_0 = 2\pi\rho l^2/\epsilon. \qquad (27)$$

From Eqs. (24) and (25), we substitute for values of l and find

$$V_0 = 2\pi\rho N_s^2/\epsilon n_e^2(l), \qquad (28)$$

$$V_0 = 2\pi\rho N_s^2/\epsilon n_h^2(0). \qquad (29)$$

Substituting for ρ from Eqs. (22) and (23),

$$V_0 = 2\pi e N_s^2 / \epsilon n_e(l), \tag{30}$$

$$V_0 = 2\pi e N_s^2 / \epsilon n_h(0), \tag{31}$$

or

$$N_s = [V_0 \epsilon n_e(l)/2\pi e]^{1/2}, \tag{32}$$

$$N_s = [V_0 \epsilon n_h(0)/2\pi e]^{1/2}. \tag{33}$$

If we now assume that every donor defect level in an n-type semiconductor has yielded its electron to the conduction band, then in Eq. (32), $n_e(l) \rightarrow n_d$, the total number of defects per unit volume. We may now rewrite Eq. (32) for the number of adsorbed atoms,

$$N_s = [V_0 \epsilon n_d / 2\pi e]^{1/2}. \tag{34}$$

Assuming V_0 about 1 V, ϵ about 10, and n_d about $10^{18}/cm^3$, we find for depletive chemisorption

$$N_s \sim 2.5 \times 10^{12}/cm^2.$$

Since the number of surface sites on a metal oxide is of the order of 10^{15}, the coverage for depletive chemisorption is less than 1 %.

For the p-type semiconductor, $n_h(0)$ must be eliminated from Eq. (33) using Eq. (21) with $x = 0$. Equation (33) becomes

$$N_s = \{[V_0 \epsilon n_h(l)/2\pi e] \exp(eV_0/kT)\}^{1/2}. \tag{35}$$

If we assume that every acceptor defect level has been filled, then $n_h(l) \rightarrow n_d$, and Eq. (35) becomes

$$N_s = [(V_0 \epsilon n_d / 2\pi e) \exp(eV_0/kT)]^{1/2}. \tag{36}$$

Comparing Eqs. (34) and (36), we see that the extra exponential term in the latter should permit greater coverage just as we would expect for cumulative chemisorption, that is, for C^- on a p-type semiconductor.

It should be pointed out that there are some objections to treating cumulative chemisorption by the boundary layer theory (*12*). In this case, electrons are lost to the conduction band of an n-type semiconductor or gained from the full band of a p-type semiconductor. However, the level densities in both types of cumulative chemisorption are so great that an extremely thin boundary layer can accept or supply all the electrons required for adsorption. In some cases the boundary layer may be vanishingly small ($l \rightarrow 0$) and it would be better to describe the situation at the surface as a double layer rather than a boundary layer resulting from a space charge. Such chemisorption might better be compared with chemisorption on metals than with depletive adsorption. Even covalent bonding may be possible.

We note the form of the isotherm which may be derived. Taking the n-type semiconductor as an example, negative ion adsorption proceeds by

$$C + e \rightleftharpoons C^-, \tag{37}$$

where the electrons come from the boundary layer. The net rate of adsorption at equilibrium may be expressed

$$dN_s/dt = 0 = k_a n_e(0) p_C - k_d N_s, \tag{38}$$

where k_a and k_d are the rate constants of adsorption and desorption, respectively. Substituting $n_e(l) \exp(-V_0/kT)$ for $n_e(0)$ from Eq. (20), by means of Eq. (30) eliminating V_0, and finally replacing $n_e(l)$ by n_d, we find

$$N_s = K n_d p_C \exp[(-2\pi e^2/\epsilon kT)(N_s^2/n_d)], \tag{39}$$

which may be approximated by

$$N_s = (1/\sqrt{2e})[(\epsilon kT/\pi) n_d \ln(K n_d p_C)]^{1/2}, \tag{40}$$

where $K = k_a/k_d$. This is far different from the Langmuir mechanism, except at low coverage when $N_s \to 0$.

We also note that this mechanism of adsorption inherently involves a decrease in energy of adsorption with increasing coverage. When C^- is formed on an n-type semiconductor, the initial energy of adsorption is $Ae - \varphi e + \beta$, where A is the electron affinity of C, φ is the work function of the solid and β is the interaction energy (electrostatic attraction, interaction integral, etc.). As adsorption proceeds, the potential barrier V_0 builds up and the work function increases by this amount. At equilibrium, we have $Ae - \varphi e - V_0 e + \beta = 0$, that is, the energy of adsorption has dropped to zero and further adsorption is precluded. A similar argument follows when C^+ is formed on a p-type semiconductor, the initial energy of adsorption in this case being $\varphi e - Ie + \beta$, where I is the ionization potential of C. Chemisorption in this case continues until $\varphi e - Ie - V_0 e + \beta = 0$.

Furthermore, we note some of the details, both theoretical and experimental, of the considerations about surface states which have been made in the light of the boundary layer theory. Grimley (15), for example, finds from theoretical considerations that, contrary to expectations, localized states for the combined system, foreign atom (anion or cation) plus crystal (n-type or p-type) are not always necessary. Morrison (16) has devised ingenious experiments for measuring the energy levels of surface states. He has studied the surface states of oxygen anions on zinc oxide. The procedure consists in placing a ZnO crystal in an oxygen atmosphere and initiating a corona discharge through the oxygen. The negative oxygen ions formed by the discharge migrate to the ZnO surface. Since chemisorption of oxygen on ZnO is the depletive type, a negative layer vastly beyond normal equilibrium for the adsorption of neutral oxygen is formed by the procedure. Thus, the depression of the Fermi level is very great (~ 3 eV), and the surface states of adsorbed oxygen anions lie in the forbidden zone at an energy E below the bottom of the conduction band and far above the bulk Fermi level of the semiconductor. The decay of the surface charge takes place by electrons moving from the surface to the interior. In order for this process to occur, an electron must gain the activation energy E before it can reach the conduction band levels. The quantity E represents the energy level of the surface state and it may be obtained by measuring the rate of decay at two or more temperatures. Morrison found that the $(000\bar{1})$ orientation of the ZnO crystal exhibits an apparently monoenergetic surface state associated with adsorbed oxygen anions having energy $E = 1.0$ eV. On the other hand, the (0001) orientation shows a complex behavior, having three separate levels, $E_1 = 1.6$, $E_2 = 5.8$, and $E_3 = 1.0$ eV. Gray and Amigues (42) and Gray and Cichowski (43) have obtained similar results (see Section 15.4). Finally, the theoretical work of Mark (44, 45), Tomasek and Koutecky (46), and Levine and Davison (47) on surface states of intrinsic semiconductors should be cited.

The boundary layer theory is consistent with many experimental data. For example, in many cases it predicts the direction of change of conductivity accompanying adsorption (17). There are also examples of depletive and cumulative adsorption, that is, adsorption limited to very low coverage and adsorption which proceeds essentially to a monolayer (18). Relative to the ability to adsorb oxygen, Stone (18) has studied the decomposition of nitrous oxide on a variety of n-type, p-type, and insulator oxides. This reaction proceeds by the formation of oxygen anions on the catalyst surfaces. The n-type oxides, which can adsorb oxygen only to a limited extent, were the least active and p-type oxides which adsorb more oxygen were the most effective. The effect of incorporation of foreign ions on conductivity and adsorption has also been shown in many cases to be in accord with the theory (19–21).

One criticism (22) which has been leveled at the theory as presented by Hauffe and Engell (7, 8) is the artificial manner of approximating the distribution of free electrons and holes in the boundary layer (see Fig. 8.6). Krusemeyer and Thomas (22) have found that the Hauffe and Engell treatment holds for donor-type surface states on an n-type and acceptor states on a p-type semiconductor if the voltage drop across the space-charge layer is greater than $\approx 2kT/e$ and no degeneracy or saturation effects exist. For acceptor-type states on an n-type and donor states on a p-type semiconductor the lower limit of applicability is $\approx 4kT/e$.

The most serious fault of the theory, however, is that it considers only those types of adsorption which result from a transfer of charge. There are examples of systems in which adsorption occurs with no change in conductivity at all, thus, presumably, with no transfer of charge. Such cases fall outside the realm of the boundary-layer theory. We shall consider these in the next section.

8.4   Electron Theory of the Russian School

A great deal of work has been done by scientists of the U.S.S.R. in the area of chemisorption and catalysis on semiconductors, especially by Wolkenstein *et al.* (*23–37*). Wolkenstein's concepts are broader than the boundary-layer theory and include it. He considers adsorption in which a transfer of charge occurs with resulting change in conduction, and also adsorption in which no transfer of charge occurs. He assumes that two or more types of adsorption may occur simultaneously and calculates the equilibrium existing between them. There are many examples of gases adsorbed in different ways on the same support. For instance, hydrogen may be adsorbed on ZnO and $ZnO·Cr_2O_3$ by two mechanisms (*38*), one of which predominates at high temperatures and the other at low temperatures. From a comparison of the conductivity curves, it is concluded that chemisorption of the high-temperature type is solely responsible for the observed increase in conductivity. Therefore, chemisorption at low temperature does not involve the conduction band. High-temperature adsorption proceeds by Eq. (19) and desorption by the equation

$$2OH^- \rightarrow O^{2-} + H_2O, \tag{41}$$

and is irreversible, in contrast to low-temperature adsorption, since only water can be desorbed (*39*). In general, reversible adsorption is weaker than irreversible adsorption. Carbon monoxide also adsorbs both reversibly and irreversibly. At room temperature, CO adsorption is completely reversible on ZnO (*39*), partly reversible and partly irreversible on Cu_2O (*17*) and $ZnO·Cr_2O_3$ (*39*), and completely irreversible on Mn_2O_3 (*40*). Irreversible adsorption of CO is considered to take place according to the equation

$$CO + 2O^{2-} \rightarrow CO_3^{2-} + 2e, \tag{42}$$

producing an anion vacancy containing two electrons. With such a mechanism, we would expect subsequent oxygen adsorption to the extent of one half the CO adsorption in order to fill the anion vacancies,

$$\tfrac{1}{2}O_2 + 2e \rightarrow O^{2-}. \tag{43}$$

This has been observed experimentally (*17*). With a *p*-type oxide, electrons liberated by CO adsorption will neutralize positive holes and with an *n*-type oxide will enter the conduction band. Reversible adsorption of CO on ZnO and Cu_2O also appears to affect conductivity and is not well understood. Oxygen chemisorption decreases the conductivity of n-type oxides and increases the conductivity of p-type oxides. These results indicate that oxygen is adsorbed as negative ions, which may be O_2^-, O^-, or O^{2-}. No evidence for adsorption as O_2^- has been found. In all cases, a change in conductivity would be expected. Barry and Stone (*41*) show that oxygen is adsorbed by at least two mechanisms on ZnO. Their evidence points to adsorption as O^- at low temperatures and as O^{2-} at high temperatures. An even more complex mechanism for adsorption of oxygen on the p-type semiconductor Cu_2O (*14*) has been suggested. The experimental results are explained by assuming that in addition to forming a reactive surface layer (O^{2-}), oxygen is gradually built into the crystal probably by diffusion of metal ions to the surface. Positive holes are thus formed in the interior with consequent increase in positive hole conduction. Not all of these mechanisms can be explained by Wolkenstein's theories, but at least he goes beyond the simple boundary-layer theory.

Wolkenstein defines "weak" chemisorption and "strong" chemisorption as follows:

(1) In "weak" chemisorption, no change occurs in the number of electrons in the conduction band or positive holes in the valence band of the crystal. The chemisorbed particle with its adsorption center remains electrically neutral.

(2) In "strong" chemisorption, either donation or capture of an electron by the chemisorbed particle occurs. Thus the number of electrons in the conduction band or positive holes in the valence band of the crystal will change.

"Weak" and "strong" chemisorption as defined above are not necessarily equivalent to reversible and irreversible chemisorption. Reversible adsorption, though weak in comparison to irreversible, may sometimes involve changes in conduction as we have seen. It is, of course, possible to have only reversible adsorption, and more than one type, one representing "weak" chemisorption and another "strong" chemisorption. There are two types of "strong" chemisorption bonds: donor bonds when the chemisorbed particles donate electrons (capture positive holes) to the crystal, and acceptor bonds when the particles capture free electrons from the crystal. Each of these bonds may be purely ionic, purely covalent, or intermediate. Following Wolkenstein, we illustrate in Fig. 8.7 the various forms of chemisorption of particle C on an ionic crystal composed of the ions M^+ and R^-. "Weak" chemisorption is shown in Figs. 8.7a and d. In the former, C is chemisorbed on M^+ and an example is the adsorption of Na($\equiv$C) on Na^+($\equiv M^+$) in a NaCl crystal. This is essentially

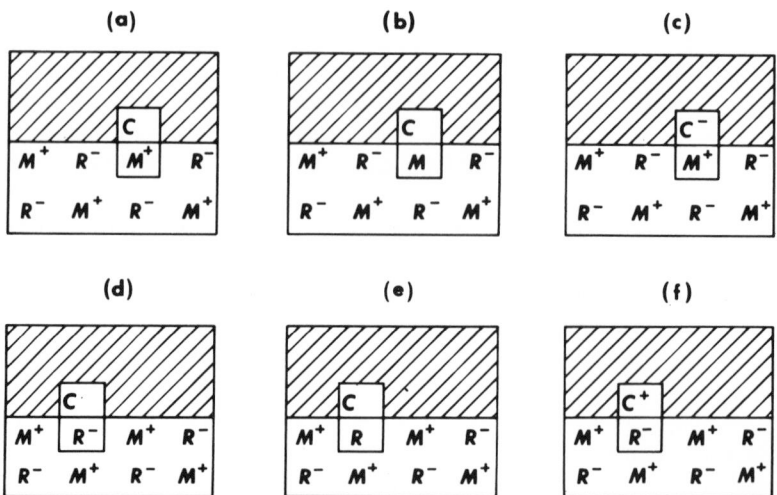

FIG. 8.7. "Strong" and "weak" bonding in chemisorption: (a) and (d) "weak" bonding; (b) and (c) "strong" acceptor bonding; (e) and (f) "strong" donor bonding (*36, 37*).

a one-electron bond (*28–30*), using only the valence electron of the chemi-sorbed Na atom and none from the crystal. As explained in Section 7.5, this bond may be completely localized between a chemisorbed Na atom and a Na^+ lattice ion directly beneath it, with a discrete energy level lying below the normal crystal band (also below any surface bands). On the other hand, only delocalized states may exist, with energies lying within the normal crystal band or surface band, or both. Whether localized or delocalized states are formed depends on the interaction parameters of the system. The greater the separation of a discrete energy level from the center of the energy band of the crystal, the more localized the band will be. As the separation decreases, the probability of the bonding electron being found on more and more distant Na^+ ions increases. The localized bond is comparable to the one-electron bonds of the molecular ions H_2^+ or Na_2^+. A similar "weak" bond is depicted in Fig. 8.7d. and, as an example, may be formed when C is a Cl atom centered over a Cl^- ion ($=R^-$) in a NaCl crystal. It is comparable to the bond in a Cl_2^- molecular ion. Both localized systems acquire dipole moments as a result of the distortion of the electron cloud about the Na atom, which, in the isolated atom, is spherically symmetrical. The moments are in opposite directions.

  "Strong" bonding is depicted in Figs. 8.7b, c, e, and f. Here again, the formation of localized bonds depends on the interaction parameters of the system (*24, 26–29*). Two extreme types exist: the purely covalent and the purely ionic. The purely covalent bond is shown in Figs. 8.7b and e. An example of the former (b) is the adsorption of $Na(\equiv C)$ on $Na(\equiv M)$ in a NaCl lattice. A free electron has been captured from the lattice and the bond is called a "strong" acceptor bond. As a result of the electron capture, the Na^+ ion of the crystal loses its positive charge. Since the Na–Na pair is homonuclear, the bond is considered to be purely covalent. It is comparable to the bonds in Na_2 or H_2. Another covalent bond type is shown in Fig. 8.7e and may be illustrated by the adsorption of a Cl atom ($\equiv C$) on a Cl atom ($\equiv R$) in the NaCl lattice. In this case, an electron has been donated to the lattice, and the bond is called a "strong" donor bond. It is comparable to the bond in a neutral Cl_2 molecule. Purely ionic bonds are shown in Figs. 8.7c and f. We may illustrate the former, Fig. 8.7c, by setting C^- equal to Cl^- and M^+ equal to Na^+ in a NaCl lattice. This is a "strong" acceptor bond, an electron has been captured from the lattice. The heteronuclear nature of the bond components gives it a strong ionic character. Figure 8.7f is an example of the "strong" donor bond with ionic character and may be illustrated by the adsorption of $Na^+(\equiv C^+)$ on $Cl^-(\equiv R^-)$.

Wolkenstein also divides chemisorbed species into radical and valence-saturated forms. We see that the "weak" adsorption of a Na atom on Na^+ (Fig. 8.7a) is an example of the radical form. When the chemisorption of a Na atom involves the capture of an electron from the crystal (Fig. 8.7b) or the donation of one to it (Fig. 8.7f), the resulting particle becomes valence-saturated. In general, the participation of a free electron or hole in chemisorption leads to the transformation of a radical into a valence-saturated particle and vice-versa. The radical forms of chemisorption presumably are more reactive toward other chemisorbed particles or toward particles in the gas phase.

The various forms of "strong" and "weak" chemisorption are considered to interconvert, and Wolkenstein (*37*) has developed the equilibrium calculations on an intrinsic semiconductor. The electron transitions may be written down as follows with their energy designations:

(1)   Transfer of an electron from the conduction band to the valence band $\cdots u$.

(2)   "Weak" bond + conduction band electron $\leftrightarrows$ "strong" acceptor bond $\cdots v^-$.

(3)   "Weak" bond + valence-band hole $\leftrightarrows$ "strong" donor bond $\cdots w^+$.

(4)   "Strong" acceptor bond + valence-band hole $\leftrightarrows$ "weak" bond $\cdots v^+$.

(5)   "Strong" donor bond + conduction band electron $\leftrightarrows$ "weak" bond $\cdots w^-$.

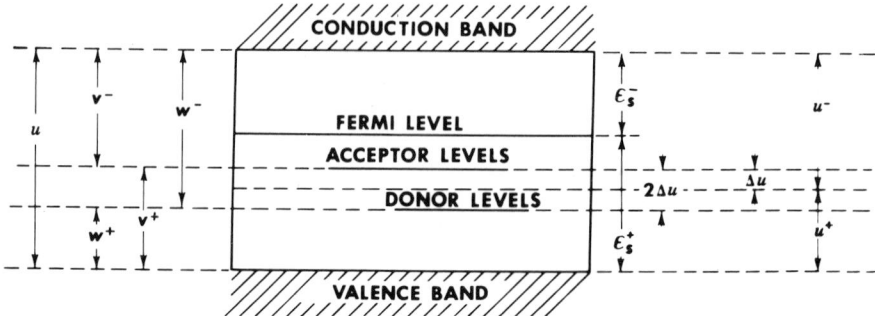

FIG. 8.8. Electron transitions, "strong" bonding $\rightleftarrows$ "weak" bonding, (36, 37). Picture at the surface, $x = 0$.

The transitions are exothermic when read from left to right. In Fig. 8.8, the various levels and the transition energies are shown including the electron accepting and donating levels of a chemisorbed particle (not to be confused with acceptor and donor impurity levels of extrinsic n- and p-type semiconductors). Taking a hole from the valence band to the acceptor or donor level is equivalent to taking an electron from the acceptor or donor level to the valence band.

Let there be a total of N_s chemisorbed particles of a given kind on the surface, of which N^0 are in a state of "weak" bonding, N^- in a state of "strong" acceptor bonding, and N^+ in a state of "strong" donor bonding. Then introduce the quantities

$$\eta^0 = N^0/N_s, \qquad \eta^- = N^-/N_s, \qquad \eta^+ = N^+/N_s, \qquad (44)$$

which give the probabilities that a chemisorbed particle will be in a given state. Thus,

$$\eta^0 + \eta^+ + \eta^- = 1. \qquad (45)$$

Now the probability that an acceptor level is occupied by an electron at equilibrium depends on its relative position with respect to the Fermi level, and is given by Eq. (13), excluding degeneracy of energy levels. This probability is the ratio of the number of occupied acceptor levels (N^-) to the total number of acceptor levels, occupied and unoccupied. The total number of acceptor levels is all the levels, $(N^0 + N^- + N^+)$, less the unoccupied donor levels (N^+), that is, $(N^0 + N^-)$. Therefore, we may rewrite Eq. (13) as

$$N^-/(N^0 + N^-) = \{1 + \exp[(\epsilon_s^- - v^-)/kT]\}^{-1}. \qquad (46)$$

Similarly, the probability that a donor level is unoccupied at equilibrium is

$$N^+/(N^0 + N^+) = \{1 + \exp[(\epsilon_s^+ - w^+)/kT]\}^{-1} \qquad (47)$$

From Eqs. (44)–(47) and Fig. 8.8, we get the probabilities

$$\eta^0 = \frac{1}{1 + 2\exp(-\Delta u/kT)\cosh[(\epsilon_s^+ - u^+)/kT]}, \tag{48}$$

$$\eta^- = \frac{\exp[-(\Delta u/kT) + (\epsilon_s^+ - u^+)/kT]}{1 + 2\exp(-\Delta u/kT)\cosh[(\epsilon_s^+ - u^+)/kT]}, \tag{49}$$

$$\eta^+ = \frac{\exp[-(\Delta u/kT) - (\epsilon_s^+ - u^+)/kT]}{1 + 2\exp(-\Delta u/kT)\cosh[(\epsilon_s^+ - u^+)/kT]}. \tag{50}$$

Referring to Fig. 8.8 and Eqs. (48)–(50), we see that, as the Fermi level moves upward from near the top of the valence band toward the bottom of the conduction band, η^- increases monotonically, η^+ decreases monotonically, and η^0 passes through a maximum at $\epsilon_s^+ = u^+$ (measured from the top of the valence band). When the Fermi level lies close to the top of the valence band, $\eta^+ \gg \eta^-$; and when the Fermi level lies close to the bottom of the conduction band, $\eta^- \gg \eta^+$. At equilibrium, it is evident that the position of the Fermi level determines the ratio of the different forms of chemisorption on the surface.

At low coverages, the adsorption isotherm is given by

$$ap = (1 + \exp(-\epsilon_s^-/kT) + \exp(-\epsilon_s^+/kT))\eta^0 N_s \exp(-q^0/kT), \tag{51}$$

where a is a constant and q^0 is the heat of desorption of a particle held in a state of "weak" chemisorption. It is important to note that the limitation on coverage in depletive chemisorption according to the boundary-layer theory (see Table 8.1) is the result of neglecting "weak" bonding. When "weak" bonding is admitted, this limitation disappears.

REFERENCES

1. Bloch, F., *Z. Physik* **52**, 555 (1928).
2. Wigner, E., and Seitz, F., *Phys. Rev.* **43**, 804 (1933).
3. Wigner, E., and Seitz, F., *Phys. Rev.* **46**, 509 (1934).
4. Mott, N. F., and Jones, H., "The Theory of the Properties of Metals and Alloys," Oxford Univ. Press, London and New York, 1936.
5. Wilson, A. H., "Semiconductors and Metals," Cambridge Univ. Press, London and New York, 1939.
6. Aigrain, P., and Dugas, C., *Z. Elektrochem.* **56**, 363 (1952).
7. Hauffe, K., and Engell, H. J., *Z. Elektrochem.* **56**, 366 (1952).
8. Engell, H. J., and Hauffe, K., *Z. Elektrochem.* **57**, 762 (1953).
9. Weisz, P. B., *J. Chem. Phys.* **20**, 1483 (1952).
10. Weisz, P. B., *J. Chem. Phys.* **21**, 1531 (1953).

11. Mott, N. F., and Gurney, R. W., "Electronic Processes in Ionic Crystals," Oxford Univ. Press, London and Washington, D.C., 1950.
12. Stone, F. S., "Chemistry of the Solid State" (W. E. Garner, ed.), p. 367. Butterworth, London and Washington, D.C., 1955.
13. Garner, W. E., Gray, T. J., and Stone, F. S., *Proc. Roy. Soc. (London), Ser. A* **197**, 294 (1949).
14. Garner, W. E., Stone, F. S., and Tiley, P. F., *Proc. Roy. Soc. (London), Ser. A* **211**, 472 (1952).
15. Grimley, T. B., *Advan. Catalysis* **12**, 24 (1960).
16. Morrison, S. R., Private communication (1966).
17. Garner, W. E., Gray, T. J., and Stone, F. S., *Proc. Roy. Soc. (London), Ser. A* **197**, 294, (1949).
18. Stone, F. S., *Advan. Catalysis* **13**, 31 (1962).
19. Molinari, E., and Parravano, G., *J. Am. Chem. Soc.* **75**, 5233 (1953).
20. Parravano, G., *J. Am. Chem. Soc.* **75**, 1452 (1953).
21. Heckelsberg, L. F., Clark, A., and Bailey, G. C., *J. Phys. Chem.* **60**, 559 (1956).
22. Krusemeyer, H. J., and Thomas, D. G., *J. Phys. Chem. Solids* **4**, 78 (1958).
23. Wolkenstein, Th., *J. Phys. Chem. (USSR)* **22**, 311 (1948).
24. Wolkenstein, Th., *J. Phys. Chem. (USSR)* **26**, 1462 (1952).
25. Wolkenstein, Th., *J. Phys. Chem. (USSR)* **28**, 422 (1954).
26. Wolkenstein, Th., and Riginsky, S. Z., *J. Phys. Chem. (USSR)* **29**, 485 (1955).
27. Wolkenstein, Th., *Advan. Phys. Sci. (USSR)* **60**, 249 (1956).
28. Wolkenstein, Th., *J. Phys. Chem. (USSR)* **21**, 1317 (1947).
29. Wolkenstein, Th., *Bull. Acad. Sci. USSR Div. Chem. Sci.* 916, (1957).
30. Wolkenstein, Th., *J. Chim. Phys.* **54**, 175 (1957).
31. Wolkenstein, Th., *Bull. Acad. Sci. USSR, Div. Chem. Sci.* 143 (1957).
32. Wolkenstein, Th., *Bull. Acad. Aci. USSR, Div. Chem. Sci.* 924 (1957).
33. Wolkenstein, Th., *J. Chim. Phys.* **54**, 181 (1957).
34. Wolkenstein, Th., and Bonch-Bruevich, V. L., *J. Exptl. Theoret. Phys. (USSR)* **20**, 624 (1950).
35. Wolkenstein, Th., *J. Exptl. Theoret. Phys. (USSR)* **22**, 184 (1952).
36. Wolkenstein, Th., "Theorie Electronique de la Catalyse Sur les Semi-Conducteurs," Masson, Paris, 1961.
37. Wolkenstein, Th., *Advan. Catalysis* **12**, 189 (1960).
38. Kubokawa, Y., and Toyama, O., *J. Phys. Chem.* **60**, 833 (1956).
39. Garner, W. E., and Veal, F. J., *J. Chem. Soc.* 1487 (1935).
40. Ward, T., *J. Chem. Soc.* 1244 (1947).
41. Barry, T. J., and Stone, F. S., *Proc. Roy. Soc. (London), Ser. A* **255**, 124 (1960).
42. Gray, T. J., and Amigues, P., *Surface Sci.* **13**, 209 (1969).
43. Gray, T. J., and Cichowski, R. S., Ph.D. Thesis, New York State College of Ceramics at Alfred University (1968).
44. Mark, P., *Catalysis Rev.* **1**, 165 (1967).
45. Mark, P., *J. Phys. Chem. Solids* **29**, 689 (1968).
46. Tomasek, M., and Koutecky, J., *Int. J. Quantum Chem.* **3**, 249 (1969).
47. Levine, J. D., and Davison, S. G., *Phys. Rev.* **174**, 911 (1968).

THE KINETICS OF CHEMISORPTION

9.1 Introduction

In this chapter, we shall discuss the kinetics of activated and nonactivated chemisorption. Activated adsorption, that is, adsorption occurring with a measurable activation energy, was first proposed by Taylor and Williamson (*1*). In the interpretation of nonactivated chemisorption, a series of adsorption steps leading from nonactivated physical adsorption to an intermediate chemisorbed state and then to a final chemisorbed state is postulated and supported with experimental evidence. The potential energy curves of physical and chemisorption intersect in such a manner that transitions occur without the requirement of an activation energy. Absolute rate theory of chemisorption processes is discussed. It is pointed out that the theory is so general that if disagreement between theory and experiment exists, it is difficult to determine the physical causes. Electronic theories following the boundary-layer theory and the more general theory of the Russian school are discussed in some detail. Neither theory can be checked satisfactorily, for their expressions contain electronic quantities which cannot be measured. Both theories are of interest, however, for the insight they provide into electronic mechanisms of adsorption. Simple expressions are derived for the kinetics of chemisorption on nonuniform surfaces. The greatest difficulty in these cases is the uncertainty as to the relationship between heats of adsorption and activation energies of adsorption. Chemisorption data are best correlated by means of the Elovich equation. Although many models for the rate of chemisorption lead to equations of the Elovich form, it is probably best to consider the Elovich equation as an empirical equation useful in correlating adsorption data.

9.2   Activated and Nonactivated Chemisorption

Langmuir (2) was the first to suggest a similarity between adsorption forces and chemical forces. Later, in the work of Benton and White (3), Garner and Kingman (4), and Taylor and Williamson (1), two forms of adsorption were distinguished—physical adsorption and chemisorption. Taylor proposed that chemisorption, which often occurs at higher temperatures and slower rates than physical adsorption, is associated with an activation energy. But the work of Roberts (5) in particular showed that in some cases strong adsorption could occur almost instantaneously at low temperatures, e.g., hydrogen on a clean tungsten filament, and therefore that chemisorption is not always associated with an activation process.

In the present section, we restrict the discussion to the Langmuir model of adsorption. For a discussion of adsorption kinetics for other models, the reader is referred to the work of Baret (61).

The kinetics of rapid chemisorptions are difficult to measure experimentally and the mechanism of the slow step in activated adsorption is often difficult to interpret. For these reasons, less is known about the dynamics of chemisorption than its statics.

The adsorption and desorption process may be expressed by

$$A + \Sigma \rightleftharpoons (A{-}\Sigma), \tag{1}$$

where an atom or molecule of A is adsorbed on a vacant site Σ. The rate of adsorption may be given by

$$r_a = [\sigma p_A/(2\pi m k T)^{1/2}]f(\theta) \exp(-E_a/kT), \tag{2}$$

where $p_A/(2\pi m k T)^{1/2}$ is the number of molecules of A striking unit surface area per second, p_A being the gas-phase pressure of A, and m the mass of A. The term $\exp(-E_a/kT)$ is the fraction of molecules of A with energies in excess of the activation energy E_a, $f(\theta)$ the fraction of the collisions of molecules of A taking place on available sites, and σ (condensation coefficient) the fraction of the molecules striking available sites with energies in excess of E_a which adsorbs. It is often found that E_a and σ are functions of coverage θ as a result of interactions or nonuniformity. If we assume that a molecule must collide directly with a vacant site in order to be adsorbed, then,

$$f(\theta) = (1 - \theta), \tag{3}$$

the fraction of vacant sites. If the adsorption process is dissociative, we have

$$A_2 + \Sigma_2 \leftrightarrows 2(A\text{—}\Sigma), \tag{4}$$

where A_2 is a diatomic molecule which collides with a dual site, that is, with a pair of vacant nearest-neighbor sites, Σ_2. In order to evaluate $f(\theta)$ in this case, we use a procedure similar to that employed in deriving the Bragg–Williams approximation (see Section 3.4) in which adsorption is considered immobile. Each site has on the average $c(B - N_s)/B = c(1 - \theta)$ nearest-neighbor unoccupied sites, assuming a random distribution, where c is the number of nearest-neighbors, B the total number of sites, and N_s the number of occupied sites. For a total of $(B - N_s)$ unoccupied sites, the number, N_{Σ_2}, of vacant dual sites is

$$N_{\Sigma_2} = c(1 - \theta)(B - N_s)/2, \tag{5}$$

where division by 2 is necessary to exclude counting each site twice. The total number of dual sites is $cB/2$, so that $f(\theta)$, the fraction of vacant dual sites, is

$$N_{\Sigma_2}/cB/2 = f(\theta) = (1 - \theta)^2. \tag{6}$$

For the desorption process, we read Eqs. (1) and (4) from right to left. The corresponding rate expressions are, for nondissociative adsorption,

$$r_d = k_d \theta \exp(-E_d/kT), \tag{7}$$

and, for dissociative adsorption,

$$r_d = k_d \theta^2 \exp(-E_d/kT). \tag{8}$$

The activation energy of desorption E_d can be expressed in terms of the activation energy of adsorption plus the heat of adsorption $(q_d = -\Delta E)$,

$$E_d = E_a + q_d. \tag{9}$$

Rapid, essentially nonactivated, chemisorption has been observed frequently on both metals and semiconductors. For example, rapid chemisorption of oxygen (6) and nitrogen (7) on tungsten wire and hydrogen (8) on nickel wire occurs with activation energies of 0.6, 0.4, and 0.4 kcal mole^{-1}, respectively and with high heats of adsorption. Initial chemisorption of oxygen on NiO and ZnO–Cr$_2$O$_3$, carbon monoxide on ZnO, and ethylene on Cu$_2$O takes place rapidly with activation energies essentially zero and heats of adsorption of 54, 42.9, 9–13, and 20 kcal mole^{-1}, respectively (9). Langmuir (10) suggested as early as 1929 that chemisorption might involve weak precursor states. In the case of oxygen adsorption on tungsten, he visualized that molecules of oxygen are physically adsorbed on top of chemisorbed oxygen atoms. Later work of Taylor and Langmuir (11) on the adsorption of cesium on tungsten supported this suggestion. They found that the sticking coefficient (fraction of incident molecules that stick) remained at unity even at coverages as high as 0.98 of a monolayer, when most molecules colliding with the surface would strike occupied sites. Note that this finding is in contradiction to the assumption made in deriving Eqs. (3) and (6). Using molecular beam techniques, Smith and Fite (12) showed that at 300°K, the angular distribution of hydrogen molecules desorbing from a tungsten surface is random, indicating adsorption for a short time before evaporation. At this temperature, the primary chemisorbed layer should be nearly complete, and therefore the process must involve a weakly held layer with a sticking coefficient close to unity. Farnsworth and Madden (13) using low-energy electron-diffraction techniques for studying the adsorption of oxygen on nickel also showed the presence of a physically adsorbed precursor state. Finally, the theoretical work of Lennard-Jones (14) shows that the potential energy of a molecule as a function of its separation from a surface has two minima—a shallow one due to physical forces and a deeper one closer to the surface resulting from chemical forces. In Fig. 9.1, we indicate schematically the possible relationships between a precursor state of physical adsorption and a chemisorbed state (Type C state). Depending on the energy relationships, adsorption from the physically adsorbed state into the Type C chemisorbed state may be activated or nonactivated.

Frequently on metals and semiconductors it is found that rapid nonactivated chemisorption is followed by adsorption at a slower rate which is easily

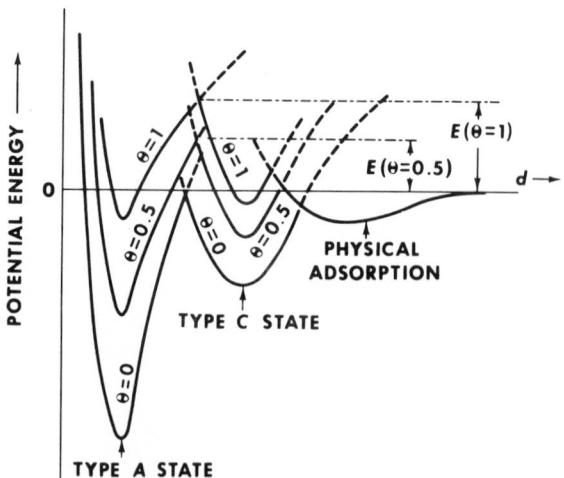

Fig. 9.1. Appearance of an activation energy between Type C and Type A states and the transition from the state of physical adsorption to Type C states. [After Gundry and Tompkins, Trans. Faraday Soc. 52, 1609 (1956). By permission of The Faraday Society.]

measured. In some cases, the kinetics at constant pressure may be expressed by

$$dq/dt = kp^n e^{-bq}, \qquad (10)$$

where q is the amount adsorbed and k, n, and b are constants (see the Elovich equation in Section 9.6). The constant n is often less than unity. For oxygen on W, Mo, and Rh (15), Si (16), and Ge (17, 18), for hydrogen on Fe (19) and Ni (20, 21), and for nitrogen on Ta, Cr, and Fe (22), n is approximately 0.5. For CO on Fe (20, 21), n has a value of 0.85, and on Ni (20, 21) a value of 0.65. If activated adsorption is direct from the gas-phase or from a precursor state of physical adsorption, a value of unity for n would be expected. The dependence on the square root of pressure for the rates of hydrogen, oxygen, and nitrogen adsorption can be explained by assuming a state of chemisorption intermediate between a physically adsorbed state and a final chemisorbed state in which the molecules are dissociated into atoms (20). Figure 9.1 illustrates schematically the nonactivated transition from a physically adsorbed state to a chemisorbed state, Type C, and subsequent transition from Type C state into a final chemisorbed state, Type A. The transition from Type C to Type A is represented as initially nonactivated, and the diagram shows how the activation energy may increase with increasing coverage while the heat of adsorption drops.

Gundry and Tompkins (20) derive a rate expression based on this model. The slow step is assumed to be the transition from Type-C state to Type-A state, while the physically adsorbed state and the Type-A state are assumed to be at equilibrium. Equilibrium between gas-phase molecules and physically adsorbed molecules P is expressed by the constant

$$K_1 = [A_2]_P/p_{A_2}, \qquad (11)$$

and equilibrium between physically adsorbed molecules and Type C adsorbed atoms is expressed by the constant

$$K_2 = [A]_C^2/[A_2]_P, \qquad (12)$$
$$[A]_C = \{K_2[A_2]_P\}^{1/2} = (K_1 K_2)^{1/2} p_{A_2}^{1/2}.$$

The rate of the slow step, that is, the transition from Type C to Type A is given by

$$r_a = k_2(K_1 K_2)^{1/2} p_{A_2}^{1/2} = k_1 p_{A_2}^{1/2} \exp(-E_a/kT),$$
$$k_2 = A \exp(-E_a/kT), \qquad k_1 = A(K_1 K_2)^{1/2}. \qquad (13)$$

Thus dependence on the square root of pressure is explained. Up to this point, no variation in activation energy with coverage has been assumed. If, however, it is assumed that the activation energy varies linearly with the amount of gas adsorbed, then

$$E_a = \gamma(q - q_0), \tag{14}$$

where γ is a constant and q_0 is the amount adsorbed when the activation energy first appears. The rate becomes

$$r_a = kp_{A_2}^{1/2}e^{-bq}, \qquad k = k_1 e^{bq_0}, \qquad b = \gamma/kT, \tag{15}$$

which, at constant pressure, according to Eq. (10), is frequently the experimentally observed rate for slow adsorption. A great deal more experimental work is necessary before the general applicability of this model can be determined. Nondissociative adsorption in Type C state need not be excluded, but no examples are available. As we shall see, there are several models that will fit Eq. (10), the Elovich form.

There are examples of systems with activation energies over the entire range of surface coverage. In some cases, such systems may not be fundamentally different from those systems which develop an activation energy after appreciable coverage has been attained. The difference may be due simply to the relative positions of the potential energy curves in Fig. 9.1. There are a few cases which do appear to involve an important difference. But this difference is not concerned so much with the point of onset of an activation energy as with the pressure dependence of the rate. For example, the adsorption of oxygen on Cu_2O (*23, 24*) proceeds over the range of monolayer coverage according to the equation

$$r_a = kp(1 - \theta)^2 \exp(-E_a/kT), \tag{16}$$

with E_a constant at 7 kcal/mole. The equation corresponds to dissociative adsorption. But the dependence on the first power of the pressure could indicate that chemisorption occurs either directly from the gas phase or from a state of molecular physical adsorption. Thus, the intermediate state of chemisorption may be absent in this case. Usually, the activation energy varies with coverage; and if, as in the derivation of Eq. (15), it is assumed that the variation is linear, $E_a = E_0 + \gamma q$, then Eq. (16) becomes

$$r_a = k'(1 - \theta)^2 pe^{-bq}, \qquad b = \gamma/kT. \tag{17}$$

In the lower range of coverage, where variation of $(1 - \theta)^2$ is overshadowed by that of the exponential term, we may write

$$r_a = k'p \exp(-bq), \tag{18}$$

which is similar in form to Eq. (15) and, therefore, at constant pressure, corresponds to an Elovich form, Eq. (10).

Another example is the adsorption of nitrogen on promoted iron (*25, 26*) which also is activated over the entire range of coverage and is proportional to the first power of the pressure. This is in contrast to nitrogen adsorption on an iron film where the rate is proportional to the square root of the pressure (*22*).

9.3   Absolute Rate Theory

For a process to occur, a system must pass through an activated state on its reaction path, and the system in this state is described as an activated complex. This activated complex is considered to be at the top of an energy barrier lying between the normal states of reactants and products. The rate of reaction is controlled by the rate at which molecules in the activated complex pass over the top of the barrier. A simplifying assumption made by Eyring (*27–29*) allows the formulation of a method of calculating the reaction rates. He assumed that the activated complexes are in statistical equilibrium with the reactants.

The method has been applied to rates of adsorption (*30*). We consider immobile adsorption expressed by the reaction

$$A + \Sigma \leftrightharpoons (A\text{---}\Sigma)^{\ddagger} \rightarrow (A\text{---}\Sigma), \tag{19}$$

where $(A - \Sigma)^{\ddagger}$ represents the activated complex. Molecules at the top of the energy barrier are in statistical equilibrium with molecules in the gas phase and with vacant surface sites. The activated complexes are assumed to vibrate along the reaction coordinate (i.e., perpendicular to the surface) with frequency v. The molecules in activated complexes are at a potential maximum, and, therefore, they enter a more stable state of lower energy as they move toward or away from the surface. With no restoring force, each vibration results in adsorption or desorption. The frequency of decomposition of the complexes is given therefore by v, and if $c^{\ddagger} = N^{\ddagger}/\alpha$ is the number of complexes per unit area of surface, then the rate of adsorption is

$$r_a = vc^{\ddagger}. \tag{20}$$

The object now is to derive expressions for $vc^{\ddagger}$.

Let $c_{\Sigma}(= N_{\Sigma}/\alpha)$ and $c_G(= N_G/V)$ be the number of vacant sites per unit area and gas molecules per unit volume, respectively. Then, according to the assumption of equilibrium between reactants and activated complexes, we have for the equilibrium constant

$$K^{\ddagger} = \frac{c^{\ddagger}}{c_G \, c_{\Sigma}} = \frac{(q^{\ddagger}/\alpha)}{(q_G/V)(q_{\Sigma}/\alpha)} = \frac{q^{\ddagger}}{(q_G/V)q_{\Sigma}}. \tag{21}$$

The last two members of the equality give the usual statistical mechanical expression for the equilibrium constant in terms of molecular partition functions per unit area for complexes ($q^{\ddagger}/\alpha$) and vacant sites (q_{Σ}/α) and per unit volume for the gas (q_{G}/V). Just as in Chapter II, Sections 2.2 and 2.3, we may separate the respective zero-point potential energies from each of these partition functions, $\exp(-E^{\ddagger}/kT)$, $\exp(-E_{\Sigma}/kT)$, and $\exp(-E_{G}/kT)$ to give

$$K^{\ddagger} = \frac{c^{\ddagger}}{c_{G}\,c_{\Sigma}} = \frac{q^{\ddagger}}{(q_{G}/V)q_{\Sigma}}\exp\frac{-E_{a}}{kT}, \tag{22}$$

where the q's will now represent the molecular partition functions with zero-point energies removed and $E_{a} = E^{\ddagger} - E_{G} - E_{\Sigma}$ is the zero-point energy of the complexes with reference to the reactants and therefore the activation energy of adsorption at absolute zero. As we have mentioned, one of the degrees of freedom of the activated complex is a vibration along the reaction coordinate of frequency ν. Since there is no restoring force in this vibration, its frequency will be very low. It is expressed by the usual vibrational partition function which we expand and retain the first term

$$\lim_{\nu \to 0} [1 - \exp(-h\nu/kT)]^{-1} = [1 - (1 - h\nu/kT)]^{-1} = kT/h\nu. \tag{23}$$

Separating this term from $q^{\ddagger}$, we obtain

$$K^{\ddagger} = \frac{c^{\ddagger}}{c_{G}\,c_{\Sigma}} = \frac{q_{\ddagger}(kT/h\nu)}{(q_{G}/V)q_{\Sigma}}\exp\frac{-E_{a}}{kT}, \tag{24}$$

which gives, in view of Eq. (20), the rate of adsorption

$$r_{a} = \nu c^{\ddagger} = c_{G}\,c_{\Sigma}\frac{kT}{h}\frac{q_{\ddagger}}{(q_{G}/V)q_{\Sigma}}\exp\frac{-E_{a}}{kT}. \tag{25}$$

Slight modifications of this equation give the equation for dissociative adsorption. The terms on the right of Eq. (25) may be evaluated specifically as follows:

(1) For a perfect gas, $c_{G} = p/kT$.

(2) $c_{\Sigma} = (B/\alpha)f(\theta)$, where B/α is the number of sites per unit area and $f(\theta)$ is the fraction of sites available for chemisorption.

(3) $q_{\ddagger}$ for immobile adsorption has no translational or rotational factors, only vibrational factors represented by $b_{\ddagger}$. With high-frequency vibrations, $b_{\ddagger}$ is close to unity.

(4) q_{Σ} consists only of high-frequency vibrational terms and may be taken equal to unity.

(5) $(q_{G}/V) = (2\pi mkT)^{3/2}b_{G}/h^{3}$, where b_{G} includes vibrational and rotational factors and the rest is the translational factor.

Making these substitutions in Eq. (25), we find

$$r_a = [Bf(\theta)/\alpha][h^2/(2\pi mkT)^{3/2}](b_{\ddagger}/b_G) \exp(-E_a/kT)p. \tag{26}$$

This equation can be rearranged and compared with Eq. (2) and an expression for σ, the condensation coefficient, obtained,

$$\sigma = Bh^2 b_{\ddagger}/2\pi mkT b_G. \tag{27}$$

For nonlocalized adsorption, in which it is presumed that the activated complex as well as molecules in the final adsorbed state are mobile, equilibrium may be represented by

$$K^{\ddagger} = \frac{c^{\ddagger}}{c_G} = \frac{q^{\ddagger}/\alpha}{q_G/V}. \tag{28}$$

Proceeding just as in the immobile case, we obtain for the rate

$$r_a = c_G \frac{kT}{h} \frac{(q_{\ddagger}/\alpha)}{(q_G/V)} \exp \frac{-E_a}{kT}. \tag{29}$$

If we assume that $(q_{\ddagger}/\alpha)$ and (q_G/V) have the same vibrational and rotational factors, differing only in that the activated complex has two modes of translational freedom whereas the gas molecules have three; and, if we further assume that complexes form only above vacant sites, then the rate expression becomes

$$r_a = [p/(2\pi mkT)^{1/2}]f(\theta) \exp(-E_a/kT), \tag{30}$$

where again we have put $c_G = p/kT$. Comparing Eq. (30) with Eq. (2), we find that $\sigma = 1$. From calculations of σ for the immobile case using Eq. (27), values appear to be about 10^{-5}. Thus the theory predicts that rates are much faster for mobile than for immobile adsorption.

For desorption, we have the reverse process

$$A - \Sigma \leftrightarrows (A - \Sigma)^{\ddagger} \to A + \Sigma. \tag{31}$$

The rate expression becomes

$$r_d = c_s(kT/h)(q_{\ddagger}/q_s) \exp(-E_d/kT), \tag{32}$$

where $c_s(= N_s/\alpha)$ is the number of adsorbed molecules per unit area, $q_{\ddagger}$ and q_s the molecular partition functions of the activated complex and adsorbed A, respectively with zero-point energies removed in each case, and E_d the zero-point energy of the complex with reference to adsorbed A and, therefore, the activation energy of desorption at absolute zero. If the activated complex and adsorbed A are assumed to be immobile, then the ratio $q_{\ddagger}/q_s$ should be close to unity. The concentration of adsorbed A may be expressed by $c_s =$

$(B/\alpha)\theta$ where B/α is the total number of adsorption sites per unit area and θ is the fraction of sites occupied. Eq. (32), with these substitutions, becomes

$$r_d = (B/\alpha)\theta(kT/h)\exp(-E_d/kT). \qquad (33)$$

Comparing Eqs. (32) and (33) with Eq. (7), $r_d = k_d\,\theta\exp(-E_d/kT)$, we find

$$k_d = (B/\alpha)(kT/h)(q_{\ddagger}/q_s) \cong (B/\alpha)(kT/h). \qquad (34)$$

Similar expressions for associative desorption are easily derived by replacing c_s and q_s by their squares.

There are many examples where the agreement of absolute rate equations with experiment is satisfactory in view of the necessary approximations of partition functions and the sensitivity to small changes in activation energy. Since *a priori* calculations of reliable activation energies are difficult, it is customary to insert experimental activation energies into the theoretical expressions and compare rates so obtained with experimental rates. The insertion of experimental activation energies determined at temperature T in place of activation energies referred to zero-point levels is justifiable to within a variation of about RT, which is usually within the limits of experimental accuracy. Experimental and calculated rates for adsorption of hydrogen on doubly promoted iron (*31*) and on copper powder (*32*), and nitrogen on doubly promoted iron (*33, 34*), singly promoted iron (*26*) and tungsten powder (*35*) are in fair agreement. But rates of adsorption of hydrogen on tungsten powder (*5*) and nickel wire (*8*), which have approximately zero activation energy, are in disagreement by a factor of about 10^5, suggesting that a mechanism more complex than direct adsorption may be occurring. Fair agreement has been obtained for the rates of desorption of carbon monoxide and oxygen from platinum and tungsten, respectively (*36*). In this example, rates of desorption were calculated using experimental values of the activation energy at $\theta \cong 1$ and calculated values of $(B/\alpha)\,(kT/h)$ in Eq. (33). The value of B/α was estimated to be about 10^{15} sites per square centimeter.

The application of absolute-rate theory to adsorption and desorption problems is disappointing overall. One reason is that absolute-rate equations contain no surface parameters except as they operate collectively through the activation energy. If mathematical agreement is attained, the desired information about mechanism is missing.

9.4   Electronic Theories

Hauffe (*37*) and Engell and Hauffe (*38*) derive an expression for the rate of chemisorption based on the boundary-layer theory, assuming that the rate-determining step is electron transfer between adsorbent and adsorbate and neglecting the reverse reaction. Thus, the rate is governed by the potential barrier V_0, which increases with increasing coverage. If E_0 is the height of the energy barrier at zero coverage ($N_s = 0$, $t = 0$), a is the distance between the surface and the centers of charge of the chemisorbed atoms, and other symbols have their previous meanings, it is found for negative-ion adsorption on a p-type semiconductor,

$$dN_s/dt = [ken_h(l)/2\pi eN_s^2]V_0 \exp[-(eE_0 + 4\pi e^2 aN_s)/\epsilon kT]. \quad (35)$$

The exponential term is assumed to be of primary importance in determining the rate although this is a rather forced relationship in view of the $1/N_s^2$ dependence of the pre-exponential term. Therefore, Eq. (35) may be considered to be approximately the Elovich form. Integrating Eq. (35) with the pre-exponential term considered constant, we find

$$N_s = (1/b) \ln(t + t_0) - (1/b) \ln t_0,$$
$$b = 4\pi e^2 a/\epsilon kT, \qquad t_0 = \text{constant}. \quad (36)$$

Engell and Hauffe (38) obtained a straight-line plot of $\ln(t + t_0)$ versus volume adsorbed for oxygen on NiO. Actually, the plot consisted of two straight-line segments, the first interpreted to represent the chemisorption process and the second the formation of new oxide molecules on the surface.

Wolkenstein and Peshev (39, 40) have considered the kinetics of chemisorption in the light of the electron theory (see Section 8.4), including both charged and neutral forms of chemisorbed particles, which represent " strong " and " weak " bonding to the surface. They assumed that chemisorbed particles are of the acceptor type and exist on the surface in the negatively charged or neutral state. They further assumed that only neutral particles desorb from the surface, that is, charged particles do not exchange with the gas phase. With these assumptions, the kinetics of chemisorption involves the following individual steps:

(1) Rate of adsorption of neutral particles, number of molecules per second per square centimeter,

$$r_a = [\sigma p/(2\pi mkT)^{1/2}](1 - \theta) \exp(-E_a/kT)$$

[see Eq. (2)].

(2) Rate of desorption of neutral particles,

$$r_d = k_d N^0 \exp(-E_d/kT),$$

where N^0 is the concentration of neutral chemisorbed particles.

(3) Rate of electron transitions, number per second per square centimeter, designated as 1, 2, 3, 4 in Fig. 9.2,

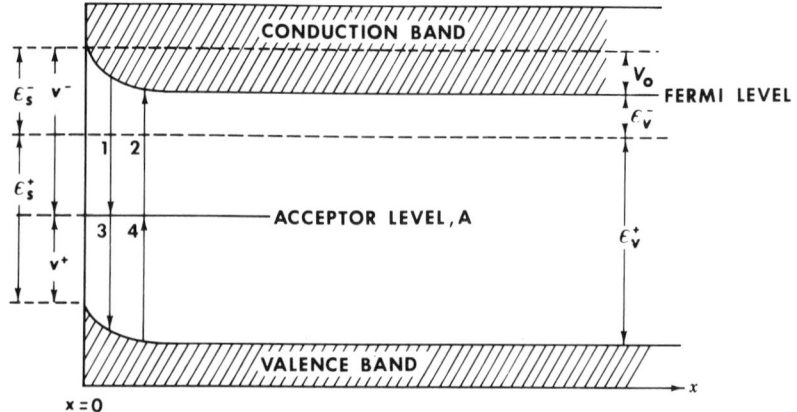

FIG. 9.2. Energy relationships for acceptor-type adsorption on a semiconductor (39, 40).

$$C_1 = k_1 n_e N^0, \qquad C_2 = k_2 N^- \exp(-v^-/kT),$$
$$C_3 = k_3 n_h N^-, \qquad C_4 = k_4 N^\circ \exp(-v^+/kT),$$

where, as before, n_e and n_h refer to concentrations of free electrons in the conduction band and holes in the valence band, respectively at the surface, $x = 0$; and N^- is the concentration of negatively charged, chemisorbed particles.

Rates of change of concentration of neutral and charged chemisorbed species are expressed by

$$dN^0/dt = (r_a - r_d) - [(C_1 - C_2) - (C_3 - C_4)], \qquad (37)$$
$$dN^-/dt = (C_1 - C_2) - (C_3 - C_4). \qquad (38)$$

Let τ denote the average lifetime of a particle in the chemisorbed state, τ^0 and τ^- the average lifetimes of a chemisorbed particle in the electrically neutral and charged states, respectively. They are given by

$$1/\tau = k_d \exp(-E_d/kT),$$
$$1/\tau^0 = k_1 n_e + k_4 \exp(-v^+/kT) = (C_1 + C_4)/N^0, \qquad (39)$$
$$1/\tau^- = k_3 n_h + k_2 \exp(-v^-/kT) = (C_2 + C_3)/N^-.$$

At low coverage, to which the derivation will be limited, $\theta \to 0$ and the rate of adsorption of neutral particles r_a becomes a constant at constant pressure. At adsorption equilibrium ($t \to \infty$), we write

$$r_a = r_d = k_d N_\infty^0 \exp(-E_d/kT) = N_\infty^0/\tau, \qquad (40)$$

where $N_\infty{}^0$ is the concentration of chemisorbed neutral particles at final adsorption equilibrium. This holds for r_a at any t at low coverage. Using Eqs. (39) and (40), the rate expressions in Eqs. (37) and (38) become

$$dN^0/dt = (N_\infty{}^0 - N^0)/\tau - (N^0/\tau^0 - N^-/\tau^-), \tag{41}$$

$$dN^-/dt = N^0/\tau^0 - N^-/\tau^-. \tag{42}$$

The problem is to solve the simultaneous Eqs. (41) and (42).

In general, $\tau^0 = \tau^0(N^0)$ and $\tau^- = \tau^-(N^-)$. But in order to solve the simultaneous equations, it is assumed that τ^0, $\tau^- = $ constant. With this assumption, the solutions are

$$N^0(t) = [N_\infty{}^0/(\tau_2 - \tau_1)\tau]$$
$$\times \{\tau_2(\tau - \tau_1)[1 - \exp(-t/\tau_2)] + \tau_1(\tau_2 - \tau)[1 - \exp(-t/\tau_1)]\}, \tag{43}$$

$$N^-(t) = [N_\infty{}^-/(\tau_2 - \tau_1)]\{\tau_2[1 - \exp(-t/\tau_2)] - \tau_1[1 - \exp(-t/\tau_1)]\}, \tag{44}$$

where $N_\infty{}^-$ is the concentration of charged chemisorbed particles at adsorption and electron equilibrium, dN^0/dt, $dN^-/dt = 0$, and where

$$1/\tau_1 = \lambda\{1 + [1 - (\mu/\lambda^2)]^{1/2}\}, \quad \lambda = \tfrac{1}{2}[(1/\tau^0) + (1/\tau^-) + (1/\tau)],$$
$$1/\tau_2 = \lambda\{1 - [1 - (\mu/\lambda^2)]^{1/2}\}, \quad \mu = (1/\tau^-)(1/\tau). \tag{45}$$

The solution is vastly simplified if it is assumed that either $\tau^0, \tau^- \ll \tau$ or $\tau \ll \tau^0, \tau^-$. For these conditions both lead to $\mu/\lambda^2 \ll 1$, and, therefore, to

$$1/\tau_1 = 2\lambda = (1/\tau^0 + 1/\tau^- + 1/\tau), \quad \tau_1\tau_2 = \tau\tau^-,$$
$$1/\tau_2 = \mu/2\lambda = (1/\tau^-)(1/\tau)\tau_1, \quad \tau_1/\tau_2 = \mu/4\lambda^2 \ll 1. \tag{46}$$

The following additional restrictive assumptions are also made: Either $\tau^0 \ll \tau^-$ or $\tau^- \ll \tau^0$. In summary, the following four cases are examined

$$\tau_0, \tau^- \ll \tau \begin{cases} \tau^0 \ll \tau^- \\ \tau^- \ll \tau^0, \end{cases}$$
$$\tau \ll \tau^0, \tau^- \begin{cases} \tau^0 \ll \tau^- \\ \tau^- \ll \tau^0. \end{cases} \tag{47}$$

From Eq. (46) for $1/\tau_1$ it is evident that with the assumptions of Eq. (47) that τ_1 equals whichever is the least of the three quantities τ_0, τ^-, and τ. Each of the four cases is examined in three time regions of adsorption: $t \ll \tau_1$, $t \gg \tau_1$, and $\tau_1 \ll t \ll \tau_2$.

In Region I, $t \ll \tau_1$, it is found, for the four cases given by Eq. (47), that Eqs. (43) and (44) yield

$$N^0(t) = N_\infty{}^0(t/\tau) \ll N_\infty{}^0, \tag{48}$$

$$N^-(t) = N_\infty{}^-(t^2/2\tau_1\tau_2) = N_\infty{}^-(t^2/2\tau\tau^-) \ll N_\infty{}^-. \tag{49}$$

The ratio of neutral to charged chemisorbed particles is

$$N^-/N^0 = (N_\infty{}^-/N_\infty{}^0)(t/2\tau^-) = t/2\tau^0 \ll 1, \qquad (50)$$

where the relationship $N_\infty{}^-/N_\infty{}^0 = \tau^-/\tau^0$ derived from Eq. (42) at equilibrium has been used. Since $t/2\tau^- \ll 1$, Eq. (50) shows that $N^-/N^0 \ll N_\infty{}^-/N_\infty{}^0$, and therefore in this region the system is far from electron equilibrium. The overall rate expression is

$$\frac{dN_s}{dt} = \frac{dN^0}{dt} + \frac{dN^-}{dt} = \frac{N_\infty{}^0}{\tau} + \frac{N_\infty{}^- t}{\tau\tau^-}$$

$$= \frac{N_\infty{}^0}{\tau} + \frac{N_\infty{}^0}{\tau}\frac{t}{\tau^0} = \frac{N_\infty{}^0}{\tau} = \frac{\sigma p}{(2\pi mkT)^{1/2}} \exp \frac{-E_a}{kT}. \qquad (51)$$

In Region II, $t \gg \tau_1$, it is found for $\tau^0, \tau^- \ll \tau$ that

$$N^0(t) = N_\infty{}^0[1 - \exp(-t/\tau_2)] \qquad N^-(t) = N_\infty{}^-[1 - \exp(-t/\tau_2)], \qquad (52)$$

and for $\tau \ll \tau^0, \tau^-$,

$$N^0(t) = N_\infty{}^0, \qquad N^-(t) = N_\infty{}^-. \qquad (53)$$

Thus in both cases,

$$N^-/N^0 = N_\infty{}^-/N_\infty^0 \qquad (54)$$

and electron equilibrium exists. In the first case, $\tau^0, \tau^- \ll \tau$, electron equilibrium exists at any time $t(\gg\tau_1)$, while in the second case, $\tau \ll \tau^0, \tau^-$, electron equilibrium is established only after adsorption equilibrium sets in. The overall rate is

$$dN_s/dt = (N_\infty{}^0/\tau) \exp(-t/\tau_2). \qquad (55)$$

In Region III, $\tau_1 \ll t \ll \tau_2$, all cases in Eq. (47) give

$$N_0(t) = N_\infty{}^0\tau_1(\tau_2 - \tau)/\tau_2\tau = N_\infty{}^0,$$
$$N^-(t) = N_\infty{}^-(t/\tau^-) \ll N_\infty{}^-. \qquad (56)$$

The fraction of neutral chemisorbed particles remains constant in this time interval while that of the charged chemisorbed particles increases with time.

The system is far from electron equilibrium. The overall rate is

$$dN_s/dt = N_\infty{}^0/\tau^0 = N_\infty{}^-/\tau^-. \qquad (57)$$

If the Fermi level is not too low, Eq. (57) can be shown to take the form

$$dN_s/dt = dN^-/dt = k_1 n_e N_\infty{}^0 = k_0 p \exp(-E/kT), \qquad (58)$$

where

$$E = E_a - E_d + \epsilon_s{}^- = E_a - E_d + \epsilon_v{}^- + V_0. \qquad (59)$$

It is possible to derive Eq. (57) with the restriction of τ^0, $\tau^- =$ constant lifted. Furthermore, if the Fermi level is not too low, Eq. (58) is again obtained, while V_0 in Eq. (59) is now a function of coverage, $V_0 = V_0(N^-)$. Therefore, the activation energy E becomes a function of coverage and Eq. (58) takes the general form of the boundary-layer equation, Eq. (35), that is, essentially the Elovich form, Eq. (10).

9.5 Theories Based on a Nonuniform Surface

It is presumed that a surface having a distribution of adsorption potentials among its sites will also have a distribution of activation energies of adsorption in one-to-one correspondence. On this basis, Peers (*41*) has derived an expression for coverage as a function of time. Following Halsey (*42*), Porter and Tompkins (*19*), and Stone (*43*), he assumes that the distribution function $N = (1/B)[dB(E_a)/dE_a]$ is a constant, where E_a is the activation energy, B the number of adsorption sites per unit area, and $0 \lesssim B(E_a) \lesssim B$, that is, $B(E_a)$ is the integral number of sites with activation energies up to E_a. If E_1 is the lower limit of E_a and E_2 the upper limit, then $B(E_1) = 0$ and $B(E_2) = B$. Desorption rates are assumed to be negligible. As in Section 9.2, the rate of adsorption on sites of given E_a is

$$d\theta_{E_a}/dt = A(1 - \theta_{E_a}), \qquad (60)$$

where

$$A = \lambda p \exp(-E_a/kT), \qquad (61)$$

and λ and p, the pressure, are constant. On integration, Eq. (60) gives

$$\theta_{E_a} = 1 - e^{-At}. \qquad (62)$$

The fraction of the total surface covered is obtained by summing over all activation energies from the lower limit E_1 to the upper limit E_2.

$$\theta = \int_{E_1}^{E_2} (1/B)[dB(E_a)/dE_a]\theta_{E_a}\, dE_a = N \int_{E_1}^{E_2} \theta_{E_a}\, dE_a. \qquad (63)$$

To simplify the integration of Eq. (63), it is assumed that sites are filled strictly in the order of increasing values of E_a, rather than filling simultaneously at different rates determined by E_a. This approximation is reasonable if the variation of E_a over the surface is sufficiently large. Thus sites of a given activation energy are either completely filled or completely empty and $\theta_{E_a} = 1$, so that Eq. (63) becomes

$$\theta = N \int_{E_1}^{E_a} dE_a = N(E_a - E_1), \qquad (64)$$

where E_a is the activation energy of sites at the limit of coverage. To obtain E_a as a function of time (*19, 42, 43*), it is assumed that each set of sites fills instantaneously at $t = 1/A$, so that from Eq. (61), $E_a = kT \ln \lambda pt$. Assuming another value for t merely changes the value of the constant λ. Substituting in Eq. (64),

$$\theta = N(kT \ln \lambda pt - E_1), \tag{65}$$

or, multiplying through by B, we obtain an expression for the amount adsorbed

$$B\theta = q = k_1 \ln t + k_2, \tag{66}$$

where $k_1 = BNkT$ and $k_2 = BN(kT \ln \lambda p - E_1)$, both constants. Equation (66) may be compared with the Elovich form, usually written

$$dq/dt = ae^{-bq}, \qquad a, b = \text{constant} \tag{67}$$

and after integration ($q = 0$, $t = 0$) gives

$$q = (1/b) \ln(t + t_0) - (1/b) \ln t_0, \qquad t_0 = 1/ab \tag{68}$$

The significance of t_0 will be discussed in Section 9.6. From Eq. (64) we find

$$E_a = E_1 + (1/N)\theta, \tag{69}$$

that is, the activation energy increases linearly with coverage.

Brunauer et al. (44) made the assumption, as above, that sites are filled strictly in the order of increasing E_a, and then assumed directly that E_a increases linearly with coverage (see Section 9.2). For low coverage and constant pressure, they obtained approximate agreement with the Elovich form, Eq. (68). In addition to the assumptions made in the above derivation of Peers (41), it was further assumed by Halsey (42), Porter and Tompkins (19), and Stone (43) that E_a is linearly proportional to the energy of adsorption. They obtained a relationship of the same form for the amount adsorbed versus time, Eq. (66).

The assumption of the linear relationship between the activation energy and the energy of adsorption, $E_a = r\chi$, is difficult to justify. Since E_a increases with coverage, the relationship $E_a = r\chi$ leads to increasing energy of adsorption with increasing coverage for immobile adsorption. Such a variation of χ contradicts nearly all available experimental data. However, if adsorption energies are measured for increments of coverage at a series of increasing pressures, the lower limit of the occupied region would move to successively lower values of χ according to the site-filling procedure described above, and, consequently, χ would decrease with increasing coverage. Nevertheless, the relationship implies that in each small region filled at constant pressure, sites of larger χ are the last to be occupied. Other workers (45–47) have suggested a relationship indicating a decrease of E_a with increasing χ, for example, $E_a = E_2 - r\chi$ (47).

An expression has been derived by Kubokawa (48) and corrected with respect to final integration limits by Peers (41) without the use of arbitrary site-filling procedures. Kubokawa differentiates Eq. (63) with respect to the parameter t,

$$d\theta/dt = N \int (d\theta_{E_a}/dt)\, dE_a = N \int (d\theta_{E_a}/dt)(dE_a/dA)\, dA, \tag{70}$$

where A is given by Eq. (61). An expression for $(d\theta_{E_a}/dt)$ is obtained by differentiating Eq. (62),

$$d\theta_{E_a}/dt = Ae^{-At}, \tag{71}$$

and for (dE_a/dA) by differentiating Eq. (61).

$$dE_a/dA = -kT/A. \tag{72}$$

Substituting into Eq. (70), we find

$$d\theta/dt = -NkT \int_{A_1}^{A_2} e^{-At}\, dA,$$

where the limits A_1 and A_2 are the values of A corresponding to the lower and upper limits of E_a, respectively, E_1 and E_2. Performing the integration

$$d\theta/dt = (NkT/t)(e^{-A_2 t} - e^{-A_1 t}). \tag{73}$$

This expression may be simplified by assuming the limit E_2 is very large, in which case $A_2 \to 0$, and Eq. (73) becomes

$$d\theta/dt = (NkT/t)(1 - e^{-A_1 t}). \tag{74}$$

Peers (41) gives an analysis of conditions under which Eq. (74) approximates the Elovich form, Eq. (68).

9.6 The Elovich Equation

Many experimental data at constant pressure and even at constant volume (49, 50) with pressures falling to as little as 10 or 20% of their initial values fit the Elovich equation (51, 52), Eq. (67). The integrated form is given by Eq. (68) when the boundary conditions are $q = 0$, $t = 0$. In many cases, an initial volume of gas is adsorbed rapidly. Assuming that the initial adsorption is instantaneous, we find for the integrated expression of the Elovich equation.

$$q = (1/b)\ln(t + k) - (1/b)\ln t_0, \tag{75}$$

where $k = t_0 \exp(bq_0)$ and $t_0 = 1/ab$, or

$$q - q_0 = (1/b)\ln(t + k) - (1/b)\ln k. \tag{76}$$

The value of k (or t_0) is found by trial plotting of $\ln(t + k)$ against q until a linear plot is obtained, $1/b$ is the slope of the linear plot, and a is found from $t_0 = 1/ab$. The value of q_0 is obtained by extrapolation of the linear plot to $t = 0$; it equals approximately the amount of initial rapid adsorption. If $k \rightarrow 0$ (i.e., $t_0 \rightarrow 0$, and $a \rightarrow \infty$), then the equation becomes

$$q = (1/b) \ln t - (1/b) \ln t_0. \tag{77}$$

In this case, there is obviously an initial rapid adsorption, but it goes over continuously into the slow adsorption and cannot be separated easily by extrapolation to $t = 0$. An example is the adsorption of hydrogen on ZnO·Cr$_2$O$_3$ (52).

In preceding sections of this chapter, we have derived various theoretical expressions for rates of adsorption based on widely differing models. In all cases, the resulting expressions are approximately in the Elovich form. With only one exception of limited application (53), all theoretical interpretations of Elovich kinetics are based on the Langmuir kinetic formulation. Further, desorption is considered to be negligible so that the overall rate expression is (see Section 9.2),

$$dq/dt = kpn, \tag{78}$$

where n is the instantaneous number of unoccupied sites. Models may be divided into two classes depending on whether n or k in Eq. (78) is considered to be an exponential function of q.

First we consider n as an exponential function of q. If dq/dt is proportional to the instantaneous number of unoccupied sites in accordance with Eq. (78), then the Elovich equation, Eq. (67), requires

$$n = n_0 e^{-bq}, \tag{79}$$

where n_0 is the number of sites at $q = 0$. This differs from the conventional relation, $n = n_0 - q$. Differentiation gives

$$- dn/dq = bn, \tag{80}$$

and

$$- dn/dt = bn(dq/dt). \tag{81}$$

From Eqs. (67) and (79) we then obtain

$$- dn/dt = (ab/n_0)\, n^2. \tag{82}$$

This equation expresses a bimolecular rate of adsorption-site decay. Taylor and Thon (50) postulate that the initial event which sets off the decay is the sudden creation of sites suggested by Wolkenstein (54) to occur on initial contact of gas and solid. Other site-generating mechanisms have been considered by Landsberg (*55*), Meller (*53*), and Cook and Oblad (*56*). Taylor and Thon justify subsequent decay by analogies taken from the phenomena of luminescence and photoconductivity. Such arguments, are, at the moment tenuous.

We now consider the second class of models in which k is an exponential function of q in Eq. (78) and $n = n_0 - q$ with q small enough that n may be considered constant. The models of this class appear more reasonable than the site-generating models. Since k includes the activation energy, we find that k becomes an exponential function of q when we allow the activation energy to increase linearly with coverage. An equation based on the boundary-layer model and approximating the Elovich form has already been given in Eq. (35). Another model leading to linear dependence of activation energy on coverage requires an *a priori* heterogeneity of the surface with respect to activation energy of adsorption. Examples have been given in Sections 9.2 and 9.5. It is obvious that with certain approximations any model invoking a potential barrier to adsorption which increases linearly with coverage will lead to an equation of the Elovich form.

In all the models proposed so far, the interpretations are subject to certain restrictions:

(1) They are restricted without exception to adsorption at constant pressure and thus do not account for those instances where Elovich kinetics are observed under constant-volume conditions.

(2) They do not explain complex systems in which $q - \ln(t + t_0)$ plots with two or more linear segments are obtained.

(3) In all cases, desorption is considered negligible.

(4) They cannot account for the frequently observed tendency (*38, 51, 57, 58*) for b to increase with increasing initial pressure.

(5) In many instances (see Section 9.5), the derived equations are of the form, $q = k_1 \ln t + k_2$, where t_0 of the Elovich form is missing in the first term on the right.

Concerning the last point, Stone (*43*) and Kubokawa (*48*) conclude that t_0 is missing from Eq. (66) because desorption has been neglected. However, Peers (*41*) shows convincingly that this is not the case. Equation (66) leads to an infinite rate of adsorption as $t \to 0$ while the Elovich equation leads to a finite value due to the presence of the term t_0. But in the derivation of the theoretical equation, it was shown that the initial rate on any set of sites of given E_a is equal to the finite value A from Eq. (71). Thus, Eq. (66), which should be the sum of a series of finite initial rates, is inconsistent in this respect with the model from which it was derived, and the discrepancy arises from the approximate method of integration used by Stone *et al.* in arriving at the final result, and not from the neglect of desorption.

Peers (59) and Sutherland and Winfield (60) propose a model that is free from the above restrictions. The model is distinct from the previous models in that gaseous diffusion and not the act of adsorption is the rate-limiting process. The Knudsen flow mechanism employed leads to Elovich kinetics under constant-volume conditions as well as constant-pressure conditions, and also predicts an inverse relation between b and the initial pressure. The model also allows a theoretical evaluation of b. Further testing of this model is needed.

At present, it is perhaps best to consider the Elovich equation as an empirical equation useful in correlating adsorption data. So far, it has not been particularly valuable in settling questions of mechanism.

REFERENCES

1. Taylor, H. S., and Williamson, A.T., *J. Am. Chem. Soc.* **53**, 2168 (1931).
2. Langmuir, I., *J. Am. Chem. Soc.* **40**, 1361 (1918).
3. Benton, A. F., and White, T. A., *J. Am. Chem. Soc.* **52**, 2325 (1930).
4. Garner, W. E., and Kingman, F. E. T., *Trans. Faraday Soc.* **27**, 322 (1931).
5. Roberts, J. K., *Proc. Roy. Soc. (London),* Ser. A **152**, 445, 464, 477 (1925).
6. Morrison, J. L., and Roberts, J. K., *Proc. Roy. Soc. (London),* Ser. A **173**, (1939).
7. Becker, J. A., and Hartman, C. D., *J. Phys. Chem.* **57**, 153 (1953).
8. Matsuda, A., *J. Res. Inst. Catalysis Hokkaido Univ.* **5**, 71 (1957).
9. Trapnell, B. M. W., and Hayward, D. O., "Chemisorption," p. 85. Butterworth, London and Washington, D.C., 1964.
10. Langmuir, I., *Chem. Rev.* **6**, 451 (1929).
11. Taylor, J. B., and Langmuir, I., *Phys. Rev.* **44**, 423 (1933).
12. Smith, J. N., and Fite, W. L., *J. Chem. Phys.* **37**, 898 (1962).
13. Farnsworth, H. E., and Madden, H. H., Solid Surfaces and the Gas–Solid Interface, *in* "Advances in Chemistry," Vol. 33, p. 114, American Chemical Society, Washington, D.C. (1961).
14. Lennard-Jones, J. E., *Trans. Faraday Soc.* **28**, 333 (1932).
15. Lanyon, M. A. H., and Trapnell, B. M. W., *Proc. Roy. Soc. (London),* Ser. A **227**, 387 (1955).
16. Low, J. T., *J. Phys. Chem. Solids* **4**, 91 (1958).
17. Green, M., Kofalas, J. A., and Robinson, P. H., "Semiconductor Surface Physics," (R. H. Kingston, ed.), p. 349. Univ. of Pennsylvania Press, Philadelphia, Pennsylvania, 1957.
18. Bennett, M. F., and Tompkins, F. C., *Proc. Roy. Soc. (London),* Ser. A **259**, 28 (1960).
19. Porter, A. S., and Tompkins, F. C., *Proc. Roy. Soc. (London),* Ser. A **217**, 529, 554 (1953).
20. Gundry, P. M., and Tompkins, F. C., *Trans. Faraday Soc.* **52**, 1609 (1956).
21. Gundry, P. M., and Tompkins, F. C., *Trans. Faraday Soc.* **53**, 218 (1957).
22. Greenhalgh, E., Slack, N., and Trapnell, B. M. W., *Trans. Faraday Soc.* **52**, 865 (1956).
23. Rudham, R., and Stone, F. S., "Chemisorption" (W. E. Garner, ed.), p. 205. Butterworth, London and Washington, D.C., 1957.
24. Jennings, T. J., and Stone, F. S., *Advan. Catalysis* **9**, 441 (1957).
25. Scholten, J. J. F., and Zwietering, P., *Trans. Faraday Soc.* **53**, 1363 (1959).

26. Scholten, J. J. F., Zwietering, P., Konvalinka, J. A., and de Boer, J. H., *Trans. Faraday Soc.* **55**, 2166 (1959).
27. Eyring, H., *J. Chem. Phys.* **3**, 107 (1935).
28. Eyring, H., *Chem. Rev.* **17**, 65 (1935).
29. Eyring, H., *Trans. Faraday Soc.* **34**, 41 (1938).
30. Laidler, K. J., *Catalysis* **1**, 75, 195 (1954).
31. Emmett, P. H., and Harkness, R. W., *J. Am. Chem. Soc.* **57**, 1631 (1935).
32. Kwan, T., *J. Res. Inst. Catalysis Hokkaido Univ.* **3**, 16, 109 (1953).
33. Emmett, P. H., and Brunauer, S., *J. Am. Chem. Soc.* **56**, 35 (1939).
34. Brunauer, S., Love, K. S., and Keenan, R. G., *J. Am. Chem. Soc.* **64**, 751 (1942).
35. Davis, R. T., *J. Am. Chem. Soc.* **68**, 1395 (1946).
36. Laidler, K. J., *Catalysis* **1**, 204, (1954).
37. Hauffe, K., *Advan. Catalysis* **7**, 232 (1955).
38. Engell, H. G., and Hauffe, K., *Z. Elektrochem.* **57**, 762 (1953).
39. Wolkenstein, Th., and Peshev, O., *Kinetika i Kataliz* **6**, 95 (1965).
40. Wolkenstein, Th., and Peshev, O., *J. Catalysis* **4**, 301 (1965).
41. Peers, A. M., *J. Catalysis* **4**, 499 (1965).
42. Halsey, G. D., *J. Phys. Colloid. Chem.* **55**, 21 (1951).
43. Stone, F. S., "Chemistry of the Solid State" (W. E. Garner, ed.), p. 367. Butterworth, London and Washington, D.C., 1955.
44. Brunauer, S., Love, K. S., and Keenan, R. G., *J. Am. Chem. Soc.* **64**, 751 (1942).
45. de Boer, J. H., *Advan. Catalysis* **8**, 18 (1956).
46. Ashmore, P. G., "Catalysis and Inhibition of Chemical Reactions," p. 165. Butterworth, London and Washington, D.C., 1963.
47. Scholten, J. J. F., Zwietering, P., Konvalinka, J. A., and de Boer, J. H., *Trans. Faraday Soc.* **55**, 2166 (1959).
48. Kubokawa, Y., *Bull. Chem. Soc. Japan* **33**, 734 (1960).
49. Low, M. J. D., *Chem. Rev.* **60**, 267 (1960).
50. Taylor, H. A., and Thon, N., *J. Am. Chem. Soc.* **74**, 4169 (1952).
51. Roginsky, S., and Zeldovich, Ya., *Acta Physicochim.* **1**, 554, 595 (1934).
52. Elovich, S. Yu., and Zhabrova, G. M., *Zh. Fiz. Khim.* **13**, 1761 (1939).
53. Meller, A., *Monatsh. Chem.* **87**, 491 (1956).
54. Wolkenstein, Th., *Zh. Fiz. Khim.* **23**, 917 (1949).
55. Landsberg, P. T., *J. Chem. Phys.* **23**, 1079 (1955).
56. Cook, M. A., and Oblad, A. G., *Ind. Eng. Chem.* **45**, 1456 (1953).
57. Cimino, A., Molinari, E., and Cipollini, E., *Actes Congr. Intern. Catalyse, 2ᵉ, Paris* **1**, 263 (1960).
58. Kubokawa, Y., *Bull. Chem. Soc. Japan* **33**, 550 (1960).
59. Peers, A. M., *J. Catalysis* **4**, 672 (1965).
60. Sutherland, K. L., and Winfield, M. E., *Australian J. Chem.* **6**, 221 (1953).
61. Baret, J. F., *J. Colloid Interface Sci.* **30**, 1 (1969).

CATALYSIS

X

ADSORPTION AND CATALYSIS

10.1 Introduction

We have seen that chemisorption involves phenomena which are considerably more complex than those of physical adsorption. When adsorption, physical or chemical, is accompanied by catalytic reaction, the complexity becomes even greater. Broadly, we can say that adsorption is the precursor of catalysis. However, the variety of interrelationships between adsorption and catalysis are almost limitless. Few general theories are satisfying. Catalysis on the surface of solids is most likely to involve localized defects. These localized adsorption centers can be as specific as normal chemical bonds and no purely physical theories can give an adequate explanation of them. Each of the vast number of catalytic systems has its own catalytically important peculiarities not completely amenable to generalization. In the following chapters, we shall discuss a few general theories of catalysis and many specific catalytic systems. In this chapter, we want to sketch some very general interrelationships between adsorption and catalysis.

10.2 Adsorption and Reaction Rate

To show how reaction rate may be affected by the presence of a catalyst, consider the special case of a bimolecular reaction, $A + B \rightarrow$ product. If the reaction proceeds homogeneously in the gas phase, we may represent it according to absolute-rate theory by

$$A + B \leftrightarrows (A\text{---}B)^{\ddagger} \rightarrow product \qquad (1)$$

where $(A\text{---}B)^{\ddagger}$ is the activated complex. If it proceeds heterogeneously on the surface of a solid catalyst, we write

$$A + B + \Sigma_2 \leftrightarrows \left(\begin{array}{c} A\text{---}B \\ | \quad | \\ \overline{-\Sigma\text{---}\Sigma-} \end{array} \right)^{\ddagger} \rightarrow product. \qquad (2)$$

where Σ_2 is a vacant pair of adjacent sites. For the homogeneous reaction, the rate is given by

$$r_{hom} = c_A c_B \frac{kT}{h} \frac{q_{\ddagger}/V}{(q_A/V)(q_B/V)} \exp \frac{-E_{hom}}{RT}, \qquad (3)$$

where c_A and c_B are the gas-phase concentrations of reactants A and B, respectively; (q_A/V), (q_B/V), and $(q_{\ddagger}/V)$ are the partition functions per unit volume of A, B, and the activated complex; and E_{hom} is the activation energy of the homogeneous reaction. As in Section 9.3, the partition function of vibration in the direction of the reaction coordinate kT/h has been separated from the total partition function for the activated complex. For the heterogeneous reaction, the rate is expressed by

$$r_{het} = c_A c_B c_{\Sigma_2} \frac{kT}{h} \frac{q_{\ddagger}/\alpha}{(q_A/V)(q_B/V)(q_{\Sigma_2}/\alpha)} \exp \frac{-E_{het}}{RT}, \qquad (4)$$

where c_{Σ_2} is the concentration of vacant pairs of sites, $(q_{\ddagger}/\alpha)$ and (q_{Σ_2}/α) are the partition functions per unit area for the activated complex and a vacant pair of sites respectively, E_{het} is the activation energy of the heterogeneous reaction, and other terms have their previous meaning and units. Equation (4) is the reaction rate for a unit volume of gas in the presence of a unit area of catalyst. Since $q_{\ddagger}$ and q_{Σ_2} often contain only high frequency vibrations, they are set equal to unity as an approximation so that we can write

$$r_{het} \cong c_A c_B c_{\Sigma_2} \frac{kT}{h} \frac{1}{(q_A/V)(q_B/V)} \exp \frac{-E_{het}}{RT}. \qquad (5)$$

Dividing Eq. (5) by Eq. (3), we find the ratio of the heterogeneous to the homogeneous rate per unit volume of gas and per unit area of catalyst surface,

$$\frac{r_{het}}{r_{hom}} = \frac{c_{\Sigma_2}}{(q_{\ddagger}/V)} \exp \frac{\Delta E}{RT}, \qquad \Delta E = E_{hom} - E_{het}. \qquad (6)$$

A reasonable value for c_{Σ_2} is 10^{15} $(cm^2)^{-1}$; and for $(q_{\ddagger}/V)$, which contains a translational term, is 10^{24} to 10^{30}, say 10^{27} $(cm^3)^{-1}$. Then

$$r_{het}/r_{hom} = 10^{-12} \exp(\Delta E/RT) \quad cm, \qquad (7)$$

which indicates that the heterogeneous reaction is much slower than the homogeneous reaction for the same activation energy and per unit area of catalyst surface. Porous catalysts may have surface areas as high as 10^7 cm^2/gm. Thus, using 1 gm of such catalyst per cm^3 of gas phase, the ratio in Eq. (7) for $\Delta E = 0$ is still small, 10^{-5}. It is concluded that the activation energy of the heterogeneous reaction must be considerably lower than that of the homogeneous reaction in order that a significant advantage in rate be attained through the use of a catalyst. For example, ΔE at 300°K must be 16.5 kcal/mole when the catalyst surface is 1 cm^2 and 7 kcal/mole when it is 10^7 cm^2 in order that $r_{het} = r_{hom}$. At 500°K, the corresponding figures are 27.6 and 11.5 kcal/mole. Surface activation energies are in fact considerably lower than gas-phase activation energies in many systems as the data in Table 10.1 illustrate (*1*).

TABLE 10.1

ACTIVATION ENERGIES FOR HOMOGENEOUS AND HETEROGENEOUS
REACTIONS

Decomposition of	Catalyst	E_{het}, kcal/mole	E_{hom}, kcal/mole
HI	Au (2)	25.0	44.0
	Pt (3)	14.0	
N_2O	Au (4)	29.0	58.5
	Pt (5)	32.5	
NH_3	W (6)	39.0	>80.0
	Os (7)	47.0	
CH_4	Pt (8)	55–60	~ 80.0

Another important role which the catalyst surface frequently plays in accelerating the rate of reaction is the "preparation" of a reactant. For example, hydrogenations may be speeded up by the dissociation of hydrogen on a metal surface, a task which is far more difficult to perform in the absence of a catalyst. Another example is the orienting of a reactant by the surface in such a way as to facilitate the formation of a transition state. Both of these examples stress the importance of the geometric properties of the surface in catalysis, for the ease of dissociation depends strongly on the distance of separation of adjacent sites, and a surface in general restricts the direction of approach of molecules.

10.3 Strength of Adsorption Bond and Catalysis

Electronic properties of the surface are also important (*2–8*) for they affect the energy of the adsorption bond. In general, the energy of adsorption must be low in order for catalysis to take place (*9*). Reactions cannot be sustained unless the energy of desorption of product is sufficiently low that the product can leave the surface. Bimolecular surface reactions may be further restricted by low surface mobility of the reactants at high adsorption energies. Both physical adsorption and chemisorption may be involved in catalysis. Physical adsorption may be involved directly, for example, in the polymerization of ethylene to high polymers on chromium oxide–silica–alumina catalysts (*10*), where ethylene molecules physically adsorbed on silica–alumina sites may feed into growing polymer chains on chromium sites. Physical adsorption may also be involved indirectly in catalysis through such a mechanism as shown in Fig. 9.1 where physical adsorption leads to an intermediate state of chemisorption, Type C, and then to a final state of chemisorption, Type A, in which catalysis may proceed.

Catalysis on nonuniform surfaces is not well understood. Different sites may have different rate-limiting steps in which case the reaction cannot be considered to have an overall rate-limiting step. In some cases, a broad range of adsorption sites may be catalytically effective, while in other cases only a narrow range may be effective. In the latter case, the surface will behave kinetically as though it were uniform. Thus, the relationships between adsorption and catalysis may be extremely complex on nonuniform surfaces, and rate theories based on overall adsorption data are likely to be in error. For sites contributing strongly to catalysis may contribute negligible amounts to the adsorption isotherm.

10.4   Adsorption Equilibrium and Catalysis

If a reaction is taking place on a surface, the following adsorption conditions are possible (*11*):

(1)   equilibrium adsorption,
(2)   steady state nonequilibrium adsorption,
(3)   unstable nonequilibrium adsorption.

Equilibrium adsorption will exist essentially if the rate of surface reaction is very slow compared to rates of adsorption and desorption of reactants and a negligible fraction of the surface is covered by adsorbed product. We illustrate this case for a unimolecular reaction in which the Langmuir adsorption equilibrium for reactant is

$$\theta = Kp/(1 + Kp), \tag{8}$$

where $K = k_a/k_d$ is the equilibrium constant of adsorption. The rate of reaction is then

$$r = k_r \theta = k_r Kp/(1 + Kp), \tag{9}$$

where k_r is the surface-reaction velocity constant. For a bimolecular surface reaction, A + B, a similar expression may be derived using the mixed Langmuir isotherm to express surface concentrations, θ_A and θ_B, in terms of gas-phase pressures, p_A and p_B, and substituting in the rate expression, $r = k_r \theta_A \theta_B$.

If surface reaction rates cannot be neglected in comparison to rates of adsorption and desorption, then a steady state nonequilibrium treatment may be employed. For a unimolecular reaction, we have

$$d\theta/dt = k_a p(1 - \theta) - k_d \theta - k_r \theta = 0, \tag{10}$$

where we have again assumed no build-up of product on the surface. In this case, the reaction rate is

$$r = k_r \theta = k_r k_a p/(k_d + k_r + k_a p) = k_r K'p/(1 + K'p) \qquad K' = k_a/(k_d + k_r), \tag{11}$$

which has the same form as Eq. (9) to which it reduces if $k_d \gg k_r$, while Eq. (10) reduces to the adsorption equilibrium condition. For a bimolecular reaction, A + B, we can write

$$d\theta_A/dt = k_{Aa}(1 - \theta_A - \theta_B)p_A - k_r \theta_A \theta_B - k_{Ad} \theta_A = 0,$$
$$d\theta_B/dt = k_{Ba}(1 - \theta_A - \theta_B)p_B - k_r \theta_A \theta_B - k_{Bd} \theta_B = 0. \tag{12}$$

With these equations, expressions for θ_A and θ_B in terms of p_A and p_B may be obtained and substituted in the surface rate expression $r = k_r \theta_A \theta_B$ as before for the equilibrium case.

If, however, removal of adsorbed material is by reaction alone, that is, A and B do not desorb, then the steady-state equations become

$$d\theta_A/dt = k_{Aa}(1 - \theta_A - \theta_B)p_A - k_r \theta_A \theta_B = 0,$$
$$d\theta_B/dt = k_{Ba}(1 - \theta_A - \theta_B)p_B - k_r \theta_A \theta_B = 0, \tag{13}$$

which represents a condition of unstable nonequilibrium adsorption. When $k_{Aa} p_A > k_{Ba} p_B$, the only solution is $\theta_A = 1$, $\theta_B = 0$; and when $k_{Aa} p_A < k_{Ba} p_B$, the only solution is $\theta_B = 1$, $\theta_A = 0$. That is, ultimately the surface is completely covered either with A or with B, neither of which desorbs. The reaction rate in both cases is zero. These results show the wide variation in surface conditions that are obtained depending on whether equilibrium or nonequilibrium conditions are considered.

REFERENCES

1. Laidler, K. J., *Catalysis* **1**, 234 (1961).
2. Eyring, H., *J. Chem. Phys.* **3**, 107 (1935); *Chem. Rev.* **17**, 65 (1935); *Trans. Faraday Soc.* **34**, 41 (1938).

3. Laidler, K. J., Glasstone, S., and Eyring, H., *J. Chem. Phys.* **8**, 659, 667 (1940); Glasstone, S., Laidler, K. J., and Eyring, H., "The Theory of Rate Processes," Chapter VII, McGraw-Hill, New York, 1941.

4. Topley, B., *Nature* **128**, 115 (1931).

5. Kimball, G. E., *J. Chem. Phys.* **6**, 447 (1938).

6. Temkin, M., *Acta Physicochim. U.S.S.R.* **8**, 141 (1938).

7. Emmett, P. H., and Brunauer, S., *J. Am. Chem. Soc.* **56**, 35 (1934); Emmett, P. H., and Harkness, R. W., *J. Am. Chem. Soc.* **57**, 1631 (1935).

8. Roberts, J. K., *Proc. Roy. Soc. (London), Ser. A.* **152**, 445 (1935).

9. de Boer, J. H., *Advan. Catalysis* **8**, 149 (1956).

10. Clark, A., Polymerization and Polycondensation Processes, Symp. Natl. Am. Chem. Soc., San Francisco, California, April 1–5, 1968.

11. Halsey, G. D., Jr., *J. Chem. Phys.* **67**, 2038 (1963).

XI

KINETICS OF HETEROGENEOUS CATALYSIS
—DIFFUSION STEPS NEGLECTED

11.1 Introduction

Kinetic studies are hampered by empiricism. Only in the simplest homogeneous gas reactions is it possible through activated-complex theory, collision theory, or more general reaction-rate theories (*1*) to calculate kinetic constants independently with any assurance whatsoever. Most heterogeneous catalytic reactions are too complex to be treated satisfactorily at present by such theories, and it is therefore necessary to rely heavily on classical, empirical methods. Even though it is assumed that diffusion is not controlling, as in this chapter, heterogeneous catalytic reactions may be complex, consisting of at least three single steps: adsorption, surface reaction, and desorption. Considerable simplification results when there is a single rate-determining step. However, only when surface reaction is the rate-limiting step are overall kinetics governed by the kinetics of the reaction step. If desorption of product is the rate-limiting step, for example, then overall kinetics as measured in the gas phase do not reflect the surface-reaction rate. In homogeneous reactions, partial pressures of reactants are often directly measurable. In heterogeneous reactions, however, the important quantities are surface concentrations, for at least one reactant will be adsorbed, and such concentrations usually cannot be measured except indirectly through partial pressures. [Under certain conditions, Tamaru and others (2) have measured surface concentrations during reaction.] Thus, the kinetics of heterogeneous reactions are far less accessible than the kinetics of homogeneous reactions. Heterogeneous reactions are usually confined to a thin surface layer and are far less reproducible than homogeneous reactions because the nature of the catalyst surface, which has a profound effect on the reaction, is often extremely sensitive to methods of preparation. Furthermore, the active centers on a catalyst surface are not necessarily distributed in a random manner as are reacting centers in a homogeneous system. The order of a heterogeneous reaction may be sensitive to temperature. For example, in a bimolecular reaction at low temperature, reactant A, may be adsorbed considerably more strongly than reactant B, so that overall gas-phase kinetics may be approximately first order. At higher temperatures, reactants A and B may be adsorbed in approximately equal quantities, and the kinetics may become second order. Finally, overall kinetics involving partial pressures often lead to peculiar Arrhenius plots. It is quite common to find a system (3) for which a plot of ln k versus $1/T$ gives a maximum (see Section 11.2).

Because of the great difficulties in the kinetics of heterogeneous catalytic reactions, it has become popular (36, 37) in devising models to rely heavily on mathematical statistics. But it is important to realize that physically oriented approaches are still essential in identifying realistic rate models.

11.2   Unimolecular Reactions

In heterogeneous catalytic reactions which involve at least three mechanistic steps, adsorption, surface reaction, and desorption, a molecule is held on the surface for a time τ_d, and may react while on the surface with a characteristic time τ_s (4). If τ_d/τ_s ($=\beta$) is small, most molecules will leave the surface before they react. Consequently, the reactant is essentially in adsorption equilibrium between the gas phase and the surface, and the rate-controlling step is the surface reaction. If β is large, most molecules will undergo reaction before they desorb, and surface reaction is not the rate-controlling step. The parameter β measures the competition between surface reaction and desorption. These steps lead to intrinsic kinetics, which have a characteristic time τ_k. When the catalytic reaction takes place within a porous catalyst particle, another sequence competes in the determination of the overall rate process. A molecule wanders in a pore for a time proportional to the Einstein time $\tau_E = R^2/4D$. If τ_E/τ_k ($=\varphi$) is small, the kinetics are not influenced by diffusion. If φ is large, the diffusion step becomes rate controlling. As $\beta \to 0$ and $\varphi \to 0$, surface reaction becomes rate controlling.

In Section 10.4, Eq. (9), we described briefly the case where β and φ are very small, that is, when there is no diffusion limitation, surface reaction is controlling, adsorption equilibrium exists for the reactant, and product adsorption is negligible. We consider the specific case of a system of fixed volume V containing N_A molecules of reactant A at time t and M gm of catalyst having B catalytic sites per gram. Reactant A is equilibrated between the gas phase, N_G molecules, and the adsorbed phase, N_S molecules, with $N_A = N_S + N_G$. The rate in molecules of A disappearing from the system per second per gram of catalyst is

$$r = -(1/M)(dN_A/dt) = k_r{'}B\theta_A = k_r\theta_A, \tag{1}$$

where $B\theta_A = N_S$ and $k_r{'}$ is the surface reaction velocity constant. If $N_S \ll N_G$, that is, V is very large compared to the volume of catalyst, then, just as in a flow system where N_s is constant in each small increment of reaction zone at steady state conditions,

$$r = -(1/M)(dN_A/dt) \cong -(1/M)(dN_G/dt) = -(1/M)(V/kT)(dp_A/dt) \tag{2}$$

where p_A is the partial pressure of A. As in Section 10.4, using the Langmuir adsorption equilibrium expression for θ, we obtain

$$r = k_r K_A p_A/(1 + K_A p_A) \tag{3}$$

At very low pressure, when $Kp_A \ll 1$, we have

$$r = k_r K_A p_A, \tag{4}$$

a first-order reaction; and when $Kp_A \gg 1$, we find

$$r = k_r, \tag{5}$$

a zero-order reaction.

When a reaction product (or extraneous poison) x is adsorbed to a significant extent, we have two components in adsorption equilibrium. Thus

$$r = k_r\theta_A = k_r K_A p_A/(1 + K_A p_A + K_x p_x) \tag{6}$$

Obviously, approximations can be made depending on the relative magnitudes of 1, $K_A p_A$, and $K_x p_x$, and various ratios of inhibited to uninhibited reaction rates can be found. When inhibition occurs, initial rate measurements will not give the whole picture. It is customary to check for inhibitory effects by addition of product in various amounts to the reactant.

On the basis of equations developed above, relationships between apparent and true activation energies can be derived. For example, Eq. (4) shows a first-order relationship with respect to pressure, but the rate constant is $k_r K_A$ compared to k_r for the surface reaction expressed in terms of the surface concentration. The constant k_r is associated with the true activation energy E

of the surface reaction, and K_A, the equilibrium constant of adsorption is associated with the entropy and enthalpy of adsorption, so that

$$k_r K_A = [A' \exp(-E/RT)][\exp(\Delta S/R) \exp(-\Delta H/RT)] \tag{7}$$
$$= A \exp[-(E + \Delta H)/RT] = A \exp(-E_a/RT),$$

where A and A' are constants and

$$E_a = E + \Delta H_A \tag{8}$$

is defined as the apparent activation energy. The energy E_a is thus lower than the true activation energy E (>0) by the heat of adsorption ΔH_A (<0). The correction arises because desorption occurs, that is, θ decreases at constant partial pressure p_A as the temperature is increased. If saturation adsorption, $\theta = 1$, exists over the entire temperature range studied, then the reaction rate is given by Eq. (5) and

$$E_a = E. \tag{9}$$

In this case, the apparent activation energy is the true activation energy, for there is no desorption to consider. When the reaction is strongly inhibited by a product, so that $K_x p_x \gg 1 + K_A p_A$, we have, from Eq. (6),

$$r = k_r K_A p_A / K_x p_x, \tag{10}$$

and the apparent activation energy is

$$E_a = E + \Delta H_A - \Delta H_x. \tag{11}$$

We mentioned in Section 11.1 that many catalytic reactions are first order with respect to pressure, obeying Eq. (4). The polymerization of ethylene to high polymers over chromium oxide–silica–alumina and the disproportionation of propylene to ethylene and butylene are examples (3). Furthermore, each of these reactions exhibits a maximum in the Arrhenius plot of rate constant versus reciprocal temperature. If these reactions were *elementary* first-order reactions, they would not possess rate–temperature maxima. However, it is possible, as noted above, for catalytic reactions to exhibit first-order kinetics over wide pressure and temperature ranges and to possess rate–temperature maxima in the same ranges. The causes of these phenomena are specific to each system and are not well understood. It should be noted that expressions such as Eq. (3) can show a maximum in rate versus temperature and at the same time give an approximately linear plot of rate versus pressure in the temperature region where the maximum occurs. In this case, a plot of the logarithm of the experimentally determined, first-order rate constant versus the reciprocal of the absolute temperature would give a curve with a maximum.

The lack of equilibrium in adsorption during a catalytic reaction can completely alter the picture of a catalytic process as we have seen in Section

10.4. Yet kinetic data are commonly analyzed in terms of Langmuir equilibrium theory, even though a large body of evidence exists which indicates that the use of this theory is unjustified. For engineering correlations, the consequences are not serious, although mathematically simpler expressions can be obtained with other approaches (*5*). But caution must be exercised in mechanistic studies. We restate here the two chief shortcomings of the Langmuir approach (see Chapter II for complete discussion):

(1)   Interaction between adsorbed molecules is ignored.
(2)   Heat of adsorption is assumed not to change with coverage.

A discussion of the attempts to overcome these and other shortcomings will be treated in later sections of this chapter and in Chapter XII.

Brief mention will be made of an interesting approach involving the principle of LFER (Linear Free Energy Relationships) which Mochida and Yoneda (*38–40*) have applied principally to the dealkylation of alkylbenzenes. For dealkylation of monoalkylbenzenes (*38*), they showed that a plot of the logarithm of the rate constant against the enthalpy $[\Delta H_{c^+} (R_1)]$ of hydride abstraction from the paraffin corresponding to the alkyl group (R_1) is linear over a broad range of R_1. The relationship appears to hold for various solid catalysts including silica–alumina, silica–magnesia, alumina, alumina–boria, etc.

11.3   A General Approach to the Unimolecular Reaction

In this section, we want to set up general equations for the Langmuir–Hinshelwood model which deal with all possible combinations and permutations of the rate-limiting steps (*6*). We consider the following reaction steps,

$$A_{gas} \xleftrightarrow[k_2]{k_1} A_{ads}, \tag{12}$$

$$A_{ad} \xleftrightarrow[k_4]{k_3} B_{ads}, \tag{13}$$

$$B_{ads} \xleftrightarrow[k_6]{k_5} B_{gas}. \tag{14}$$

Hutchinson *et al.* (*7*) have considered this sequence for the conversion of parahydrogen to ortho-hydrogen. Assuming a uniform surface and no adsorbate–adsorbate interactions, we may write the following rate equations for the three steps,

$$r_{ads} = k_1 p_A (1 - \theta_A - \theta_B) - k_2 \theta_A, \tag{15}$$

$$r_s = k_3 \theta_A - k_4 \theta_B, \tag{16}$$

$$r_{des} = k_5 \theta_B - k_6 p_B (1 - \theta_A - \theta_B). \tag{17}$$

Under steady state conditions, $r_{ads} = r_s = r_{des}$. If one of the three steps is sufficiently slow, then the other two steps will be approximately at equilibrium. If two of three steps are sufficiently slow, then the third step will be approximately at equilibrium. If none of the three steps is slow with respect to the others, then we cannot use the approximation of equilibrium for any of the steps. For each of these cases, θ_A and θ_B can be eliminated from Eqs. (15)–(17) and an expression obtained for the overall rate in terms of rate constants and p_A and p_B. We list the expressions as given by Hutchinson et al. (7),

Adsorption controlling:

$$r = (k_1 k_3 k_5 p_A - k_2 k_4 k_6 p_B)/[k_3 k_5 + (k_3 k_6 + k_4 k_6)p_B]. \qquad (18)$$

Surface-reaction controlling:

$$r = (k_1 k_3 k_5 p_A - k_2 k_4 k_6 p_B)/(k_2 k_5 + k_1 k_5 p_A + k_2 k_6 p_B), \qquad (19)$$

which is identical with Eq. (6) when there is no back surface reaction, that is, $k_4 = 0$.

Desorption controlling:

$$r = \frac{k_1 k_3 k_5 p_A - k_2 k_4 k_6 p_B}{k_2 k_4 + (k_1 k_3 + k_1 k_4)p_A} . \qquad (20)$$

Surface reaction and desorption steps slow, adsorption fast and approaching equilibrium,

$$r = \frac{k_1 k_3 k_5 p_A - k_2 k_4 k_6 p_B}{(k_2 k_4 + k_2 k_5) + (k_1 k_3 + k_1 k_4 + k_1 k_5)p_A + k_2 k_6 p_B} . \qquad (21)$$

Adsorption and desorption steps slow, surface reaction fast and approaching equilibrium,

$$r = \frac{k_1 k_3 k_5 p_A - k_2 k_4 k_6 p_B}{(k_2 k_4 + k_3 k_5) + (k_1 k_3 + k_1 k_4)p_A + (k_3 k_6 + k_4 k_6)p_B} . \qquad (22)$$

Adsorption and surface reaction steps slow, desorption fast and approaching equilibrium,

$$r = \frac{k_1 k_3 k_5 p_A - k_2 k_4 k_6 p_B}{(k_2 k_5 + k_3 k_5) + k_1 k_5 p_A + (k_2 k_6 + k_3 k_6 + k_4 k_6)p_B} . \qquad (23)$$

No rate-limiting steps,

$$r = \frac{k_1 k_3 k_5 p_A - k_2 k_4 k_6 p_B}{(k_2 k_4 + k_2 k_5 + k_3 k_5) + (k_1 k_3 + k_1 k_4 + k_1 k_5)p_A} . \qquad (24)$$
$$+ (k_2 k_6 + k_3 k_6 + k_4 k_6)p_B$$

All of these expressions are of the form,

$$r = (k_1 k_3 k_5 p_A - k_2 k_4 k_6 p_B)/(A + Bp_A + Dp_B), \tag{25}$$

where A, B, and D are different combinations of the rate constants k_1 through k_6. Eq. (25) can be rearranged to give

$$r = k(p_A - p_{Ae})/(1 + k'p_A), \tag{26}$$

where

$$k = k_1 k_3 k_5 P/[p_{Be}(A + DP)], \tag{27}$$

$$k' = (B - D)/(A + DP), \tag{28}$$

and p_{Ae} and p_{Be} are equilibrium concentrations of A and B and $P = p_A + p_B$ is the total pressure when only A and B in addition to catalyst are present.

It is evident that all permutations and combinations of the three steps give the same mathematical expression. If the Langmuir–Hinshelwood model is not valid, then one simple test can prove it so regardless of the detailed nature of rate-limiting steps. The disadvantage is that it is impossible to tell which steps are rate-controlling.

Madix (41) has examined theoretically the validity of assuming a rate-determining step on a nonuniform surface. He concludes that the validity depends critically on the values of the parameters describing the relationship between the reactivity of the intermediates and the distribution variable.

11.4 Bimolecular Reactions

When the reaction involves two molecules, A and B, then two general mechanisms can be visualized (8, 9): the Langmuir–Hinshelwood mechanism with both A and B reacting in the adsorbed state, and the Rideal mechanism with A (or B) in the gas phase reacting with B (or A) in the adsorbed phase. For the Langmuir–Hinshelwood model, it is assumed that the rate of reaction is determined by the number of pairs of unlike species on adjacent sites, which gives

$$r = k_r \theta_A \theta_B = k_r K_A K_B p_A p_B/(1 + K_A p_A + K_B p_B)^2. \tag{29}$$

We have assumed equilibrium adsorption of both A and B, so that

$$\theta_A = K_A p_A/(1 + K_A p_A + K_B p_B), \quad \text{and} \quad \theta_B = K_B p_B/(1 + K_A p_A + K_B p_B). \tag{30}$$

A more general approach as in Section 11.3 for the unimolecular reaction is possible. For the Rideal model, we find

$$r = k_r \theta_A p_B = k_r K_A p_A p_B/(1 + K_A p_A + K_B p_B), \tag{31}$$

where it is assumed that interaction occurs between adsorbed A and gaseous B, even though some B is adsorbed.

If the partial pressure of A is held constant while that of B is varied, we find for the Langmuir–Hinshelwood model that the rate goes through a maximum at $K_B p_B = K_A p_A + 1$, or at $K_A p_A = K_B p_B + 1$ when the pressure of B is held constant and that of A varied. In the Rideal model, however, the rate does not go through a maximum. Rather, it approaches a limiting value asymptotically.

Many special cases of each model exist. In the Langmuir–Hinshelwood model, if A and B are both weakly adsorbed, Eq. (29) becomes

$$r = k_r K_A K_B p_A p_B. \tag{32}$$

If A is weakly adsorbed and B strongly adsorbed,

$$r = k_r K_A K_B p_A p_B/(1 + K_B p_B)^2, \tag{33}$$

which has a maximum only when p_B is varied at constant p_A. If B is so strongly adsorbed that $K_B p_B \gg 1$, then

$$r = k_r K_A p_A/K_B p_B. \tag{34}$$

If the reaction is inhibited by x, which may be a product of the reaction, we find

$$r = k_r K_A K_B p_A p_B/(1 + K_A p_A + K_B p_B + K_x p_x)^2. \tag{35}$$

This equation can also be simplified by various assumptions about the relative magnitudes of the terms in the denominator. In the Rideal model, if B is not adsorbed at all,

$$r = k_r K_A p_A p_B/(1 + K_A p_A). \tag{36}$$

Apparent activation energies may be obtained as in the case of unimolecular reactions.

For the Langmuir–Hinshelwood model:

(1)   A and B weakly adsorbed, corresponding to Eq. (32)

$$E_a = E + \Delta H_A + \Delta H_B, \tag{37}$$

(2)   B strongly adsorbed, corresponding to Eq. (34)

$$E_a = E + \Delta H_A - \Delta H_B. \tag{38}$$

For the Rideal model:

(1) A weakly adsorbed, B not adsorbed at all, corresponding to Eq. (36) with $K_A p_A \ll 1$

$$E_a = E + \Delta H_A, \tag{39}$$

(2) A strongly adsorbed, B not adsorbed at all, corresponding to Eq. (36) with $K_A p_A \gg 1$

$$E_a = E. \tag{40}$$

11.5 Complex Reactions

When there is more than one reaction step, the question of the relative amounts of products formed in a given time interval arises (*10–13*). There are three types of complex reactions to consider:

(1)

$$A \xrightarrow{k_1} B \qquad\qquad \text{Type I}$$
$$C \xrightarrow{k_2} D$$

$$A \begin{array}{c} \xrightarrow{k_1} B \\ \xrightarrow{k_2} C \end{array} \qquad \text{Type II}$$

$$\qquad\qquad\qquad\qquad \text{Type III}$$

$$A \xrightarrow{k_1} B \xrightarrow{k_2} C$$

where k_1 and k_2 are overall rate constants for gas-phase kinetics. Type I represents two simultaneous reactions of different reactants, Type II represents two simultaneous reactions of the same reactant, and Type III represents two consecutive reactions. We follow Wheeler (*14*) in assuming first-order kinetics with respect to the gas phase, although more general forms similar to Eqs. (3) or (6) could be used. Rate constants are referred to unit area or weight of catalyst. Wheeler defines a selectivity factor $\sigma = k_1/k_2$ for all three types. A high value of σ denotes a preponderance of product B. In Type I reaction, the rates are expressed by

$$dp_A/dt = -k_1 p_A, \tag{41}$$

$$dp_C/dt = -k_2 p_C. \tag{42}$$

Dividing Eq. (41) by Eq. (42), integrating, and setting α_A and α_C equal, respectively, to the fractions of A and C reacted, we obtain

$$\alpha_A = 1 - (1 - \alpha_C)^\sigma. \tag{43}$$

Similarly for Type II,

$$\alpha_B = \sigma \alpha_C \tag{44}$$

where α_B and α_C are the fractions of A reacted to B and C respectively. And for Type III, the relative rates of change of the concentrations in terms of partial pressures of A and B are given by

$$-dp_B/dp_A = 1 - (1/\sigma)(p_B/p_A), \tag{45}$$

which integrates to

$$\alpha_B = [\sigma/(\sigma - 1)](1 - \alpha_A)[(1 - \alpha_A)^{(1-\sigma)/\sigma} - 1], \tag{46}$$

where α_A is the fraction of A reacted as before, and α_B is the fraction of A reacted to B. In Fig. 11.1, Type I and Type III plots of fractions reacted at various values of σ are given (15).

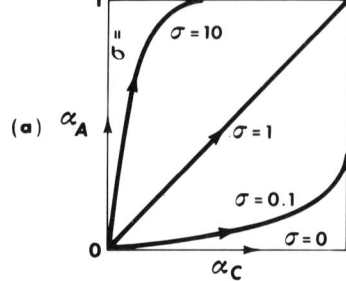

FIG. 11.1. (a) Type-I selectivity, the fractions of A and C converted to products for various selectivity factors. (b) Type-III selectivity, the fractions of A converted to B and C for various selectivity factors. Arrows denote direction of time (15).

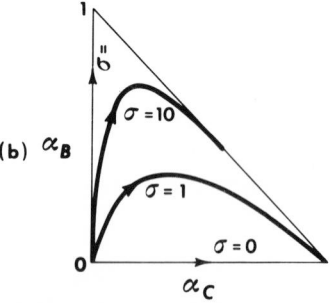

We note that in a reaction of Type I a *thermodynamic* factor plays an important role in selectivity. This may be seen when surface reactions are controlling by writing the equations for the disappearance of A and C in the general Langmuir form,

$$-\frac{dp_A}{dt} = (k_r)_A \theta_A = \frac{(k_r)_A K_A p_A}{(1 + K_A p_A + K_C p_C + K_B p_B + K_D p_D)}, \tag{47}$$

$$-\frac{dp_C}{dt} = (k_r)_C \theta_C = \frac{(k_r)_C K_C p_C}{(1 + K_A p_A + K_C p_C + K_B p_B + K_D p_D)}, \tag{48}$$

where $(k_r)_A K_A = k_1$ and $(k_r)_C K_C = k_2$ of Eqs. (41) and (42). Dividing Eq. (47) by Eq. (48), we have

$$\frac{dp_A}{dp_C} = \frac{(k_r)_A K_A p_A}{(k_r)_C K_C p_C} = \frac{(k_r)_A}{(k_r)_C} \frac{p_A}{p_C} \exp(-\delta \Delta F_a/RT), \qquad (49)$$

where the ratio of the adsorption equilibrium constants K_A/K_C has been put in terms of free energy using $\Delta F_A = -RT \ln K_A$ and $\Delta F_C = -RT \ln K_C$ with $\delta \Delta F_a = \Delta F_C - \Delta F_A$. The thermodynamic selectivity factor is also significant in Type III selectivity, for preferential formation of B in the presence of A depends on (i) the ability of B to desorb rapidly, or to be replaced by A, before undergoing further reaction and (ii) the inability of B, once desorbed, to re-adsorb in the presence of A. Selectivity in Type II reactions, however, does not depend on the thermodynamic factor. In this case, highly selective formation of the desired product depends on specific properties of the catalyst, and is thus described as *mechanistic* in origin.

Consecutive reactions, Type III, are the most common and we wish to present a few more details concerning them. For this purpose, we shall take into account adsorption and desorption of A, B, and C as well as surface reactions (*16*). We start with the complete steady-state equations for the surface and make assumptions about slow steps which are controlling and therefore may be dropped from the steady-state equations.

$$d\theta_A/dt = 0 = (k_a)_A\left(1 - \sum_i \theta_i\right)p_A - (k_r)_A \theta_A - (k_d)_A \theta_A, \qquad (50)$$

$$d\theta_B/dt = 0 = (k_a)_B\left(1 - \sum_i \theta_i\right)p_B + (k_r)_A \theta_A - (k_d)_B \theta_B - (k_r)_B \theta_B, \qquad (51)$$

$$d\theta_C/dt = 0 = (k_a)_C\left(1 - \sum_i \theta_i\right)p_C + (k_r)_B \theta_B - (k_d)_C \theta_C. \qquad (52)$$

In these material balance equations, k_a represents adsorption rate constants; $(1 - \sum_i \theta_i)$ is the fraction of vacant catalytic sites; $(k_r)_A$ and $(k_r)_B$ are reaction rate constants for $A_{ads} \rightarrow B_{ads}$ and $B_{ads} \rightarrow C_{ads}$ respectively, related to k_1 and k_2 through K_A and K_B the adsorption equilibrium constants as shown below; and k_d stands for desorption rate constants. Following de Boer and van der Borg (*16*), four cases are examined:

(i) Slow surface chemical reactions.
(ii) None of the adsorption–desorption reactions is fast with respect to the chemical reactions.
(iii) Fast adsorption and desorption with respect to chemical reactions for intermediate product but not for A.

(iv)   Rates of surface reactions comparable with rates of adsorption and desorption of intermediate product, rate of surface reaction of A slow.

CASE (i). In Eqs. (50)–(52), we may neglect terms involving surface reactions, $(k_r)_A \theta_A$ and $(k_r)_B \theta_B$. Adsorption equilibrium of A, B, and C exists and the equilibrium constants are

$$K_A = (k_a)_A/(k_d)_A = \theta_A/p_A\left(1 - \sum_i \theta_i\right), \tag{53}$$

$$K_B = (k_a)_B/(k_d)_B = \theta_B/p_B\left(1 - \sum_i \theta_i\right), \tag{54}$$

$$K_C = (k_a)_C/(k_d)_C = \theta_C/p_C\left(1 - \sum_i \theta_i\right). \tag{55}$$

The rates of formation of gaseous A, B, and C may be written as

$$-dp_A/dt = (k_r)_A \theta_A, \tag{56}$$

$$dp_B/dt = (k_r)_A \theta_A - (k_r)_B \theta_B, \tag{57}$$

$$dp_C/dt = (k_r)_B \theta_B. \tag{58}$$

We substitute expressions from Eqs. (53)–(55) for θ_A and θ_B into Eqs. (56)–(58) to obtain

$$-dp_A/dt = (k_r)_A K_A\left(1 - \sum_i \theta_i\right)p_A, \tag{59}$$

$$dp_B/dt = (k_r)_A K_A\left(1 - \sum_i \theta_i\right)p_A - (k_r)_B K_B\left(1 - \sum_i \theta_i\right)p_B, \tag{60}$$

$$dp_C/dt = (k_r)_B K_B\left(1 - \sum_i \theta_i\right)p_B. \tag{61}$$

Dividing Eq. (59) by (61), we obtain

$$\frac{(k_r)_A K_A p_A}{(k_r)_B K_B p_B} = \frac{(k_r)_A(k_a)_A(k_d)_B}{(k_d)_A(k_r)_B(k_a)_B}\frac{p_A}{p_B} = \frac{k_1}{k_2}\frac{p_A}{p_B} = \sigma\frac{p_A}{p_B}, \tag{62}$$

where $\sigma = k_1/k_2$ is the Type III selectivity factor.

Because it was assumed that adsorption equilibrium is established rapidly, and surface reactions are slow, nearly all of B desorbs before it reacts to form C. Thus the reaction A → C, in which (i) gaseous A adsorbs, reacts to form adsorbed B, (ii) B reacts without desorbing to form adsorbed C, and (iii) C desorbs to give gaseous C, is negligibly slow. That is, $k_3 \cong 0$ in the kinetic scheme

(2)

$$\begin{array}{c} \text{B} \\ {}^{k_1}\nearrow \quad \searrow^{k_2} \\ \text{A} \xrightarrow{\ k_3\ } \text{C} \end{array}$$

(63)

Therefore, in this case, two rate constants suffice and the kinetic scheme is given by

$$\text{A} \xrightarrow{\ k_1\ } \text{B} \xrightarrow{\ k_2\ } \text{C}.$$

(64)

CASE (ii). Since none of the adsorption-desorption reactions are fast with respect to the chemical reactions on the surface, all of the terms in the material balance Eqs. (50)–(52) must be retained. We find

$$\theta_\text{A} = \frac{(k_\text{a})_\text{A}\left(1 - \sum_i \theta_i\right)}{(k_\text{d})_\text{A} + (k_\text{r})_\text{A}}\, p_\text{A},$$

(65)

and

$$\theta_\text{B} = \frac{(k_\text{r})_\text{A}}{(k_\text{r})_\text{B} + (k_\text{d})_\text{B}}\, \theta_\text{A} + \frac{(k_\text{a})_\text{B}\left(1 - \sum_i \theta_i\right)}{(k_\text{r})_\text{B} + (k_\text{d})_\text{B}}\, p_\text{B}.$$

(66)

In this case, the rates of formation of gaseous B and C are:

$$dp_\text{B}/dt = (k_\text{r})_\text{A}\,\theta_\text{A} - (k_\text{r})_\text{B}\,\theta_\text{B} = (k_\text{d})_\text{B}\,\theta_\text{B} - (k_\text{a})_\text{B}\left(1 - \sum_i \theta_i\right)p_\text{B},$$

(67)

$$dp_\text{C}/dt = (k_\text{r})_\text{B}\,\theta_\text{B}.$$

(68)

Substituting expressions for θ_A ánd θ_B from Eqs. (65) and (66), we obtain

$$\frac{dp_\text{B}}{dt} = \frac{(k_\text{d})_\text{B}(k_\text{r})_\text{A}(k_\text{a})_\text{A}\left(1 - \sum_i \theta_i\right)}{[(k_\text{r})_\text{A} + (k_\text{d})_\text{A}][(k_\text{r})_\text{B} + (k_\text{d})_\text{B}]}\, p_\text{A} - \frac{(k_\text{a})_\text{B}(k_\text{r})_\text{B}\left(1 - \sum_i \theta_i\right)}{(k_\text{r})_\text{B} + (k_\text{d})_\text{B}}\, p_\text{B}$$

(69)

$$\frac{dp_\text{C}}{dt} = \frac{(k_\text{a})_\text{A}(k_\text{r})_\text{A}(k_\text{r})_\text{B}\left(1 - \sum_i \theta_i\right)}{[(k_\text{r})_\text{A} + (k_\text{d})_\text{A}][(k_\text{r})_\text{B} + (k_\text{d})_\text{B}]}\, p_\text{A} + \frac{(k_\text{a})_\text{B}(k_\text{r})_\text{B}\left(1 - \sum_i \theta_i\right)}{(k_\text{r})_\text{B} + (k_\text{d})_\text{B}}\, p_\text{B}.$$

(70)

Equation (69) gives the algebraic sum of the rates of the reactions,

$$\text{A} \to \text{A}_\text{ads} \to \text{B}_\text{ads} \to \text{B}, \qquad \text{A} \xrightarrow{\ k_1\ } \text{B}$$

$$\text{B} \to \text{B}_\text{ads} \to \text{C}_\text{ads} \to \text{C}, \qquad \text{B} \xrightarrow{\ k_2\ } \text{C}.$$

Equation (70) gives the sum of the rates,

$$\text{A} \to \text{A}_\text{ads} \to \text{B}_\text{ads} \to \text{C}_\text{ads} \to \text{C}, \qquad \text{A} \xrightarrow{\ k_3\ } \text{C},$$

$$\text{B} \to \text{B}_\text{ads} \to \text{C}_\text{ads} \to \text{C}, \qquad \text{B} \xrightarrow{\ k_2\ } \text{C}.$$

Dividing the first term of Eq. (69) by the second, we obtain

$$\frac{(k_a)_A}{[(k_r)_A + (k_d)_A]} \frac{(k_d)_B (k_r)_A}{(k_a)_B (k_r)_B} \frac{p_A}{p_B} = \frac{k_1}{k_2} \frac{p_A}{p_B} = \sigma \frac{p_A}{p_B}. \tag{71}$$

Dividing the first term of Eq. (69) by the first term of Eq. (70), we find

$$(k_d)_B/(k_r)_B = k_1/k_3. \tag{72}$$

Thus, three overall rate constants and two selectivity factors are required in this case as in reaction scheme (63). A high selectivity for B requires small values of k_2 and k_3 compared to k_1.

CASE (iii). The surface material balance for adsorbed A will be given by Eq. (50). But in the balance for adsorbed B, Eq. (51), the surface reaction term may be dropped because adsorption and desorption of B are assumed fast compared to reaction. Thus θ_A and θ_B become

$$\theta_A = \left[(k_a)_A \left(1 - \sum_i \theta_i \right) \middle/ (k_d)_A + (k_r)_A \right] p_A, \tag{73}$$

$$\theta_B = [(k_a)_B/(k_d)_B] \left(1 - \sum_i \theta_i \right) p_B = K_B \left(1 - \sum_i \theta_i \right) p_B. \tag{74}$$

Using these values in the rate expressions, we obtain

$$-dp_A/dt = (k_r)_A \theta_A = \frac{(k_r)_A (k_a)_A (1 - \sum_i \theta_i)}{(k_d)_A + (k_r)_A} p_A, \tag{75}$$

$$dp_B/dt = (k_r)_A \theta_A - (k_r)_B \theta_B, \tag{76}$$

$$dp_C/dt = (k_r)_B \theta_B = (k_r)_B K_B \left(1 - \sum_i \theta_i \right) p_B. \tag{77}$$

Dividing Eq. (75) by (77), we find,

$$\frac{(k_a)_A}{[(k_d)_A + (k_r)_A]} \frac{(k_r)_A}{K_B (k_r)_B} \frac{p_A}{p_B} = \frac{(k_a)_A}{[(k_d)_A + (k_r)_A]}$$

$$\times \frac{(k_d)_B (k_r)_A}{(k_a)_B (k_r)_B} \frac{p_A}{p_B} = \frac{k_1}{k_2} \frac{p_A}{p_B} = \sigma \frac{p_A}{p_B}, \tag{78}$$

which is identical with Eq. (71). In this case, only two rate-constants and one selectivity factor are required and the results agree with reaction scheme (64).

CASE (iv). The adsorption–desorption equilibrium is established rapidly for A, and therefore the surface reaction term in the surface material balance for A may be neglected, Eq. (50). The comparable term for B in Eq. (51) cannot be neglected according to our assumptions. As in the previous cases, rate expressions in terms of partial pressures are determined. We find the following selectivity factors:

$$\frac{(k_d)_B(k_r)_A(k_a)_A}{(k_a)_B(k_r)_B(k_d)_A} = \frac{k_1}{k_2} = \sigma, \qquad \frac{(k_d)_B}{(k_r)_B} = \frac{k_1}{k_3}, \qquad (79)$$

which are identical with Eqs. (71) and (72), respectively. As in Case (ii), three rate constants and two selectivity factors are required in agreement with the reaction scheme (63).

11.6 Absolute Reaction Rate Theory

In Section 9.3, absolute rate theory applied to chemisorption was discussed. For the adsorption reaction, we wrote

$$A + \Sigma \leftrightarrows (A - \Sigma)^\ddagger \to (A - \Sigma), \qquad (80)$$

where A is the adsorbate, Σ a vacant site, $(A - \Sigma)^\ddagger$ the activated complex, and $(A - \Sigma)$ the final adsorbed state of A. For a unimolecular chemical reaction, a similar expression is written in absolute rate theory,

$$A + \Sigma \leftrightarrows (A - \Sigma)^\ddagger \to \text{products}. \qquad (81)$$

The only difference between the expressions for adsorption and chemical reaction is that the decomposition of activated complex in adsorption leads to an adsorbed reactant whereas in a chemical reaction it leads to products. We can therefore write an expression similar to the adsorption rate equation, Eq. (25), Chapter IX, for the rate of a unimolecular reaction,

$$r = c_G \, c_\Sigma (kT/h)/[q_\ddagger/(q_G/V)q_\Sigma] \exp(-E/RT), \qquad (82)$$

where c_G is the concentration of reactant molecules in the gas phase, c_Σ the number of vacant sites per unit area, $q_\ddagger$ the partition function of a single activated complex, q_Σ the partition function of a vacant site, (q_G/V) the partition function per unit volume of a gaseous molecule, and E the activation energy of the reaction. When A is weakly adsorbed, c_Σ approaches B/α the total number sites per unit area, and the rate becomes proportional to c_G. First-order kinetics were observed also in Eq. (4) for a unimolecular reaction assuming surface reaction controlling (equilibrium adsorption) and low coverage. The general steady-state expression, Eq. (10) of Chapter X, also leads to first order kinetics at low coverage when the value for fractional surface coverage (θ) derived from it is substituted into the expression $r = k_r \theta$. When A is more strongly-adsorbed, we have from the Langmuir isotherm

$$c_s/c_\Sigma \, c_G = [q_s/(q_G/V)q_\Sigma] \exp(\epsilon/RT), \qquad (83)$$

where q_s is the partition function of an adsorbed molecule, c_s the number of adsorbed molecules per unit area, and $\epsilon \, (> 0)$ the energy of adsorption at absolute zero. Combining Eq. (83) with Eq. (82), we obtain

$$r = c_s(kT/h)(q_{\ddagger}/q_s) \exp[-(E + \epsilon)/RT]. \tag{84}$$

As an approximation for immobile adsorption, $q_{\ddagger}$ and q_s may be taken as unity and the rate law becomes

$$r = c_s(kT/h) \exp(-E_0/RT), \qquad E + \epsilon = E_0. \tag{85}$$

When the surface is nearly covered, c_s is essentially a constant. Thus the rate becomes independent of the pressure of the reactant at full coverage, and the kinetics are zero order.

A bimolecular reaction which proceeds according to the Langmuir–Hinshelwood mechanism can be written as in Eq. (2) of Chapter X,

$$A + B + \Sigma_2 \leftrightarrows \left(\begin{array}{c} \text{A---B} \\ | \quad | \\ \underline{-\Sigma-\Sigma-} \end{array}\right)^{\ddagger} \rightarrow \text{product}. \tag{86}$$

The activated complex consists of A and B adsorbed on adjacent sites. As in Eq. (4) of Chapter X, the rate expression is given by

$$r = c_A c_B c_{\Sigma_2} \frac{kT}{h} \frac{q_{\ddagger}}{(q_A/V)(q_B/V)q_{\Sigma_2}} \exp \frac{-E}{RT}. \tag{87}$$

Using the Bragg–Williams approximation [see Eq. (5), Chapter IX] we can express c_{Σ_2}, the number of adjacent pairs of vacant sites per unit area, in terms of c_{Σ}, the number of vacant sites per unit area,

$$c_{\Sigma_2} = \frac{1}{2} \frac{c(c_{\Sigma})^2}{B/\alpha}, \tag{88}$$

where c is the number of nearest-neighbors of a site and B/α is the total number of sites per unit area. If we assume that A and B are in adsorption equilibrium, c_{Σ} can be expressed according to the Langmuir isotherm by

$$c_{\Sigma} = \frac{B/\alpha}{1 + K_A c_A + K_B c_B}. \tag{89}$$

Substituting Eqs. (88) and (89) into Eq. (87), we obtain the general rate expression for a bimolecular reaction proceeding by the Langmuir–Hinshelwood mechanism according to the absolute rate theory,

$$r = \frac{\frac{1}{2}c(B/\alpha)c_A c_B}{(1 + K_A c_A + K_B c_B)^2} \frac{kT}{h} \frac{q_{\ddagger}}{(q_A/V)(q_B/V)q_{\Sigma_2}} \exp \frac{-E}{RT}. \tag{90}$$

Various limiting cases may be considered depending on the strengths of adsorption of A and B and their gas-phase concentrations. For example, if

$K_A c_A$ and $K_B c_B$ are both small compared to unity, Eq. (90) becomes

$$r = \frac{1}{2} \frac{cB}{\alpha} c_A c_B \frac{kT}{h} \frac{q_{\ddagger}}{(q_A/V)(q_B/V)q_{\Sigma_2}} \exp \frac{-E}{RT}. \qquad (91)$$

Rate expressions for bimolecular reactions proceeding by the Rideal mechanism can also be derived. In the case where adsorbed A reacts with gas phase B, the expression is

$$r = c_B c_a \frac{kT}{h} \frac{q_{\ddagger}}{(q_B/V)q_a} \exp \frac{-E}{RT}, \qquad (92)$$

where c_a is the concentration of adsorbed A and q_a its partition function. The rate expression corresponds to the process

$$\begin{array}{c} \text{B} \\ \vdots \\ (\text{A} - \Sigma) + \text{B} \leftrightarrows (\text{A} - \Sigma)^{\ddagger} \rightarrow \text{products}, \end{array} \qquad (93)$$

though the configuration of the activated complex is arbitrary. Using the Langmuir isotherm expressions

$$c_a = \frac{(B/\alpha)K_A c_A}{1 + K_A c_A + K_B c_B}, \qquad (94)$$

and

$$\frac{q_a}{(q_A/V)q_{\Sigma}} e^{\epsilon/RT} = K_A, \qquad (95)$$

in Eq. (92), we find

$$r = \frac{(B/\alpha)c_A c_B}{1 + K_A c_A + K_B c_B} \frac{kT}{h} \frac{q_{\ddagger}}{(q_A/V)(q_B/V)q_{\Sigma}} \exp \frac{-(E - \epsilon)}{RT}. \qquad (96)$$

A similar expression is obtained when adsorbed B reacts with gaseous A. Because of the similarity between the equations for the Langmuir–Hinshelwood and Rideal mechanisms, it is not possible to distinguish between the two types on the basis of absolute rate calculations. In fact, the general use of absolute reaction rate theory is as disappointing for catalytic reactions as it is for adsorption. The rate expressions are not sufficiently detailed and specific in content to allow many unequivocal conclusions to be drawn. There are cases where satisfactory agreement with experiment is obtained, and more cases where it is not. For discussions of applications to specific cases, the works of Bond (*9*) and Laidler (*8*) may be consulted.

In concluding this section, we want to discuss briefly the predictions of absolute rate theory concerning the differences encountered in bimolecular surface reactions when adsorption is mobile and when it is immobile. We observe first that the results of Section 11.5 for a complex reaction indicate that if B is the desired product, a high value of the ratio $(k_d)_B/(k_r)_B$ is advantageous [see Eqs. (62), (72), (78), and (79)]. Similarly, the high ratio will also be advantageous in the consecutive bimolecular reactions

$$A + M \to B,$$
$$B + M \to C. \tag{97}$$

Following de Boer and van der Borg (16), we will develop expressions for $(k_d)_B/(k_r)_B$ for the mobile and immobile cases. We start with the immobile case of a surface reaction between adsorbed B and adsorbed M. It will be assumed for convenience without loss of generality that the surface concentration of M is essentially constant. The rate of the reaction can be written

$$r_s = c_{bm}(kT/h)/[(q_{\ddagger})_{bm}/(q_s)_{bm}] \exp(-E_s/RT), \tag{98}$$

where c_{bm} is the number of adjacently adsorbed B–M pairs per unit area, $(q_s)_{bm}$ is the corresponding partition function and $(q_{\ddagger})_{bm}$ is the partition function of the activated complex. Using the Bragg–Williams approximation in a procedure similar to that used in deriving Eq. (88), we obtain

$$c_{bm} = \frac{c c_m c_b}{B/\alpha}, \tag{99}$$

where, as before, c is the number of nearest-neighbor sites of a site, c_m and c_b are the number of molecules of adsorbed M and adsorbed B, respectively per unit area, and (B/α) is the total number of sites per unit area. Division by two is not required in this case, for switching B and M on an adjacent pair of sites represents two different configurations. Substituting Eq. (99) into (98) gives

$$r_s = c_m c_b \frac{c}{(B/\alpha)} \frac{kT}{h} \frac{(q_{\ddagger})_{bm}}{(q_s)_{bm}} \exp \frac{-E_s}{RT}. \tag{100}$$

For the mobile case, we find

$$r_s = c_m c_b(kT/h)[q_{\ddagger}'\alpha/(q_s)_b(q_s)_m] \exp(-E_s/RT), \tag{101}$$

where $q_{\ddagger}'$ is the partition function for the mobile activated complex and in which

$$\frac{q_{\ddagger}'/\alpha}{[(q_s)_b/\alpha][(q_s)_m/\alpha]} = \frac{q_{\ddagger}'\alpha}{(q_s)_b(q_s)_m}. \tag{102}$$

For both cases, the rate of desorption of B is

$$r_d = c_b(kT/h)[q_{\ddagger}/(q_s)_b] \exp(-E_d/RT), \tag{103}$$

where $q_{\ddagger}$ is the partition function of the activated complex of B which will have different values depending on whether adsorption is mobile or immobile. The concentration of adsorbed M is assumed to remain essentially constant during the reaction, as previously stated. This condition is fulfilled if M is strongly adsorbed at the surface so that the reaction is zero order with respect to it and no desorption occurs. The ratio of the rates of desorption of B and of its reaction with M in the case of localized adsorption is from Eqs. (100) and (103),

$$\frac{(k_d)_B c_b}{(k_r)_B c_b c_m} = \frac{(B/\alpha)}{c_m c} \frac{q_\ddagger}{(q_s)_b} \frac{(q_s)_{bm}}{(q_\ddagger)_{bm}} \exp \frac{-(E_d - E_s)}{RT}. \tag{104}$$

For c_m, we write $(B/\alpha)\theta_M$. Now in immobile adsorption, $(q_s)_b$ and $(q_s)_{bm}$ may be set approximately equal to unity. And, if we set $q_\ddagger = (q_\ddagger)_{bm}$, we obtain

$$(k_d)_B/(k_r)_B c_m = (1/c\theta_M) \exp[-(E_d - E_s)/RT]. \tag{105}$$

In the case of mobile adsorption we find from Eqs. (101) and (103) similarly that

$$(k_d)_B/(k_r)_B c_m = [(q_s)_m/B\theta_M] \exp[-(E_d - E_s)/RT]. \tag{106}$$

At the same value of θ_M and $(E_d - E_s)$, we see from Eqs. (105) and (106) that the ratio $(k_d)_B/(k_r)_B$ for mobile adsorption differs from that for immobile adsorption by the multiplicative factor $(q_s)_m c/B$. Assuming that rotational and vibrational partition functions contribute a negligible amount to the total partition function $(q_s)_m$, we are left with two degrees of translational freedom. The partition function per unit area is then

$$(q_s)_m/\alpha = 2\pi m_m kT/h^2, \tag{107}$$

which is of the order of 10^{16} at 300°K. The total concentration of active sites per unit area B/α may be estimated at 10^{15} cm^{-2}; and $c = 3, 4,$ or 6 depending on the surface geometry. Thus, we find that the rate for mobile adsorption, Eq. (106) is at least 30 times the value for localized adsorption. This means that the selective formation of B is enhanced by mobile adsorption. An increase in temperature favors mobility and thus increases the selectivity for B. With $E_d > E_s$, an increase in temperature has a greater effect on selectivity than with $E_d < E_s$.

11.7   The Power Rate Law and Surface Nonuniformity

For many years, the Langmuir approach to the kinetics of catalytic reactions has been employed (*17*). At the same time, the paradox of using kinetics which assumed a uniform surface, when such a surface in reality is a rarity, was recognized. Constable (*18*) first considered this paradox and resolved it for a special case. He showed that a surface with a broad distribution of active sites would act as if only one kind of site were operating for a particular reaction under a given set of conditions. In other words, a real surface may be considered as a statistical collection of ideal surfaces of different adsorption energies, and a very limited number of these members play an active part in the reaction. However, Halsey (19) has shown that under certain circumstances the conclusions of Constable may not hold. For example, Constable tacitly assumes that the rate-controlling step is the same on all the surface sites, while Halsey claims that it is also necessary to consider the possibility of the rate-controlling step being different on different types of sites. Halsey supports his claims with a quantitative discussion.

With the paradox not completely resolved, many workers have looked with distrust on the Langmuir approach. In spite of the frequent good agreements with experiment, the assumptions of a homogeneous surface and no adsorbate interactions are not usually true, and the approach may be merely an exercise in "curve fitting." Looking closer at the Langmuir approach certain anomalies appear in spite of overall agreement. In the decomposition of stibine, SbH_3, for example, the rate may be expressed by

$$r = k_r Kp/(1 + Kp) \tag{108}$$

the simple Langmuir form assuming equilibrium adsorption of reactant (17). Indeed, Eq. (108) holds over a fairly wide range of temperature, and it was found that K remains essentially constant over this temperature range (17). However, $K = \exp(\Delta S/R) \exp(-\Delta H/RT)$ should decrease exponentially with temperature, unless $\Delta H \to 0$. For ΔH to be nearly zero is not physically acceptable; it should have a value at least equal to the heat of liquefaction of antimony hydride, 5700 cal/gm mole. The pre-exponential factor $\exp(\Delta S/R)$ was also found larger than permissible theoretically (17). Thus, the surface cannot be ideal in spite of the overall agreement in rate by the Langmuir approach. Boudart explains the results qualitatively. He assumes a distribution of heats of adsorption over the surface. When the heat of adsorption varies sufficiently rapidly with coverage, relatively small changes of pressure normally encountered in kinetic experiments, do not change surface coverage appreciably (see Section 9.5). Thus the surface appears homogeneous and ideally kinetic at a given temperature; but when the temperature is increased, the effect on surface coverage is more drastic. Surface coverage decreases and sites of higher adsorption energy become catalytically operative, revealing the nonideality of the surface.

Some investigators have suggested the use of a power rate law instead of the Langmuir form. Stock and co-workers (20) found that the rate of decomposition of stibine, SbH_3, could be expressed by a relation of the type

$$r = k_r p^n, \tag{109}$$

with $n = 0.6$. Boudart (17) showed that the power rate form and the Langmuir form Eq. (108) both fit the data equally well. There is some physical justification for Eq. (109). It may be derived by assuming adsorption equilibrium according to the Freundlich isotherm

$$\theta = kp^n. \tag{110}$$

As shown in Section 2.5, the Freundlich isotherm can be derived approximately from the Langmuir isotherm assuming an exponential distribution of adsorption site energies. Kwan (*21*) used the power rate law first to fit the rate data for the chemisorption of nitrogen on a promoted iron catalyst, which he expressed by

$$r = k'p\theta^{-\alpha} - k''\theta^\beta, \tag{111}$$

where k', α, k'', and β are constants. At equilibrium, this equation gives the Freundlich isotherm with n in Eq. (110) equal to $1/(\alpha + \beta)$ and k equal to k'/k''. Later, Kwan (*22*) used the power rate law to express the kinetics of the ammonia synthesis making the usual assumption that the chemisorption of nitrogen is the slow step. He related kinetic quantities, energy and entropy of activation, to thermodynamic quantities, heat and entropy of chemisorption. It must be recognized that the ability to obtain kinetic and thermodynamic quantities does not lessen the empiricism which characterizes the power rate law. These remarks apply also to the similar power rate law developed by Schuit and van Reijen (*23*), who combine the power rate law and absolute reaction rate theory. Perhaps the most straightforward and unpretentious development of the power rate law is by Weller (*5*) who suggests the use of the general rate expression

$$r = k(p_A)^m(p_B)^n(p_C)^q \cdots \tag{112}$$

over the Langmuir form for purely mathematical convenience. The exponents m, n, and q, ... are restricted to integral and half-integral values and do not necessarily bear a direct relationship to the stoichiometric quantities of the chemical reaction involving A, B, and C. He obtains good correlations of rate data for the oxidation of sulfur dioxide over platinum, the hydrogenation of codimer (trimethyl pentenes), the hydrogenation of carbon monoxide to methane and water over nickel on kieselguhr, and the formation of phosgene from carbon monoxide and chlorine over charcoal.

Boudart (*17*) points out that starting with an empirical rate equation does not preclude obtaining some knowledge of mechanism. He suggests the following procedure: (1) Write down a kinetic expression of the type in Eq. (112), and determine the constants; (2) find a possible mechanism which gives approximately the empirical rate law of step (1); (3) confirm the assumed mechanism by a rigid analysis of the kinetic constants previously determined, taking into account heterogeneity and interactions. The third step is rarely, if ever, feasible to carry out. There are, however, good reasons for proceeding beyond the first step. First, because of the paradox of heterogeneity and surface kinetics, there may be no theoretical reason in many cases against the use of Langmuir kinetics in devising mechanisms. Second, assumed mechanisms even though not fully substantiated may lead to at least a partial understanding of a reaction. Third, the results may lead to useful predictions concerning the behavior of new systems.

11.8 The Compensation Effect

A series of related catalysts (*9*), such as a series of different metals, or a given metal or metal oxide treated in various ways, or a series of catalysts of increasing promoter content, might be expected to exhibit kinetic factors for a given reaction which change according to some regular pattern. Changes in the velocity constant k will be the result of changes in the activation energy E or the pre-exponential factor A, or both. Experimentally, it has been found (*24*) that many series of related catalysts show the following relationship between A and E.

$$\ln A = bE + c, \tag{113}$$

where b and c are constants. In addition, we have the general Arrhenius relation,

$$\ln A = (E/RT) + \ln k. \tag{114}$$

We define a temperature T_s by the equation, $b = 1/RT_s$, where the Arrhenius plots ($\ln k$ versus $1/T$) for all catalysts in the series intersect; that is, where the k's are identical for all values of E, as in Fig. 11.2a. Then, from Eq. (114), we may write

$$\ln A = (E/RT_s) + \ln k_0, \tag{115}$$

where k_0 is constant for all A and E at T_s. When $E = 0$, we have

$$\ln A = \ln k_0 = c, \qquad E = 0. \tag{116}$$

Thus, we may write for all values of E,

$$\ln A = bE + \ln k_0 = (E/RT_s) + \ln k_0 \tag{117}$$

which requires that

$$b = (RT_s)^{-1} \tag{118}$$

in order that a temperature T_s exist. We now look at possible values of b in Eq. (113).

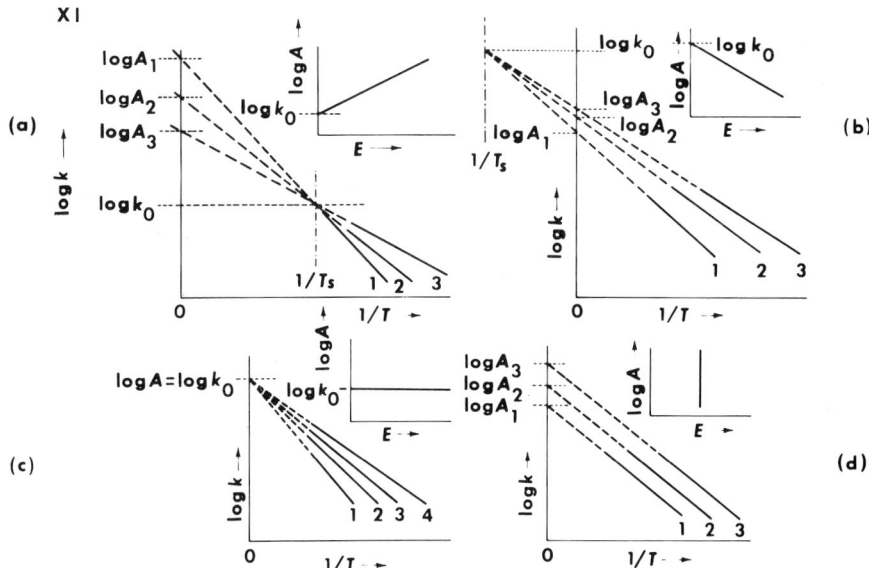

FIG. 11.2. (a) Case (i), schematic representation of the compensation effect. (b) Case (ii), schematic representation of "anticompensation" effect. (c) Case (iii), schematic representation of the "no-compensation" effect, E variable, log A constant. (d) Case (iv), schematic representation of "no-compensation" effect, E constant, log A variable (9).

CASE (i). If $0 < b < \infty$, then Arrhenius plots of k versus $1/T$ [Eq. (114)], will intersect at $1/T_s$ for all catalysts in the series, where we find from Eq. (118) that $0 < 1/T_s < \infty$. Thus, $1/T_s$, in this case, is in the range of real temperatures as shown in Fig. 11.2a. From $\ln A = bE + c$, it is evident that the changes in A and E are in the same direction and tend to compensate for each other in the equation $k = A \exp(-E/RT)$, so that the change in k is considerably restricted. This effect is known as the "compensation effect."

CASE (ii). If $0 > b > -\infty$, then Arrhenius plots will intersect in the range $0 > 1/T_s > -\infty$, which is shown by extrapolation in Fig. 11.2b. We see from $\ln A = bE + c$ that A changes in the opposite direction from E. Thus, the changes tend to augment each other in the Arrhenius equation, so that the change in k is large. This is called the "anticompensation" effect.

CASE (iii). If $b = 0$, Arrhenius plots intersect at $1/T_s = 0$ as in Fig. 11.2c. From $\ln A = c$, we find that A is constant whatever the change in E. We see from the Arrhenius equation that there is no compensation in this case.

CASE (iv). Similarly, when $b = \pm \infty$, Arrhenius plots intersect at $1/T_s = \pm \infty$ as in Fig. 11.2d. In this case, E is constant while A varies, and there is no compensation.

We are particularly interested in the physical interpretations of the compensation effect, Case (i). Of the many interpretations appearing in the literature, three will be discussed. No firm selection of the best interpretations can be made at present.

The first treatment was put forth by Constable (18) and was further developed by others (25) (see Section 11.7). Constable assumes a heterogeneous surface with B_i sites of the ith type upon which the activation energy for a particular reaction is E_i. The overall rate of reaction is given by

$$r = \sum_i A \exp(-E_i/RT)B_i, \tag{119}$$

where E_i is the activation energy of the reaction on sites of the ith kind and B_i is the number of such sites. Constable makes the assumption that sites are filled strictly in the order of their decreasing energy and thus sites of the ith kind are either completely filled or completely empty (see Section 9.5). He also assumes that the pre-exponential factor A is the same for all sites. It is now assumed that the distribution is continuous between the lower limit E_1 and the upper limit E_2, and summation is replaced by integration,

$$r = \int_{E_1}^{E_2} A \exp(-E/RT)[dB(E)/dE]\, dE, \tag{120}$$

where $dB(E)/dE$ is the distribution function. Constable presents theoretical reasons for taking the distribution function in exponential form

$$dB(E)/dE = a \exp(E/b), \tag{121}$$

where a and b are constants. The rate becomes

$$r = \int_{E_1}^{E_2} aA\, e^{-E/RT}e^{E/b}\, dE = \int_{E_1}^{E_2} aAe^{-gE}\, dE \tag{122}$$

$$g = (1/RT) - (1/b)$$

On integration, we obtain

$$r = (Aa/g)(e^{-gE_1} - e^{-gE_2}). \tag{123}$$

If the distribution of activation energies is broad, $\exp(-gE_2)$ may be neglected in comparison with $\exp(-gE_1)$. As discussed in Section 11.7, we note that Halsey (19) has considered cases in which the distribution is not sufficiently broad to make this simplification. Finally, we may write

$$r = (Aa/g)(e^{-gE_1}) = (Aa/g)e^{E_1/b}e^{-E_1/RT}. \tag{124}$$

The first significant point about this equation is that it shows the reaction taking place exclusively on the sites of highest activity E_1 (lowest value of E).

The second point is that Eq. (124) shows that the compensation effect is operating. That is, one of the terms increases and another of the terms decreases as E_1 increases, and the compensation is greater the smaller the value of g. A weak point of the interpretation is the particular form of the distribution function for the activation energy, Eq. (121). Undoubtedly, there is a distribution of activation energies, but the exact form has never been determined satisfactorily for any catalytic reaction.

A second interpretation has been made by Kemball (*26*). The interpretation is restricted to a reaction proceeding on a series of catalysts consisting either of different inhibiting gases strongly adsorbed on the same substrate or the same inhibiting gas strongly adsorbed on different substrates, for example, metals. We assume that the strongly adsorbed gas A obeys the Langmuir isotherm,

$$\theta_A = K_A p_A/(1 + K_A p_A). \tag{125}$$

The equilibrium constant of adsorption K_A, can be expressed as usual in thermodynamic form see [Eq. (110), Chapter II],

$$K_A = \exp(\Delta S/R) \exp(-\Delta H/RT). \tag{126}$$

For strong adsorption, we can write

$$(1 - \theta_A) \cong \frac{1}{K_A p_A} = \frac{\exp(-\Delta S/R) \exp(\Delta H/RT)}{p_A}. \tag{127}$$

The activation energy of any reaction which is proportional to the number of vacant sites (see absolute rate theory, Section 11.6), will be increased by $-\Delta H$ and the frequency factor by $\exp(-\Delta S/R)$. Everett (*27*) gives theoretical and experimental evidence for a linear relationship between entropies and heats of adsorption,

$$\Delta S_i/R = a\Delta H_i - b, \tag{128}$$

where the subscript i refers to either different inhibiting gases on the same catalyst or the same inhibiting gas (A) on different catalysts. With different inhibiting gases on the same catalyst, for example, we would have an activation energy

$$E_i = E_0 - \Delta H_i, \tag{129}$$

where E_0 is the activation energy of the catalyst in the absence of an inhibitor. And similarly for the frequency factor, we have

$$\ln A_i = \ln A_0 + b - a\Delta H_i, \tag{130}$$

with A_0 the frequency factor of the catalyst in the absence of an inhibitor.

Substituting for ΔH_i from Eq. (129) in Eq. (130), we find

$$\ln A_i = aE_i + c, \tag{131}$$

where $c = \ln A_0 + b - aE_0$ is a constant for the catalyst. Thus the compensation effect expressed by Eq. (131) follows from the relationship between heats and entropies of adsorption of the inhibitor A.

A third interpretation is due to Sosnovsky (28) and deals with the compensation effect observed in a series of single crystals of silver bombarded with positive argon ions. The (111), (110), and (100) faces were bombarded at voltages between 14 and 4000 V. The parameters $\ln A$ and E found from Arrhenius plots changed appreciably with bombarding ion energy for the decomposition of formic acid. For all three faces, an increase in E was always associated with an increase in $\ln A$, and the relationship was linear. Assuming that the reaction mechanism does not change with bombarding ion energy, Sosnovsky concludes that a change in the number of active sites is the predominant factor causing the change in kinetic parameters and believes that the increase in number of active sites is the result of ion bombardment producing disoriented regions which are bounded by stable arrays of dislocations. She believes that the increase in E and A which is associated with an increase in density of dislocations at the surface could be due to an interaction between the dislocations. The new dislocations are restricted to the boundaries of the small crystallite blocks so that their local density and thus their interaction is strong. The energy of each site is, therefore, lowered, giving an increase in E associated with an increase in A. For simplicity, two types of sites are assumed: the original, relatively isolated sites of lower activation energy E_1, and those of higher activation energy E_2, typical of the new dislocations crowded into boundaries. As in Eq. (119), we can express the rate by

$$r = k[B_1 \exp(-E_1/RT) + B_2 \exp(-E_2/RT)], \tag{132}$$

where B_1 and B_2 are the numbers of original and newly-created sites respectively. For rough calculations, it is assumed that B_1 is constant and equal to 10^6 sites/cm^2, that E_1 and E_2 are the maximum and minimum values obtained on each face; namely, $E_1 = 12$, $E_2 = 22$ for (111); $E_1 = 24$, $E_2 = 34$ for (110); and $E_1 = 30$, $E_2 = 35$ kcal/mole for (100). In the light of Eq. (132), it is obvious that experimental values for E and $\ln A$ will represent composite values except for the initial, undeformed crystal and for the highly deformed crystal where $B_2 \exp(-E_2/RT) \gg B_1 \exp(-E_1/RT)$. It is evident that the composite values for E and for the pre-exponential factor A will increase as B_2 increases. Fig. 11.3 gives a plot of experimentally determined values of $\log_{10} A$ versus E for the decomposition of formic acid on (111), (110), and (100) silver faces after various intensities of ion bombardment.

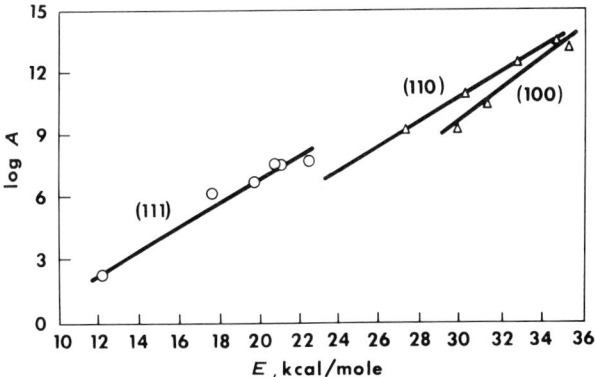

FIG. 11.3. $\text{Log}_{10} A$ versus E for the decomposition of formic acid on (111), (110), and (100) silver faces after various intensities of ion bombardment (*28*). [After Sosnovsky, *J. Phys. Chem. Solids* **10**, 304 (1959). By permission of Pergamon Press.]

Jaeger (*29*) strongly criticizes the work of Sosnovsky on the basis that dislocation concentrations were not actually measured and the topography of the surface was not investigated. He found that in comparing thin and thick (111)-oriented films with the same topography the activation energy (*E*) and pre-exponential factor (*A*) were not altered when the dislocation concentration was changed by a factor of approximately 10^2 both in annealed and unannealed films. Furthermore, in (100) films, annealing reduced the dislocation concentration but did not affect the activity. Jaeger believes that activation energy for the decomposition of formic acid on silver surfaces depends upon the stability of the chemisorbed species, which in turn is influenced by the degree of unsaturation of surface atoms.

11.9   Kinetic Studies of the Polymerization of Ethylene

There has been a vast amount of work reported in the literature on the kinetics of heterogeneous catalytic reactions. We propose to describe in this section some of the work on one reaction which has been studied in considerable detail—the formation of high polymers of ethylene over chromium oxide–silica–alumina (*3, 30–34*). Our aim is to illustrate the use of kinetics in determining mechanism rather than in correlating data.

Following Clark and Bailey (*32*), we calculate expressions for number average and weight average molecular weights and the rate of reaction according to the Rideal and Langmuir–Hinshelwood mechanisms. It is assumed that adsorption of monomer corresponds to the initiation step and the propagation rate constant, k_r, is independent of n, the number of monomer units in a polymer molecule. Also, a steady state condition is assumed to hold. This last assumption implies particular forms for the termination step. It implies a constant number of active sites. If true termination occurs with destruction of sites, then an equal number must be created. However, a true termination process occurs only in radical polymerizations, and we assume here that termination proceeds on a constant number of sites by transfer. The transfer may be spontaneous, that is, simple desorption or by reaction with monomer, as follows:

(a) Spontaneous transfer,

$$Cat \cdots CH_2—CH_2—(CH_2)_m—CH_3 \rightarrow Cat \cdots H + CH_2{=}CH—(CH_2)_m—CH_3$$

or (b) Transfer with monomer,

$$Cat \cdots CH_2—CH_2—(CH_2)_m—CH_3 + CH_2{=}CH_2 \rightarrow$$
$$Cat \cdots CH_2—CH_3 + CH_2{=}CH—(CH_2)_m—CH_3 .$$

Let $\sum N_n$ be the total number of molecules of polymer desorbed per unit catalyst surface per second. Then number average molecular weight $\bar{N}$, weight average molecular weight $\bar{W}$, and rate of reaction r may be expressed by

$$\bar{N} = M_1\left(\sum nN_n / \sum N_n\right), \tag{133}$$

$$\bar{W} = M_1\left(\sum n^2 N_n / \sum nN_n\right), \tag{134}$$

$$r = \sum nN_n, \tag{135}$$

where M_1 is the molecular weight of monomer. The rate of reaction is expressed as the number of molecules of monomer in the polymer that is desorbed from unit catalyst surface per second. The rate of reaction may be rewritten as

$$r = \sum nN_n = k_d B \sum_{n=2}^{\infty} n\theta_n \tag{136}$$

where k_d is the desorption velocity constant, B the number of sites per unit surface, θ_n the fraction of sites covered with n-mer. All sites are assumed to have the same adsorption energy, and k_d is assumed not to vary with n.

We now consider specifically the Rideal mechanism with termination by spontaneous transfer. At steady-state conditions, a balance for adsorbed n-mer gives

$$k_r c\theta_{n-1} - k_r c\theta_n - k_d \theta_n = 0 \tag{137}$$

where k_r (assumed independent of n) is the reaction-velocity constant for the reaction of unadsorbed monomer with adsorbed polymer chains, and c is the concentration of monomer in the nonadsorbed phase. It is assumed that there is no back reaction, that the concentration of polymer molecules in the nonadsorbed phase is negligible, and that readsorption of polymer molecules does not occur. Making use of balances of this type for various values of n, it can be shown that $\sum_{n=1}^{\infty} \theta_n$ is an infinite geometric series whose sum is

$$\sum_{n=1}^{\infty} \theta_n = (k_r c + k_d)\theta_1/k_d. \tag{138}$$

Therefore, we have

$$\sum_{n=2}^{\infty} \theta_n = k_r c \theta_1/k_d \tag{139}$$

$$\sum_{n=2}^{\infty} n\theta_n = k_r c \theta_1 (k_r c + 2k_d)/k_d^2 \tag{140}$$

$$\sum_{n=2}^{\infty} n^2\theta_n = k_r c \theta_1 (2k_r^2 c^2 + 5k_r c k_d + 4k_d^2)/k_d^3. \tag{141}$$

The term θ_1 may be evaluated explicitly from the balance of adsorbed monomer,

$$k_a c\left(1 - \sum_{n=1}^{\infty} \theta_n\right) - k_r c \theta_1 - k_d \theta_1 = 0 \tag{142}$$

where k_a is the adsorption velocity constant for monomer, and the term in parentheses is the fraction of unoccupied sites. Making use of Eqs. (133)–(136) and (139)–(142), we obtain the final expressions for number average molecular weight, weight average molecular weight, and the rate of polymerization,

$$\overline{N} = M_1[(k_r c/k_d) + 2], \tag{143}$$

$$\overline{W} = \frac{M_1[2(k_r c/k_d)^2 + 5(k_r c/k_d) + 4]}{(k_r c/k_d) + 2}, \tag{144}$$

$$r = Bk_a c^2 k_r (k_r c + 2k_d)/(k_a c + k_d)(k_r c + k_d). \tag{145}$$

For the Langmuir–Hinshelwood mechanism with termination by spontaneous transfer, the procedure is nearly identical with that for the Rideal mechanism. The same implicit equations for $\sum N_n$, $\sum nN_n$, and $\sum n^2 N_n$, as illustrated for $\sum nN_n$ in Eq. (136), hold. In determining explicit expressions for $\sum \theta_n$, $\sum n\theta_n$, and $\sum n^2\theta_n$, the same procedure is used except that concentration c is now replaced by $B\theta_1$. The final expressions for number average molecular weight, weight average molecular weight, and rate of reaction become,

$$\overline{N} = M_1[(K_r \theta_1/k_d) + 2], \tag{146}$$

$$\overline{W} = \frac{M_1[2(K_r \theta_1/k_d)^2 + 5(K_r \theta_1/k_d) + 4]}{[(K_r \theta_1/k_d) + 2]} \tag{147}$$

$$r = BK_r \theta_1^2(K_r \theta_1 + 2k_d)/k_d, \tag{148}$$

where $K_r = k_r B$. A monomer material balance calculation carried out using the same principles as for the Rideal case leads to the steady-state equation,

$$b^2\theta_1^3 + ab\theta_1^2 + 2b\theta_1^2 + a\theta_1 + \theta_1 - a = 0, \tag{149}$$

where $a = k_a c/k_d$ and $b = K_r/k_d$. The term θ_1 can be eliminated from each of the Eqs. (146)–(148) using Eq. (149).

For the Rideal mechanism, we find, from Eqs. (143)–(145), that $\bar{N}$, $\bar{W}$, and r all increase without limit as concentration of monomer is increased. But for the Langmuir–Hinshelwood mechanism, we see from Eqs. (146)–(148) that with increasing monomer concentration $\bar{N}$, $\bar{W}$, and r reach maximum values asymptotically which depend on the values of the rate constants. It is not difficult to show that the ratio $\bar{W}/\bar{N}$ always lies between 1 and 2 for both mechanisms. Since $\bar{W}/\bar{N}$ is a measure of the broadness of molecular weight distribution, the distribution appears to be narrow for models with all sites equal.

If termination is assumed to occur both by spontaneous transfer and transfer with monomer, then we must replace the desorption term k_d in the Rideal expressions by $(k_d + k_t c)$ and in the Langmuir–Hinshelwood expressions by $(k_d + k_t B\theta_1)$, where k_t is the reaction velocity constant of termination for transfer with monomer. With these substitutions in expressions for $\bar{N}$ and $\bar{W}$, it is easy to show that molecular weights reach maximum values asymptotically as monomer concentration is increased in *both* the Rideal and Langmuir–Hinshelwood mechanisms. But the rate expression still increases without limit for the Rideal mechanism and reaches a maximum value asymptotically for the Langmuir–Hinshelwood mechanisms as monomer concentration is increased. We summarize the various possibilities in Table 11.1. Experimental results, as shown in Table 11.1, indicate that $\bar{N}$, $\bar{W}$, and r all level out

TABLE 11.1

BEHAVIOR OF $\bar{N}$, $\bar{W}$, AND r FOR RIDEAL AND LANGMUIR–HINSHELWOOD MECHANISMS FOR TERMINATION BY SPONTANEOUS TRANSFER AND BY TRANSFER WITH MONOMER AS THE CONCENTRATION OF MONOMER IS INCREASED

	$\bar{N}$	$\bar{W}$	r
Langmuir–Hinshelwood			
Spontaneous transfer	levels out	levels out	levels out
Transfer with monomer	levels out	levels out	levels out
Rideal			
Spontaneous transfer	increases indefinitely	increases indefinitely	increases indefinitely
Transfer with monomer	levels out	levels out	increases indefinitely
Experimental	levels out	levels out	levels out

as concentration of monomer is increased. We find that apparently the poly-merization proceeds by a Langmuir–Hinshelwood mechanism. We cannot tell, however, whether termination by transfer occurs or not from the infor-mation available above.

If there is a distribution of adsorption-site energies, new expressions in-volving integrals, a distribution function, and varying values of k_r and k_d in Arrhenius form are obtained (*32*). Any decreasing monotonic distribution function, $f(q)$, where q is the heat of adsorption, will give the same relations as given in Table 11.1 for the case with all sites equal. However, there is one major difference. The ratio $\overline{W}/\overline{N}$ may now reach values far beyond the range 1 to 2 which holds for the case with all sites equal. Thus, broad distributions of polymer molecular weights appear possible when broad distributions of site energies are assumed. Experimentally, values of $\overline{W}/\overline{N}$ as high as 30 have been observed, and attributing such broad molecular weight distribution to a broad site spectrum is reasonable.

In order to find some evidence for the presence or absence of termination by transfer with monomer, we turn to the work of Ivanov *et al.* (*31*). Following Natta (*35*), they define the number average degree of polymerization of total polymer after time τ by

$$\overline{P}_n = Q \Bigg/ \left(B\theta_p + \int_0^\tau \sum r_t \, d\tau \right), \tag{150}$$

where $B\theta_p = B \sum \theta_n$ is the number of sites per unit area occupied by growing polymer chains, $\sum r_t$ represents the termination rate of growing chains by various processes, and Q is the amount of polymer produced in time τ. Assuming chain termination by spontaneous transfer and by transfer with monomer, we can write

$$\int_0^\tau \sum r_t \, d\tau = \int_0^\tau k_d \theta_p \, d\tau + \int_0^\tau k_M \theta_p c. \tag{151}$$

Note that the second integral on the right of Eq. (151) assumes a Rideal mechanism for the termination step by transfer with monomer. The amount of product, Q, expressed also in terms of the Rideal mechanism becomes

$$Q = \int_0^\tau r \, d\tau = \int_0^\tau k_p \theta_p c \, d\tau, \tag{152}$$

where k_p is the growth-rate constant. Using Eqs. (151) and (152), we can write Eq. (150) as

$$\begin{aligned}
\overline{P}_n &= \frac{Q}{B\theta_p + \displaystyle\int_0^\tau k_d \theta_p \, d\tau + \int_0^\tau k_M \theta_p c \, d\tau} \\[2mm]
&= \frac{Q}{(r/k_p c) + (k_M/k_p)Q + (k_d/k_p c)Q},
\end{aligned} \tag{153}$$

or

$$1/\bar{P}_n = (1/k_p c)(r/Q) + (k_M/k_p) + (k_d/k_p c). \tag{154}$$

Ivanov *et al.* operated for long enough times that r/Q was very small, so that approximately,

$$1/\bar{P}_n = (k_M/k_p) + (k_d/k_p)(1/c). \tag{155}$$

A plot of the reciprocal of the number average degree of polymerization $1/\bar{P}_n$ versus the reciprocal of the monomer concentration $1/c$, both experimental quantities, turns out to be linear, and thus allows determination of the ratio k_M/k_d. The result is $k_M/k_d = 19.1$ liter/mole at $60°C$. Using the value of this ratio of the rate of spontaneous transfer to that of transfer by monomer, we find

$$k_M \theta_p c/k_d \theta_p = (k_M/k_d)c = 19.1 \ c. \tag{156}$$

At $c = 1$ mole/liter and $60°C$, about 95% of the polymer is formed by transfer with monomer and only 5% by spontaneous transfer. However, the data contain a 30% uncertainty as a result of experimental difficulties and problems of deriving molecular weights from viscosity data. Furthermore, Ivanov *et al.* assume a Rideal mechanism, when the work of Clark and Bailey point to a Langmuir–Hinshelwood mechanism. For the latter mechanism, it is necessary to replace c in Eq. (155) by $B\theta_1$, the surface concentration of monomer. But this concentration cannot be determined experimentally at the present time. The results of Ivanov *et al.*, although possibly in considerable error, are believed to be sufficiently realistic to show that appreciable transfer by monomer occurs.

REFERENCES

1. Johnston, H. S., "Gas Phase Reaction Theory." Ronald Press, New York, 1966.
2. Tamaru, K., *Proc. Intern. Congr. Catalysis III* **1**, 39 (1964); Tsuchiya, S., and Shiba, T., *J. Catalysis* **6**, 270 (1966).
3. Clark, A., Polymerization and Polycondensation Processes, Symposium, Natl. Am. Chem. Soc., Meeting, San Francisco, California, March 31–April 5 (1968); Heckelsberg, L. F., Banks, R. L., and Bailey, G. C., General Papers, Division Petroleum Chemistry, Inc., Am. Chem. Soc., San Francisco, California, March 31–April 5 (1968).
4. Dwyer, F. G., Eagleton, L. C., Wei, J., and Zahner, J. C., *Proc. Roy. Soc. (London)*, *Ser. A* **302**, 253 (1968).
5. Weller, S., *Am. Inst. Chem. Eng. J.* **2**, 59 (1956).
6. Smith, J. M., "Chemical Engineering Kinetics," Chapter 9. McGraw-Hill, New York, 1956.

7. Hutchinson, H. L., Barrick, P. L., and Brown, L. F., *Chem. Eng. Progr. Symp. Ser.* **63**, 18 (1967).
8. Laidler, K. J., *Catalysis* **1**, 119 (1961).
9. Bond, G. C., "Catalysis by Metals," Chapter 7. Academic Press, New York, 1962.
10. Bischoff, K. B., and Froment, G. F., *Ind. Eng. Chem. Fundamentals* **1**, 195 (1962).
11. Weiss, A. H., *Chem. Eng.* **30**, 89 (1963).
12. Boudart, M., *Chem. Eng. Progr.* **58**, 73 (1962).
13. Kemball, C., *Intern. Congr. Catalysis 2nd, Paris* **1**, 11 (1960).
14. Wheeler, A., *Advan. Catalysis* **3**, 250 (1951).
15. Bond, G. C., "Catalysis by Metals," pp. 132, 133. Academic Press, New York, 1962.
16. de Boer, J. H., and van der Borg, R. J. A. M., *Proc. Intern. Congr. Catalysis* **1**, 919 (1961). Paris.
17. Boudart, M., *Am. Inst. Chem. Eng. J.* **2**, 62 (1956).
18. Constable, F. H., *Proc. Roy. Soc. (London) Ser. A* **108**, 355 (1925).
19. Halsey, G. D., *J. Chem. Phys.* **17**, 758 (1949).
20. Stock, A., *Ber. Deut. Chem. Ges.* **40**, 532 (1907); **41**, 1309 (1908).
21. Kwan, T., *J. Res. Inst. Catalysis* **3**, 16 (1953).
22. Kwan, T., *J. Phys. Chem.* **60**, 1033 (1956).
23. Schuit, G. C. A., and van Reijen, L. L., *Advan. Catalysis* **10**, 243 (1958).
24. Cremer, E., *Advan. Catalysis* **7**, 75 (1955).
25. Cremer, E., and Schwab, G.-M., *Z. Phys. Chem. (Leipzig)* **A144**, 243 (1929); Schwab, G.-M., *Z. Phys. Chem. (Leipzig)* **B5** 406 (1929).
26. Kemball, C., *Proc. Roy. Soc. (London), Ser. A* **217**, 376 (1953).
27. Everett, D. H., *Trans. Faraday Soc.* **46**, 957 (1950).
28. Sosnovsky, H. M. C., *J. Phys. Chem. Solids* **10**, 304 (1959).
29. Jaeger, H., *J. Catalysis* **9**, 237 (1967).
30. Yermakov, Yu. I., and Zakharov, V. A., *Intern. Congr. Catalysis 4th Moscow* Paper No. 16. (1968).
31. Ivanov, L. P., Yermakov, Yu. I., and Gel'bshtein, A. I., *Vysokomol. Soedin Siberian Acad. Sci. USSR Ser* **A9**, No. 11, 2422 (1967).
32. Clark, A., and Bailey, G. C., *J. Catalysis* **2**, 230 (1963).
33. Clark, A., and Bailey, G. C., *J. Catalysis* **2**, 241 (1963).
34. Guyot, A., *J. Catalysis* **3**, 390 (1964).
35. Natta, G., *J. Polymer Sci.* **34**, 21 (1959).
36. Happel, J., and Mozaki, R., *Catalysis Rev.* **3**, 241 (1969).
37. Kittrell, J. R., and Mozaki, R., *Ind. Eng. Chem.* **59**, 28 (1967).
38. Mochida, I., and Yoneda, Y., *J. Catalysis* **7**, 386 (1967).
39. Mochida, I., and Yoneda, Y., *J. Catalysis* **7**, 393 (1967).
40. Dunn, I. J., *J. Catalysis* **11**, 79 (1968).
41. Madix, R. J., *Chem. Eng. Sci.* **23**, 805 (1968).

XII

KINETICS OF HETEROGENEOUS
CATALYSIS—DIFFUSION CONTROLLING

12.1   The Mechanisms of Diffusion

To the steps of adsorption, surface reaction, and desorption, we must now add the steps of diffusion. In general, an overall process of chemical transformation on porous catalytic materials will involve the following series of steps:

(1)   Diffusion of reactants to the external surface of the catalyst particle.
(2)   Diffusion of reactants in the pores.
(3)   Adsorption of reactants.
(4)   Surface reaction.
(5)   Desorption of products.
(6)   Diffusion of products out of the pores.
(7)   Diffusion of reactants from the external surface of the catalyst particle to the surrounding fluid.

Thus, there are two general diffusional processes of importance in catalysis: (i) mass transfer to and from the external surface of the catalyst and (ii) mass transfer in and out of the catalyst pores. Mass transfer to the external surface is controlled in a thin boundary layer, usually less than a millimeter in thickness, next to the surface. The velocity of a fluid passing over the surface of a particle varies rapidly normal to the flow across this boundary layer. At the catalyst surface, the fluid velocity is zero, and rapidly approaches the bulk-stream velocity a short distance from the surface. Near the surface, where fluid velocity is low, little mixing of reactants and products occurs; mass transfer normal to the surface is by molecular diffusion, and may be taken as proportional to the molecular diffusion coefficient D. In the bulk-fluid stream, mass transfer is essentially independent of the molecular diffusion coefficient. As might be expected, the overall mass transfer from bulk fluid to the surface of a solid particle is found experimentally to be proportional to D^n, where n varies from zero to unity. Data on mass transfer from fluid to solid surface are often expressed in terms of mass-transfer coefficient, k_c, defined by

$$N = k_c(c_0 - c_s), \tag{1}$$

where N is the diffusion flux of the substance in question, c_s is the concentration at the surface, and c_0 the concentration in the bulk fluid. Detailed treatments of transport from fluid to solid surfaces can be found in many engineering texts and articles (*1–5*).

More important for our objectives is mass transfer within the pores. In the complex internal structure of high surface area catalysts, the nature of the individual pores and the porous structure are extremely important in determining the accessibility of the various regions of the catalyst surface to the reactants. We are interested in pore shape and average pore size, in pore-size distribution, and in how pores are interconnected. Only simple approximations to the complex patterns existing in real catalysts are available.

Some knowledge of the shapes of capillaries in porous adsorbents can be obtained from the hysteresis loops of adsorption–desorption isotherms (*6*). At least five types of hysteresis loops have been distinguished corresponding to capillaries of various shapes. McBain (*7*), for example, has attributed hysteresis effects to pores shaped like ink bottles with a narrow neck and a larger-diameter body. A good discussion of hysteresis and shapes of capillaries is given by Thomas and Thomas (*8*). The simplest expression for average pore radius assumes a cylindrical pore and is given by

$$\bar{r} = 2(V_g/\alpha_g), \tag{2}$$

where V_g is the pore volume and α_g the surface area per gram. Pore volume may be determined experimentally by the mercury–helium method (*9*) and surface area by the well-known BET method (*10*). For shapes other than cylindrical, the relationship

$$\bar{r} = (1/\gamma)(2V_g/\alpha_g) \tag{3}$$

may be used, where γ is a factor characteristic of the pore geometry. Values for γ have been given by Everett (*11*) for simple uniform structures. More

TABLE 12.1

VALUES OF γ IN $\bar{r} = (1/\gamma)(2V_g/\alpha_g)$ FOR VARIOUS STRUCTURES

Basic structure	Length parameter	Value of γ
Nonintersecting cylindrical capillaries	Capillary radius	1.000
Parallel-sided fissures	Distance apart of walls	1.000
Nonintersecting, close-packed cylindrical rods	Radius of rod	0.104
Cubic packing of spheres	Radius of sphere	0.613
Orthorhombic packing of spheres	Radius of sphere	0.433
Rhombohedral packing of spheres	Radius of sphere	0.229

complicated structures require two parameters for their description. Everett's values are given in Table 12.1. Pore-size distribution can be determined by the mercury penetration method (12) for pores down to approximately 20 Å in radius or by interpretation of low-temperature adsorption isotherms for the pore size range 0–300 Å (13). For many porous materials, e.g., silica gels and cracking catalysts, the shape of the pore size distribution curve is approximately that of a normal probability curve with a narrow spread about the average (maximum probability) value. However, many catalysts, especially pilled catalysts prepared by compacting fine powders, reveal a bimodal pore-size distribution, often called a *bidisperse* structure, or *macro–micro* distribution. The fine pore structure occurs within the small particles of the original powder and the macro structure is formed by the passageways around the original particles. Wakao and Smith (14) have developed a model to represent the bimodal distribution of a pelleted catalyst and have derived equations for the diffusion flux.

In spite of the many studies of pore shape and pore-size distribution, one is usually forced, in practice, to use approximations of average values for pore constants. Equations (2) and (3) represent the simplest approach. A refined version of Eq. (2) has been derived by Wheeler (15) with the assumptions that the sum of the surface areas of the pores is equal to the BET surface area, and the sum of the pore volumes is equal to the experimental pore volume. Let V_p be the total volume per pellet, V_g the pore volume per gram, and ρ_p the pellet density. Then $V_p \rho_p V_g = V_p \psi$ is the experimental pore volume, where $\psi = \rho_p V_g$ is termed the porosity, the fraction of the volume of the pellet which is void. The theoretical pore volume may be expressed by $(\alpha_x n_p)\pi \bar{r}^2 \bar{L}$, where α_x is the external geometrical surface area of the pellet, n_p the number of pores per unit external surface area, and $\pi \bar{r}^2 \bar{L}$ is the mean volume of a cylindrical pore with $\bar{r}$ and $\bar{L}$ the mean radius and length of pore, respectively. We have

$$V_p \rho_p V_g = V_p \psi = (\alpha_x n_p)\pi \bar{r}^2 \bar{L}. \qquad (4)$$

We now derive the corresponding equation for total surface area of the pellet. The pore walls may not be perfectly smooth so that a roughness factor f must be introduced. Furthermore, pores may interesect so that only a part of the total pore-wall should be counted in the surface area. Wheeler imagines a uniformly porous unit cube sliced into thin sections each of thickness Δx and with each section having the same area of pore mouths A_p. The total pore volume is, thus, $A_p \Delta x$ summed over all slices, which is simply A_p. However, the total pore volume per unit pellet volume is, by definition, ψ, and therefore,

$$A_p = \psi. \tag{5}$$

Since it is assumed that pore walls possess the same properties as any other surface drawn in the particle, then only a fraction $(1 - \psi)$ of the pore wall is solid surface. Corresponding to the volume expression, Eq. (4), we now write for the total surface area,

$$V_p \rho_p \alpha_g = (\alpha_x n_p) 2\pi \bar{r}(1 - \psi) f \bar{L}, \tag{6}$$

where α_g is the BET area. Dividing Eq. (4) by Eq. (6), we find

$$\bar{r} = (2V_g/\alpha_g)(1 - \psi)f, \tag{7}$$

in place of the less exact expression, Eq. (2). In order to obtain an expression for $\bar{L}$, we start with the fact that the total number of pore mouths equals the total area of pore mouths divided by the area per pore mouth. If the pores all run perpendicular to the external surface, then the number per unit area would be $n_p = A_p/\pi \bar{r}^2 = \psi/\pi \bar{r}^2$. It is more reasonable, though, to assume a random direction of pore orientation with respect to the external surface, with an average direction of 45°. An average pore of this orientation intersects the external surface in an ellipse of area $\pi \bar{r}^2/\sin 45° = \sqrt{2}\,\pi \bar{r}^2$. With this assumption, we have

$$n_p = (A_p/\pi \bar{r}^2)(1/\sqrt{2}) = (\psi/\pi \bar{r}^2)(1/\sqrt{2}). \tag{8}$$

Substituting this value of n_p into Eq. (4), we find

$$\bar{L} = (V_p/\alpha_x)\sqrt{2}. \tag{9}$$

Thus, the average pore length turns out to be $\sqrt{2}$ times the ratio of the volume to the external geometrical surface area of the pellet. For pellets in the shape of spheres, cubes, and cylinders whose diameter equals their length, V_p/α_x turns out to be $d_p/6$, where d_p is the diameter of the sphere, the side of the cube, or the length of the cylinder. Therefore, we assume for most practical shapes of catalyst pellets that the average length of a typical pore is $\sqrt{2}\,d_p/6$. As we have seen, values for $\bar{r}$ and $\bar{L}$ may be determined from experimentally measurable quantities, and they will be used in subsequent calculations of diffusion and reaction rates in pores.

Pore diffusion occurs by one or more of four mechanisms; Knudsen diffusion, ordinary bulk diffusion, surface diffusion, and forced flow. The first two are most important.

KNUDSEN DIFFUSION[1]

At low gas-density or small pore-diameter, or both, when the mean free path is large in comparison with the pore diameter, molecules will collide more frequently with the walls than with each other. If the molecules are reflected elastically from the walls, the flow will be identical with the flow through an orifice in a plane, thin wall, and all the molecules entering the pore mouth will pass through the tube regardless of its length; but the amount of gas passing through a capillary under the conditions of Knudsen flow has been found to depend on the length of the capillary. To account for this, Knudsen (16) assumed that molecules leave a surface randomly independent of the direction in which they hit. Random emission from the wall may take place as a result of rough pore surface or inelastic collisions. Some entering molecules are sent back in the direction from which they came; and the longer the capillary, the greater the fraction of molecules returned. The amount of gas flowing through the capillary decreases with length. The fraction returned is independent of gas pressure.

Let Z be the number of molecules hitting unit cross-sectional area of a cylindrical pore in unit time. The net flow of molecules through a capillary (positive x direction) has been found to be

$$\tfrac{1}{2}B(dZ/dx),\tag{10}$$

where for a cylindrical pore

$$B = 16\pi r^3/3.\tag{11}$$

The quantity Z has been shown to decrease linearly with x, so that dZ/dx may be replaced by $(Z_2 - Z_1)/L$, where Z_2 and Z_1 are the values of Z at the entrance and exit of the pore, respectively, and L is the pore length. From kinetic theory of gases

$$Z = N_A\, p/(2\pi MRT)^{1/2},\tag{12}$$

where N_A is Avogadro's number, p the pressure and M the molecular weight of the diffusing component. Thus, the number of moles of gas flowing through the capillary per second is given by

$$dn/dt = \tfrac{1}{2}B[1/(2\pi MRT)^{1/2}][(p_2 - p_1)/L],\tag{13}$$

[1] See Knudsen (16), Loeb (17), and Taylor and Glasstone (18).

where p_2 and p_1 are the inlet and outlet pressures of the diffusing component. In terms of concentrations, $c = p/RT$, we have,

$$dn/dt = \tfrac{1}{2}B(RT/2\pi M)^{1/2}[(c_2 - c_1)/L] = \pi r^2 \tfrac{2}{3}r(8RT/\pi M)^{1/2}[(c_2 - c_1)/L].$$
(14)

From Eq. (14), we see that the Knudsen diffusion coefficient is expressed by

$$D_K = \tfrac{2}{3}r(8RT/\pi M)^{1/2} = \tfrac{2}{3}r\bar{v},$$
(15)

where $\bar{v}$ is the mean Maxwellian velocity. Thus, Knudsen flow may be included in a general flow equation by using the Knudsen diffusion coefficient.

BULK DIFFUSION[2]

At higher pressures or with larger pores, when the mean free path is much smaller than the pore diameter, collisions between gas molecules will be more frequent than collisions with the wall. The rate of diffusion will then be independent of pore diameter. For example, at a pressure of 1 atm, the mean free path is about 10^{-5} cm. Therefore, in pores larger in diameter than, say, 10^{-4} cm, collisions between gas molecules will predominate. We suppose that there are two kinds of molecules, 1 and 2. In Fig. 12.1, we show the situation with respect to molecules of 1 in a cylindrical pore. We consider the number of moles of 1 crossing plane B from the direction of plane A, a distance λ, the mean free path, above it. We consider also the number of moles of 1 crossing plane B from the direction of plane C, a distance λ below it. We assume that molecules which reach plane B from above or below it have traveled on the average only a distance λ since colliding with other molecules. If there is no

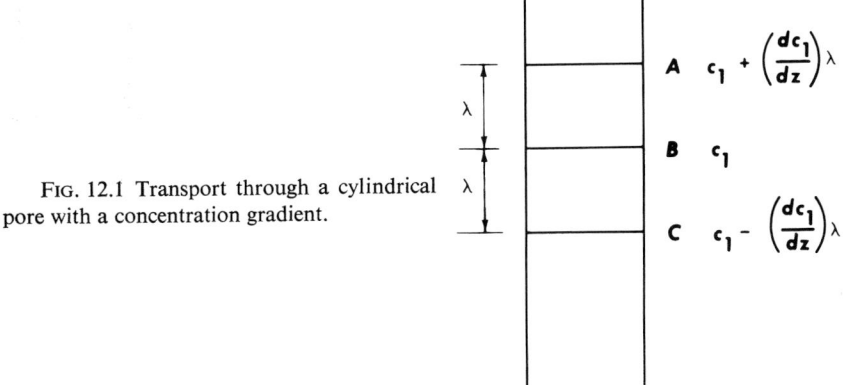

FIG. 12.1 Transport through a cylindrical pore with a concentration gradient.

---

[2] See Thomas and Thomas (8) and Moelwyn-Hughes (19).

concentration gradient down the pore, the number of moles traversing the cross-sectional area α at any distance z, in either direction, is $c_1\alpha\bar{w}$, where c_1 is the concentration (moles/cm^3) of molecules of 1 and $\bar{w}$ is their average component of molecular velocity in a direction perpendicular to a cross-sectional plane. When there is a concentration gradient, the number of moles of 1 crossing in opposite directions is different and there is a net transport of matter. But the number of moles of the two types 1 and 2, which in one second cross a plane perpendicular to the z axis in any direction must be zero; otherwise there would be a local accumulation of matter. Thus,

$$(dn_1/dt) + (dn_2/dt) = 0. \qquad (16)$$

Diffusion causes no change in total pressure. Referring to Fig. 12.1, we note that the number of moles crossing plane B from above is $\alpha\bar{w}[c_1 + (dc_1/dz)\lambda]$ and from below is $\alpha\bar{w}[c_1 - (dc_1/dz)\lambda]$. It has been assumed that the rate at which c_1 changes with respect to z, that is dc_1/dz, is constant (independent of z) over the small range 2λ. The net amount of moles of 1 per second crossing plane B is,

$$dn_1/dt = 2\bar{w}\alpha(dc_1/dz)\lambda. \qquad (17)$$

The number of moles crossing unit area per second and per unit concentration gradient is the diffusion coefficient. Hence, we find

$$D_B = 2\bar{w}\lambda. \qquad (18)$$

Now $\bar{w} = \bar{v}/4$ (*19*), where $\bar{v}$ is the mean Maxwellian velocity, so that

$$D_B = \tfrac{1}{2}\bar{v}\lambda. \qquad (19)$$

SURFACE DIFFUSION

The transport of adsorbed molecules into pores has not been studied extensively. Available data (*20–22*) show that it contributes appreciably to diffusion in pores only when adsorption is strong. Surface diffusion should exert more influence at low rather than at high temperatures. Recent considerations (*23*) stress the possibility that surface diffusion may play as significant, if not a greater role than Knudsen diffusion in pores.

FORCED FLOW

When a total pressure difference is maintained in a pore, forced flow occurs. If the mean free path is large compared to pore diameter, forced flow is indistinguishable from Knudsen flow. But, if the mean free path is small compared to pore diameter, and a total pressure differential exists, then forced flow must be superimposed on bulk diffusion. The diffusion coefficient for forced flow can be written as (*8, 24, 25*)

$$D_F = \bar{r}^2 c_T RT/8\eta, \tag{20}$$

where c_T is the total molar concentration of all molecules present and η is the viscosity coefficient.

12.2   Diffusion and Reactions in Pores

If diffusion into a pore is fast compared to reaction, the entire pore surface will promote reaction as efficiently as if it were extended into a plane surface (*1, 8, 15*). Reactants will reach all parts of the pore surface before they have time to react. The concentration gradient of reactant between the inside surface of the pore and the external surface surrounding it will be small, and as a result of this small gradient, the diffusion of reactants in and of products out will be slow. At steady state, the rate of diffusion of reactants into the pore will equal the rate of diffusion of products out, and each will be equal to the total rate of reaction on the surface of the pore.

If, on the contrary, the rate of reaction is fast compared to diffusion, reaction will set in before reactant molecules have diffused very far into the pore. Most of the reaction will be occurring on the periphery of the pore. There will be a strong concentration gradient into the pore. Reactant molecules will diffuse rapidly a short distance into the pore and product molecules will diffuse out rapidly. Thus, the interior of the pore will contribute little to reaction rate because the concentration of reactant will be low there. At steady state, the rate of diffusion of reactants in will equal the rate of diffusion of products out, and each will be equal to the rate of reaction on the periphery of the pore.

In developing a quantitative approach to reaction in pores, we follow the pioneer work of Thiele (*26*), extended by Wheeler (*15*) to catalyst pellets. We consider only one-dimensional diffusion. For a discussion of transverse as well as longitudinal diffusion see the work of Bischoff (*27*).

REACTION IN A SINGLE PORE

We select a single, open-ended cylindrical pore of radius r and length $2L$, as shown in Fig. 12.2. The concentration of reactant at the pore mouth ($z = 0$) is c_0. We consider an arbitrary section dz. Molecules will diffuse equally rapidly from either end of the pore, making the plane at $z = L$ a plane of symmetry. Therefore, we may consider one-half of the pore length, from $z = 0$

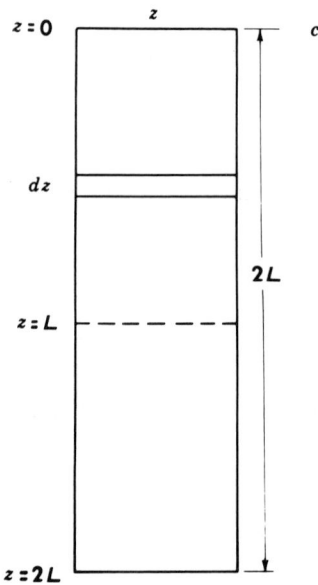

FIG. 12.2. Material balance in a cylindrical pore.

to $z = L$. No flow occurs across $z = L$, so that $(dc/dz)_{z=L} = 0$. Assuming no flow except by diffusion, the rate of diffusion of reactant molecules at steady state into any region of the pore minus the rate of diffusion of reactant molecules out of that region is simply the rate of reaction in that region. Thus, we may set the diffusion expression equal to the reaction rate expression. The moles of reactant that diffuse into the region dz per second is

$$-dn/dt = -v = \pi r^2\, D(dc/dz), \tag{21}$$

where dc/dz is the concentration gradient at distance z from the pore mouth and D is the appropriate diffusion coefficient. By differentiation we obtain the change in v in traversing dz,

$$-dv = \pi r^2\, D(d^2 c/dz^2)\, dz. \tag{22}$$

The first-order reaction rate in the surface area $(2\pi r\, dz)$ of the pore section is $2\pi r\, dz\, k_1 c$, where k_1 is the intrinsic reaction velocity constant defined as the number of moles reacted per second per square centimeter of catalyst surface per unit concentration of reactant molecules (cm/sec). Note the difference between this constant and the one usually employed in homogeneous systems, which is the number of moles *per unit volume* reacted per second per unit concentration of reactant ($\sec^{-1}$). We have

$$\pi r^2\, D(d^2 c/dz^2) = 2\pi r k_1 c. \tag{23}$$

This linear, second-order differential equation is solved with the boundary conditions $c = c_0$ at $z = 0$, and $(dc/dz) = 0$ at $z = L$, to obtain

$$c = c_0 \frac{\{\exp(h_1[1 - (z/L)]) + \exp(-h_1[1 - (z/L)])\}}{e^{h_1} + e^{-h_1}}$$

$$= \frac{c_0 \cosh\{h_1[1 - (z/L)]\}}{\cosh h_1}, \tag{24}$$

where

$$h_1 = L(2k_1/rD)^{1/2} \tag{25}$$

is a dimensionless quantity. By differentiating Eq. (24) with respect to z, we obtain an expression for the concentration gradient, dc/dz, along the pore. Now the rate of reaction per half pore, $R_{1/2}$, is equal to the rate at which reactant flows into the pore. This follows because there is no transport of matter across the plane at $z = L$. The rate per half pore is

$$R_{1/2} = - \pi r^2 D(dc/dz)_{z=0} = \pi r^2 D(h_1 c_0/L) \tanh h_1. \tag{26}$$

A simple expression for the fraction of pore area available for reaction can now be obtained. If no concentration gradient exists along the pore, that is, if the pore area is just as available as the external surface, then the rate of reaction per half pore is given by $R_0 = 2\pi r L k_1 c_0$. Dividing $R_{1/2}$ of Eq. (26) by R_0, we obtain the "effectiveness factor" for the pore,

$$f = R_{1/2}/R_0 = (1/h_1) \tanh h_1. \tag{27}$$

When $h_1 > 2$, $\tanh h_1 \cong 1$, and $f = 1/h_1$. In this case, the effectiveness factor is inversely proportional to the pore length, and, as a first approximation, pore length is proportional to pellet size.

Similar calculations have been made for a second-order reaction. The differential equation which must be solved in this case, comparable to Eq. (23), is

$$\pi r^2 D(d^2c/dz^2) = 2\pi r k_2 c^2. \tag{28}$$

The solution is complex, involving elliptical integrals. It is a relatively simple matter to derive an expression for (dc/dz) and from it an expression for $R_{1/2}$, the reaction rate in a half pore, and for f, the effectiveness factor in the pore. We give the results

$$R_{1/2} = \pi r^2 D[(2k_2/rD)\tfrac{2}{3}(c_0^3 - c_c^3)]^{1/2}, \tag{29}$$

$$f = (1/h_2)\{\tfrac{2}{3}[1 - (c_c/c_0)^3]\}^{1/2}, \tag{30}$$

where

$$h_2 = L(2k_2 c_0/rD)^{1/2}, \tag{31}$$

a dimensionless parameter, and c_c is the concentration at the center of the pore. For fast reactions, $c_c \ll c_0$, and the effectiveness factor becomes

$$f = (1/h_2)\sqrt{\tfrac{2}{3}}. \tag{32}$$

Expressions for the zero-order reaction may also be developed. The differential equation becomes

$$\pi r^2 D(d^2c/dz^2) = 2\pi r k_0, \tag{33}$$

with the boundary conditions $c = c_0$ at $z = 0$, and $(dc/dz) = 0$ at $z = L$. We write the solution

$$c = c_0\{1 - h_0^2[(z/L) - \tfrac{1}{2}(z/L)^2]\}, \tag{34}$$

where h_0, a dimensionless parameter, is defined by

$$h_0 = L(2k_0/rDc_0)^{1/2}. \tag{35}$$

If $h_0 < \sqrt{2}$, c is positive in any part of the pore, so that reaction occurs on the entire pore surface. On the other hand, if $h_0 > \sqrt{2}$, then c and, therefore (dc/dz), becomes zero at some point in the pore, say, $z_0 < L$. The boundary conditions are therefore different. For this case, Eq. (33) is integrated to give

$$c = c_0\{1 - 2[(h_0 z/\sqrt{2}L) - \tfrac{1}{2}(h_0 z/\sqrt{2}L)^2]\}, \quad z \leqslant z_0 \tag{36}$$

$$c = 0 \qquad\qquad\qquad\qquad z \geqslant z_0. \tag{37}$$

The point z_0, where $c = 0$ and $(dc/dz) = 0$, is found from Eq. (36) and is

$$z_0 = \sqrt{2}L/h_0. \tag{38}$$

When $h_0 > \sqrt{2}$, we find the effectiveness factor for the pore to be

$$f = 2\pi r k_0 z_0/2\pi r k_0 L = \sqrt{2}/h_0, \tag{39}$$

and the reaction rate in a half pore is

$$R_{1/2} = 2\pi r L k_0 f = 2\pi r (rk_0 Dc_0)^{1/2}. \tag{40}$$

The developments given above all assume behavior according to Fick's first law,

$$dn/dt = -\alpha D(dc/dz). \tag{41}$$

When there is no change in the number of moles on reaction so that there is an equivalent diffusion of reactants in and products out, this law holds. It is also an approximation for the reaction with volume change when the concentration of reactants and products is low. Furthermore, if diffusion is entirely by Knudsen flow, then a change in the number of moles on reaction has no effect on the rate of mass transfer. But when bulk diffusion occurs with a change in number of moles on reaction, a hydrodynamic flow takes place. An increase in the number of moles on reaction makes it more difficult for reactants to diffuse into the pore, thus decreasing the effectiveness factor. And a decrease in the number of moles on reaction makes it easier for reactant to diffuse into the pore, thus increasing the effectiveness factor. Thiele has solved the problem for the special case of a first-order reaction, $A \rightarrow qB$, with the diffusion coefficient D_A and the total molar concentration c_T independent of z. The fundamental differential equation to be solved with these assumptions is,

$$\frac{d}{dz}\left[\frac{\pi r^2 D_A}{1 + (q - 1)(c/c_T)}\frac{dc}{dz}\right] = 2\pi r k_1 c, \qquad (42)$$

with the boundary conditions, $c = c_0$ at $z = 0$ and $(dc/dz) = 0$ at $z = L$.

REACTION IN SPHERICAL PARTICLES

Consider a spherical particle of radius R. Within this spherical particle, select a spherical shell of inner radius r and thickness dr. We assume that the number of pores of a given radius decreases in proportion to the area of the spherical shell. Thus ψ, the fraction of superficial surface which is pore-mouth area, is constant for $0 \gtrless r \gtrless R$. The difference in mass flow into the shell at $r + dr$ and out of the shell at r is given by

$$4\pi(r + dr)^2\psi D[(dc/dr) + (d^2c/dr^2)\, dr] - 4\pi r^2\psi D(dc/dr)$$
$$= 4\pi r^2\psi D(d^2c/dr^2)\, dr + 8\pi r\psi D(dc/dr)\, dr \qquad (43)$$

in which second-order differentials are neglected. The diffusion coefficient D is averaged over all pores and is expressed in moles per second per unit concentration gradient per unit of pore mouth area on the superficial surface at any shell radius r. At steady state, we set the change in mass diffusion given by Eq. (43) equal to the rate of reaction in the elementary shell of surface area $(4\pi r^2\, dr)\alpha_v$, where α_v is the specific surface area per unit volume of the spherical particle. We replace ψD by D_e, an effective diffusion coefficient expressed in moles per second per unit concentration gradient per unit of superficial surface area at any shell radius r. The resulting differential equation is,

$$4\pi r^2 D_e(d^2c/dr^2)\, dr + 8\pi r D_e(dc/dr)\, dr = (4\pi r^2\, dr)\alpha_v k_1 c \qquad (44)$$

which simplifies to

$$(d^2c/dr^2) + (2/r)(dc/dr) = \alpha_v k_1 c/D_e = k_v c/D_e = (\varphi_s^2/R^2)c, \qquad (45)$$

where $\varphi_s = R(k_v/D_e)^{1/2}$ is called the Thiele diffusion modulus. A first-order reaction has been assumed. The rate constant k_1 expresses the rate per unit area of pore surface, while $k_v(=\alpha_v k_1)$ expresses the rate per unit volume of catalyst particle. Eq. (45) is to be solved with the boundary conditions

$$c = c_0 \quad \text{at} \quad r = R, \qquad dc/dr = 0 \quad \text{at} \quad r = 0. \tag{46}$$

Making the substitution, $c = s/r$, enables us to solve Eq. (45) by the method of factorization of operators. The solution is

$$c = (c_0 R/r)(\sinh br/\sinh bR), \tag{47}$$

where $b = \varphi_s/R$. Now, the total reaction rate inside the spherical catalyst particle is equal to the rate of diffusion into the particle and is given by

$$\text{Rate} = 4\pi R^2 D_e(-dc/dr)_{r=R} = 4\pi RD_e\,\varphi_s c_0[(1/\tanh\varphi_s) - (1/\varphi_s)]. \tag{48}$$

If the internal surface of the porous particle were all exposed to reactant concentration c_0, the rate would be

$$\text{Rate} = \tfrac{4}{3}\pi R^3 k_v c_0. \tag{49}$$

Taking the ratio of the rates given by Eqs. (48) and (49), we find the effectiveness factor,

$$f = (3/\varphi_s)[(1/\tanh\varphi_s) - (1/\varphi_s)]. \tag{50}$$

This equation tells us that the effectiveness of the internal surface of a porous catalyst particle approaches unity asymptotically as φ_s approaches zero, that is, as radius R or rate constant k_v become very small or the diffusion coefficient D_e becomes very large. Conversely, the effectiveness factor approaches zero for large particles, large rate constants, or small values of D_e. We note that very active catalyst particles (k_v large) tend to have low effectiveness factors, and inactive catalysts tend to have high effectiveness factors.

The Thiele modulus, $\varphi_s = R(k_v/D_e)^{1/2}$ contains the intrinsic rate constant, k_v, which is often difficult to obtain explicitly. Therefore, we proceed to develop a parameter which contains only quantities that can be observed or calculated. The rate of disappearance of reactant in catalyst volume V_c for a first-order reaction may be expressed by

$$-dn/dt = k_v V_c c_0 f, \tag{51}$$

or, alternatively, in terms of reactor volume V_R, by

$$-dn/dt = k_v V_R(1 - \epsilon)c_0 f, \tag{52}$$

where ϵ is the void fraction in the packed catalyst bed. The rate constant k_v may be evaluated from the Thiele modulus in terms of φ_s, R, and D_e, and substituted into Eq. (51) giving

$$\Phi \equiv (R^2/D_e)[-(1/V_c)(dn/dt)](1/c_0) = \varphi_s^2 f \qquad \text{(sphere, first order)}, \tag{53}$$

where Φ is a new dimensionless modulus expressed in terms of reaction rate per unit gross volume or catalyst, external concentration of reactant, radius, and effective diffusion constant, all of which can be obtained experimentally or estimated. Values of the effective diffusion constant D_e may be estimated from expressions in Section 12.1, or determined experimentally. From Eq. (50), we find that for $f < \sim 1/2, f = 3/\varphi_s$, and thus Eq. (53) gives f directly. For higher values of f, f may be obtained by successive approximations from Eq. (50) relating φ_s and f. Alternatively, corresponding values of φ_s and f, determined by Eq. (50), may be used to determine values of $\Phi \equiv \varphi_s^2 f$. A plot of Φ versus f can then be made, and values of f corresponding to experimental values of Φ can be read directly from the plot.

We have assumed a constant effective diffusion constant D_e in the direction of diffusion. Values of D_e may be estimated from expressions given in Section 12.1 or measured directly or determined from reaction-rate measurements. Diffusivity measurements and reaction rate measurements have been found to lead approximately to the same values of D_e provided the pore-size distribution is reasonably narrow. For catalysts having a wide pore-size distribution, such as the bimodal distribution of pelleted catalysts, values of D_e calculated from "average" pore radius can be misleading. Various methods of dealing with macro–micropore systems have been developed, especially by Mingle and Smith (*28*), Carberry (*29*), and Wakao and Smith (*14*).

REACTION IN PARTICLES OF OTHER SHAPES

The differential equation for first-order reaction in a flat plate of porous catalyst with reactants on one side, but sealed on the other side and on the ends, is easily shown to be

$$D_e(d^2c/dz^2) = k_v c, \tag{54}$$

where z is the depth of penetration into the plate. The Thiele modulus is defined as

$$\varphi_L = L(k_v/D_e)^{1/2}, \tag{55}$$

where L is the thickness of the plate. The effectiveness factor turns out to be (*15, 25*)

$$f = (1/\varphi_L) \tanh \varphi_L. \tag{56}$$

Aris (*30*) compares values of f calculated for a flat plate, and a sphere. For this purpose, he defines L for each shape as the ratio of volume to the geometric surface area through which reactants flow into the volume. Thus, with this definition of L, we write

$$\varphi_L = L(k_v/D_e)^{1/2} = (R/3)(k_v/D_e)^{1/2} = \varphi_s/3. \tag{57}$$

Using this relation and the expressions for effectiveness factors for the flat plate, Eq. (56), and for the sphere, Eq. (50), we find that for high values of

φ_L, the values of f approach identity. Aris further shows that at high values of φ_L the value of f also approaches the common value for the sphere and flat plate. Similarly, at low values of φ_L, all three functions again approach identity.

Solutions to the diffusion equation cannot be obtained analytically except for simple geometric forms. A simpler approach for arbitrary shapes has been devised by Wheeler (15). He assumes cylindrical pores. If a pellet is composed of N pores of average length $\bar{L}$, the rate of reaction per catalyst pellet is N times the rate per average pore. The latter quantity, $R_{1/2}$, is given by Eq. (26). The number of pores N is $(n_p \alpha_x)$ where n_p is the number of pores per unit external surface area, α_x [see Eq. (4)]. The rate per pellet becomes

$$R_p = \alpha_x n_p R_{1/2} = (\alpha_x \psi / \pi \bar{r}^2 \sqrt{2})(\pi \bar{r}^2 D / \bar{L}) c_0 h_p \tanh h_p, \tag{58}$$

where we have used the expression for n_p given in Eq. (8), and $\bar{L}$ is given by Eq. (9). The term h_p is an empirically modified value of h_1, Eq. (25), as employed in Eq. (26).

$$h_p = \bar{L}[2k_1(1 - \psi)\tau / \bar{r} D]^{1/2} = \sqrt{2}(V_p / \alpha_x)(k_1' / \bar{r} D)^{1/2}. \tag{59}$$

The modifications, $(1 - \psi)$ and τ, correct for pore intersections and tortuosity, respectively. The term $(1 - \psi)$ represents the fraction of pore wall not interrupted by pore intersections. As before, the effectiveness factor is [Eq. (27)],

$$f = (1/h_p) \tanh h_p. \tag{60}$$

Other special cases have been worked out. For example, Satterfield et al. (44, 45) have developed a generalized method of predicting the catalyst effectiveness factor when the intrinsic reaction kinetics are of the Langmuir–Hinshelwood type. They have assumed various catalyst shapes: (1) a catalyst mass infinite in two directions and of thickness L in the third, (2) spherical particles and (3) a slab particle. Schmalzer (46) has discussed the kinetic effects of a broad pore size distribution.

EFFECT OF DIFFUSION ON REACTION ORDERS AND ACTIVATION ENERGY

We write the reaction rate for an mth order reaction in a half pore as [see similar Eqs. (51) and (40)]:

$$R_{1/2} = 2\pi r L k_m c_0^m f. \tag{61}$$

First consider Knudsen diffusion where the diffusion constant D is independent of c_0 [see Eq. (15)]. For a fast first-order reaction ($m = 1$), $f \to 1/h_1$ [see Eq. (27)], where $h_1 = L(2k_1/rD)^{1/2}$ is independent of c_0. For a slow reaction, $f \to 1$. Thus, the order of the reaction is the true kinetic order whether diffusion is present (fast reaction) or absent (slow reaction). On the other hand, a second-order reaction is affected by diffusion. In this case the effectiveness factor for a fast reaction is $f = (1/h_2)(2/3)$, where $h_2 = L(2k_2 c_0/rD)^{1/2}$, Eqs. (31) and (32). Thus, $R_{1/2}$ is proportional to $c_0^{3/2}$. When the reaction is slow and diffusion is essentially absent, Eq. (30) shows that $f \to 1$, and the reaction order becomes 2, i.e., $R_{1/2} \sim c_0^2$. Similarly, Eqs. (35) and (39) show that a fast zero-order reaction is one-half order, $R_{1/2} \sim c_0^{1/2}$, but approaches true zero-order for a slow reaction. In general, we say that for fast reactions or slow diffusion, the observed order is $(m + 1)/2$, where m is the true kinetic order. Now consider bulk diffusion where D is proportional to the mean free path λ [see Eq. (19)] and, thus, inversely proportional to total pressure. When the bulk gas-phase consists of a single gaseous reactant, then D is inversely proportional to c_0. Using Eq. (61) and proceeding as in the case of Knudsen diffusion, we find that for bulk diffusion the kinetic orders become $0, 1/2,$ $1, \ldots, m/2$ for true zero, first, second, $\ldots$, and mth orders, respectively.

Similar arguments show that in fast reactions of true kinetic order, $m = 0, 1, 2, \ldots, R_{1/2} \sim (k_m D)^{1/2}$. Since $k_m = A_m \exp(-E_m/RT)$, the activation energy is one-half the true value. For slow reactions, when diffusion effects are absent, $R_{1/2} \sim k_m$, and the activation energy approaches the true value. Thus, for a given pore size and at low temperature (k_m small, $f \to 1$), the activation energy is essentially E_m. As the temperature is increased, k_m and D increase while f decreases, so that the activation energy approaches $E_m/2$. Similarly, at constant temperature and decreasing pore size, diffusion effects increase, and the transition $E_m \to E_m/2$ is to be expected.

TEMPERATURE GRADIENTS IN POROUS PELLETS

So far, we have assumed isothermal conditions in pores where reaction and diffusion are occurring. In practice, significant temperature gradients can occur. We follow the methods of Weisz and Hicks (*31*) in obtaining equations for a spherical particle which are modified to include heat effects. The rate of reaction per unit volume in a volume element at concentration c and temperature T is

$$-(1/V_c)(dn/dt) = -(dn_v/dt) = k_v c = A \exp(-E/RT)c. \qquad (62)$$

If we introduce $T = T_0 + \Delta T$ and $k_0 = k_v(T_0) = A \exp(-E/RT_0)$ into Eq. (62), where T_0 is the temperature on the surface of the particle, we obtain

$$-\frac{dn_v}{dt} = Ac \exp\left[-\frac{E}{R(T_0 + \Delta T)}\right] = k_0 c \exp\left(\frac{E}{RT_0} \frac{\Delta T}{T_0} \frac{1}{1 + \Delta T/T_0}\right). \qquad (63)$$

A relationship between the temperature increase ΔT in an element of volume, the molar heat of reaction ΔH, and concentrations has been derived by

Damköhler (*32*) from the diffusion equation for mass transport and the analogous equation for heat diffusion. The relationship is

$$\Delta T = T - T_0 = -(\Delta H D_e/\lambda)(c_0 - c),\tag{64}$$

where λ is the thermal conductivity. If this expression for ΔT is substituted into Eq. (63), there is obtained

$$-\frac{dn_v}{dt} = k_0 c \exp \frac{\gamma\beta[(c_0 - c)/c_0]}{\{1 + \beta[(c_0 - c)/c_0]\}},\tag{65}$$

where

$$\gamma = E/RT_0,\qquad \beta = -\Delta H D_e c_0/\lambda T_0.\tag{66}$$

We see that when temperature gradients occur, two additional parameters, β and γ, are required to describe the system. Using the rate expression of Eq. (65) in place of the isothermal expression of Eq. (45), we find

$$\frac{d^2c}{dr^2} + \frac{2}{r}\frac{dc}{dr} = \frac{k_0 c}{D_e} \exp \frac{\gamma\beta[(c_0 - c)/c_0]}{\{1 + \beta[(c_0 - c)/c_0]\}}.\tag{67}$$

This equation has been solved numerically by Weisz and Hicks subject to the boundary conditions $c = c_0$ at $r = R$ and $dc/dr = 0$ at $r = 0$. They constructed graphs for effectiveness factor f versus the Thiele modulus $\varphi_s = R(k_0/D_e)^{1/2}$, for values of γ of 10, 20, 30, and 40, and values of β ranging from -0.8 to $+0.8$. In this case, the effectiveness factor is defined as the ratio of the actual rate to that which would occur if the particle interior were all exposed to reactant at the same concentration (c_0) and temperature (T_0) as that existing at the outside surface of the particle. Positive values of β refer to exothermic reactions. A typical family of curves is shown in Fig. 12.3 for $\gamma = 20$. The curve $\beta = 0$ is for the isothermal case. The effectiveness factor can exceed unity. This is because the increase in rate caused by the temperature rise in the particle more than compensates for the decrease in rate caused by the negative concentration gradient which brings about a decrease in concentration towards the center of the particle. The overall rate is thus greater than it would be if the concentration and temperature in the particle were the same as on the periphery. The calculations show that the temperature gradient is highest at the periphery and flattens out as the center is approached.

As in the isothermal case, Φ is a more useful parameter than φ_s, for it may be obtained without knowledge of the intrinsic rate constant k_0 in terms of quantities which can be observed or predicted. Starting with the rate expression

$$(1/V_c)(dn/dt) = dn_v/dt = k_0 c_0 f,\tag{68}$$

we eliminate k_0 using the expression which defines the Thiele modulus, $\varphi_s = R(k_0/D_e)^{1/2}$. The expression for Φ is given in Eq. (53).

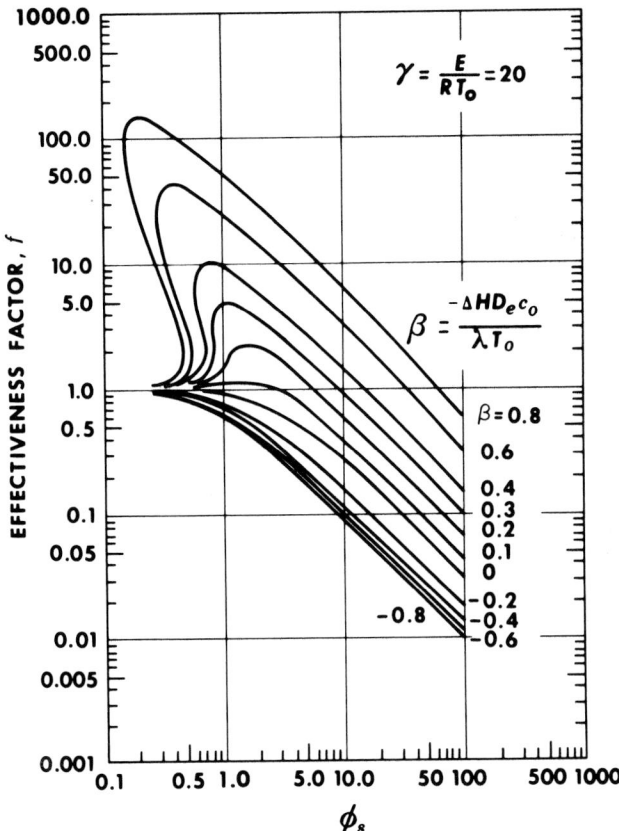

FIG. 12.3. Effectiveness factor f as a function of $\varphi_s = R(k_0/D_e)^{1/2}$ for $\gamma = 20$, first-order reaction in sphere (31). [After Weisz and Hicks, *Chem. Eng. Sci.* **17**, 265 (1962). By permission of Pergamon Press.]

CRITERIA FOR INSIGNIFICANT DIFFUSION EFFECTS

If any reaction occurs at all in a porous structure, *some* concentration gradient will exist. The effectiveness factor will never reach exactly unity except as a mathematically limiting case. For all practical purposes, one can assume that diffusion effects will be negligible when f is greater than about 0.95 (33). Weisz states, for the isothermal case ($\beta = 0$), that diffusion effects are essentially absent when

$$\Phi \leqslant 1.0 \quad \text{(first order)},$$
$$\Phi \leqslant 0.3 \quad \text{(second order)},$$
$$\Phi \leqslant 6.0 \quad \text{(zero order)}.$$

These systems cover the bulk of actual kinetic situations for the isothermal case. Diffusion effects in a given reaction will definitely be present when $\Phi >$

6.0, or absent if $\Phi < 0.3$. From Eq. (53), it is evident that the diffusion criterion for a first-order reaction, $\Phi \leqslant 1.0$, may be replaced by $\varphi_s \leqslant 1.0$, as a good approximation.

12.3   Catalyst Poisoning

Two types of poisoning action are distinguished: (a) homogeneous adsorption of the poisoning molecule, and (b) selective adsorption. In homogeneous adsorption, the poison is uniformly distributed throughout the pore and occurs when the poison is weakly adsorbed or when it is produced by a reaction whose effectiveness factor is close to unity. Selective adsorption occurs when the poison is so strongly adsorbed that the pore mouth becomes almost completely poisoned before the pore interior is affected. Our discussion is limited to first-order reactions and to time-independent poisoning. In flow systems, catalysts are usually poisoned progressively with time, but all of the general principles are developed in the time-independent equations [see Carberry and Gorring (*34*) for a good discussion of time dependent pore-mouth poisoning].

Homogeneous poisoning causes each increment of pore area to decrease in intrinsic activity k (activity per unit area per unit concentration of reactant), to the same extent. Thus, when a fraction of the total surface or sites σ is poisoned, the intrinsic activity becomes $k_1(1 - \sigma)$. In Eq. (26) for $R_{1/2}$, the reaction rate in a half pore, $h_1 = (2k_1/rD)^{1/2}$ is replaced by $h_1(1 - \sigma)^{1/2}$, so that the rate becomes

$$R_{1/2} = \pi r^2 D[h_1(1 - \sigma)^{1/2}/L]c_0 \tanh[h_1(1 - \sigma)^{1/2}]. \tag{69}$$

As before, the rate for an unpoisoned pore is $(\pi r^2 Dh_1 c_0 \tanh h_1)/L$. Dividing the rate in the poisoned pore by that in the unpoisoned pore, we find the ratio F,

$$F = h_1(1 - \sigma)^{1/2} \tanh[h_1(1 - \sigma)^{1/2}]/(h_1 \tanh h_1). \tag{70}$$

For slow reactions, h_1 is small and $\tanh h_1$ and $\tanh[h_1(1 - \sigma)^{1/2}]$ approach h_1 and $h_1(1 - \sigma)^{1/2}$, respectively, while $F \rightarrow (1 - \sigma)$. Activity decreases linearly with the amount of poison added. For fast reactions in small pores, h_1 is large and $F \rightarrow (1 - \sigma)^{1/2}$. Curves A and B in Fig. 12.4 show how the rates of slow and fast reactions are affected by the amount of homogeneous poisoning.

In selective poisoning, often called "pore-mouth" poisoning, poison is adsorbed initially about the pore mouth and extends progressively into the pore as more poison is added. An increasing length of pore becomes inactive. When a fraction σ of the pore surface is poisoned, the catalytically active length of pore is $(1 - \sigma)L$. No reaction, only diffusion occurs in the poisoned length of pore, L. Under steady state conditions, the concentration gradient

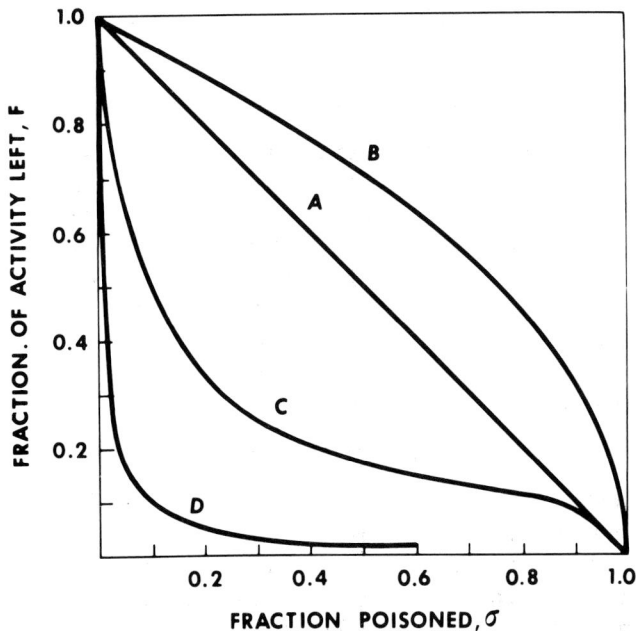

FIG. 12.4. Poisoning of catalyst pores: (*A*) the effect of homogeneous or selective poisoning on a slow reaction, $F = (1 - \sigma)$; (*B*) the effect of homogeneous poisoning on a fast reaction, $F = (1 - \sigma)^{1/2}$; (*C*) selective poisoning on a moderately fast reaction, Eq. (73), $h_1 = 10$; (*D*) selective poisoning on a fast reaction, $F = 1/(1 + \sigma h_1)$, $h_1 = 100$ (*35*). [After Wheeler, in *Catalysis* 2 by Paul H. Emmett, Copyright 1955 by Reinhold Publishing Corp., by permission of Van Nostrand Reinhold Company.]

in this poisoned length will be linear and, therefore, equal to $(c_0 - c_1)/\sigma L$, where c_0 is the concentration of reactant at the pore mouth and c_1 is the concentration at the end of the poisoned length, σL. Here we have assumed Knudsen diffusion [see Eq. (14)]. Now the rate of diffusion through the poisoned length must equal the rate of reaction in the unpoisoned length $(1 - \sigma)L$. In the reaction rate expression, Eq. (26), the intrinsic rate constant k_1, remains the same as for the unpoisoned pore, but L is replaced by $(1 - \sigma)L$. Thus, we have

$$\pi r^2 D(c_0 - c_1)/\sigma L = \pi r^2 Dc_1[h_1(1 - \sigma)/(1 - \sigma)L]\tanh[h_1(1 - \sigma)] \qquad (71)$$
$$\equiv \pi rc_1(2rk_1D)^{1/2}\tanh[h_1(1 - \sigma)],$$

where h_1, as before, is the dimensionless diffusion parameter for the unpoisoned pore of length L. Equation (71) may be solved for c_1 and the value put back into the left-hand side of Eq. (71) to give the reaction rate in the poisoned pore,

$$R_{1/2} = \pi rc_0(2rk_1D)^{1/2}\tanh[h_1(1 - \sigma)]/(1 + \sigma h_1). \qquad (72)$$

Dividing Eq. (72) by the rate in an unpoisoned pore, $(\pi r^2 c_0\, Dh_1\, \tanh h_1)/L$, we obtain

$$F = \{\tanh[h_1(1 - \sigma)]/\tanh h_1\}[1/(1 + \sigma h_1)]. \tag{73}$$

For very slow reactions, h_1 is very small, and $F \to (1 - \sigma)$, which is identical with the limit approached in the adsorption of poisons homogeneously. For fast reactions, however, $[h_1(1 - \sigma) \gtrsim 2]$, $F \to 1/(1 + \sigma h_1)$, which differs from that for homogeneous adsorption. Curve C in the poisoning curves of Fig. 12.4 represents the case for selective poisoning with an intermediate value of h_1 ($= 10$), corresponding to Eq. (73); and Curve D is the case for very fast reaction when $F = 1/(1 + \sigma h_1)$ with $h_1 = 100$.

At sufficiently high temperatures, when k_1 is very large, the reaction rate, Eq. (72), reduces to

$$R_{1/2} = \pi r^2 D(c_0/\sigma L), \tag{74}$$

which is simply the rate of diffusion through the poisoned pore-mouths at maximum gradient $c_0/\sigma L$. The temperature coefficient of such catalysts is small, since $D \sim T^{1/2}$. Thus the apparent activation energy on highly active catalysts with poisoned pore mouths will be essentially zero. On the other hand, when the temperature is sufficiently low that k_1 is very small, the reaction rate, Eq. (72), reduces to

$$R_{1/2} = 2\pi r L(1 - \sigma)k_1 c_0, \tag{75}$$

which is the maximum reaction rate for the unpoisoned portion of the pore. In this case, diffusion effects are completely absent and the measured activation energy will be the true activation energy.

12.4   Selectivity and Diffusion

In Section 11.5, we have considered selectivity in three types of complex reactions in the absence of diffusion effects. In this section, we consider the changes in selectivity in the same three types of reactions in the presence of diffusion effects ($1, 15, 35$–38). It was shown in Section 11.5 for Type I reactions,

$$\begin{aligned} A &\overset{k_{v_1}}{\to} B, \\ C &\overset{k_{v_2}}{\to} D, \end{aligned} \tag{76}$$

where the selectivity factor is $\sigma = k_{v_1}/k_{v_2}$, that the relationship between the fraction of A reacted α_A and the fraction of C reacted α_C is $\alpha_A = 1 - (1 - \alpha_C)^\sigma$. When diffusion effects are present, we may write the first-order reaction rate in a half pore [see Eq. (61)] as $2\pi r L k_1 c_0\, f$, where k_1 is the reaction velocity constant per unit area of pore surface. The ratio of reaction rates in a Type I reaction is then,

$$\frac{R_A}{R_C} = \frac{2\pi r L k_{1A} f_A c_{0A}}{2\pi r L k_{1B} f_C c_{0C}} = \frac{k_{1A} f_A c_{0A}}{k_{1B} f_C c_{0C}}, \tag{77}$$

where

$$f_A = (1/h_{1A}) \tanh h_{1A}, \qquad f_C = (1/h_{1C}) \tanh h_{1C},$$
$$h_{1A} = L(2k_{1A}/rD_A)^{1/2}, \qquad h_{1C} = L(2k_{1C}/rD_C)^{1/2}, \tag{78}$$

and c_{0A} and c_{0C} are the concentrations of A and C at the pore mouth. The terms $k_{1A} f_A$ and $k_{1C} f_C$ may be taken as the true rate constants modified by the effectiveness factors when diffusion effects are present. Their ratio may be defined as the selectivity factor. For large pores or small k_{1A} and k_{1C}, or both, h_{1A} and h_{1C} are small, and $f_A = f_C \to 1$. In this case, the selectivity factor is $\sigma = k_{1A}/k_{1C}$ as in Section 11.5, and diffusion effects are absent. For small pores or large k_{1A} and k_{1C}, or both, h_{1A} and h_{1C} are large, and the observed selectivity factor is

$$\sigma_{obs} = (k_{1A} D_A/k_{1C} D_C)^{1/2}. \tag{79}$$

Equation (79) shows that σ_{obs} may be greater or less than it would be on a plane surface in the absence of diffusion effects. Usually, however, the ratio of the diffusivities is less than the ratio of the intrinsic rate constants, and the selectivity ratio drops as the effectiveness factors decrease. Whichever reaction is favored on a plane surface will, therefore, in most cases, be less favored in a pore where diffusion effects are present. If, for example, $\sigma = 9$ for the reactions on a plane surface, the value will drop to $\sigma_{obs} = 3$ in a small pore with $D_A = D_C$.

For reactions of Type II,

$$A \overset{k_{v_1}}{\underset{k_{v_2}}{\Big\langle}} \begin{array}{l} B \\ C \end{array} \tag{80}$$

diffusion effects have no effect on the selectivity factor, which remains $\sigma = k_{v_1}/k_{v_2}$, so long as the orders of the two reactions are the same. No matter what concentration changes take place in the pore, the ratio of A going to B and to C remains constant. However, if the reaction to form B were first order and the reaction to form C were second order or higher, then the observed selectivity factor would increase with decreasing effectiveness factor. As the concentration of A drops, the reaction to form B would fall less rapidly than the higher order reaction to form C. Thus, if B is the desired product, its yield should be increased by operating with pellets of larger size or with smaller pores.

The most important complex reaction is Type III,

$$A \overset{k_{v_1}}{\to} B \overset{k_{v_2}}{\to} C, \qquad \sigma = k_{v_1}/k_{v_2}. \tag{81}$$

For reaction in a half pore, we find, from Eq. (23), that the diffusion equation for A is

$$\pi r^2 D_A(d^2 c_A/dz^2) = 2\pi r k_{1A} c_A \tag{82}$$

and, in a similar manner, for B,

$$\pi r^2 D_B(d^2 c_B/dz^2) = 2\pi r(k_{1B} c_B - k_{1A} c_A). \tag{83}$$

Equation (82) is solved, as before, giving the expression for c_A, Eq. (24). Substituting the value for c_A into Eq. (83) and solving the second-order differential equation, we find

$$\frac{c_B}{c_{0B}} = \left(1 + \frac{c_{0A}}{c_{0B}} \frac{\sigma}{\sigma - 1}\right) \frac{\cosh\{(h_{1A}/\sqrt{\sigma})[1 - (z/L)]\}}{\cosh(h_{1A}/\sqrt{\sigma})}$$

$$- \frac{c_{0A}}{c_{0B}} \frac{\sigma}{\sigma - 1} \frac{\cosh h_{1A}[1 - (z/L)]}{\cosh h_{1A}} \tag{84}$$

where σ is defined as k_{1A}/k_{1B}, the ratio of the intrinsic rate constants per unit of pore area. In this expression, it is assumed that $D_A = D_B = D$, so that $h_{1A}/\sqrt{\sigma} = h_{1B} = L(2k_{1B}/rD)^{1/2}$. The net rate at which B flows out of the pore mouth is equal to the rate of formation of B, and the net rate at which A flows into the pore mouth is the total rate of reaction of A. The ratio of these two rates is the number of moles of B formed in the pore per mole of A reacted,

$$-\frac{\pi r^2 D(dc_B/dz)_{z=0}}{\pi r^2 D(dc_A/dz)_{z=0}} = -\frac{dc_{0B}}{dc_{0A}} = \frac{\sigma}{\sigma - 1} - \left(\frac{c_{0B}}{c_{0A}} + \frac{\sigma}{\sigma - 1}\right) \frac{1}{\sqrt{\sigma}} \frac{\tanh(h_{1A}/\sqrt{\sigma})}{\tanh h_{1A}}. \tag{85}$$

Now Eq. (85) may be looked upon as the change in c_B per unit change in c_A at the pore mouth. Thus, it is comparable to Eq. (45) of Chapter XI for the same change in the absence of diffusion. In fact, when diffusion effects are absent, Eq. (85) applies at any point in the pore because c_{0A} and c_{0B} are then nearly constant throughout the pore and h_{1A} and h_{1B} are small so that

$$-dc_{0B}/dc_{0A} = 1 - (1/\sigma)(c_{0B}/c_{0A}). \tag{86}$$

Equation (86) is identical with Eq. (45) of Chapter XI. As in Eq. (46) of Chapter XI, we find by integration the relationship between c_{0A} and c_{0B}, which can be put easily in terms of α_A, the fraction of A reacted and α_B, the fraction of A reacted to form B. We repeat Eq. (46) of Chapter XI,

$$\alpha_B = [\sigma/(\sigma - 1)](1 - \alpha_A)[(1 - \alpha_A)^{(1-\sigma)/\sigma} - 1]. \tag{87}$$

If diffusion effects are present, and h_{1A} is large ($\gtrsim 3$), then Eq. (85) becomes

$$- dc_{0B}/dc_{0A} = \sqrt{\sigma}/(1 + \sqrt{\sigma}) - (1/\sqrt{\sigma})(c_{0B}/c_{0A}), \qquad (h_{1A} \gtrsim 3), \qquad (88)$$

which gives, for the relationship between c_{0A} and c_{0B} in terms of α_A and α_B,

$$\alpha_B = [\sqrt{\sigma}/(\sqrt{\sigma} - 1)](1 - \alpha_A)[(1 - \alpha_A)^{(1 - \sqrt{\sigma})/\sqrt{\sigma}} - 1]. \qquad (89)$$

Let us examine the case when $\sigma > 1$, that is, when B is favored. A comparison of Eqs. (87) and (89) shows that α_B will be larger in the absence of diffusion effects than in their presence. Thus, for a given conversion of A, the fraction of A converted to B will be less in catalysts with small pores.

It is instructive to consider the effect of pellet size on activity and selectivity. There are three possible cases. First, if activity and yields of B are independent of pellet size at constant total conversion of A, then obviously pore structure is completely available to the reaction and therefore has no influence on selectivity. Secondly, if activity and yields of B are both greater on small pellets than on large ones, pore structure must have an effect on selectivity. Equation (85) shows that selectivity is most sensitive to changes in h_{1A} and h_{1B} ($= h_{1A}/\sqrt{\sigma}$) in the region $0.3 \lesssim h_{1A}, h_{1B} \lesssim 3.0$. When h_{1A} and h_{1B} are less than 0.3, the ratio of the hyperbolic tangent terms approaches $1/\sqrt{\sigma}$. When h_{1A} and h_{1B} are greater than about 3.0, the hyperbolic tangent terms approach the constant value 1.0. Thus, if pellet size effects both activity and selectivity, then either h_{1A} or h_{1B} or both must lie in the transition range 0.3 to 3.0. Thirdly, if activity depends on pellet size but selectivity is independent of pellet size, then the pore structure must have decreased the selectivity severely to the point where it has reached a constant low value corresponding to h_{1A} and h_{1B} greater than about 3.0. Only a small fraction of the pore surface is available to the reaction. In this case, yields should be improved by drastic reduction of pellet size (e.g., fluid-bed operation), bringing either h_{1A} or h_{1B} or both into the transition region of 0.3 to 3.0.

Ostergaard (*39*) has studied the effect of temperature gradients, caused by exothermic reactions, on selectivity in Type II reactions. He adopted a flat-plate model and showed that selectivities for first-order reactions are significantly altered by temperature gradients. It will be recalled that Type II reactions operating isothermally showed no changes in selectivity as a result of diffusion effects unless the kinetic orders of the two reactions are different.

12.5 Polyfunctional Catalysis and Diffusion

In Section 11.5, we discussed the reaction sequence

$$
\begin{array}{ccc}
A & B & C \\
\updownarrow & \updownarrow & \updownarrow \\
A_{ads} \rightarrow & B_{ads} \rightarrow & C_{ads}.
\end{array}
\qquad (I)
$$

We showed that it could be represented by the gas-phase analog

$$A \xrightarrow{k_{v_1}} B \xrightarrow{k_{v_2}} C \tag{II}$$

when surface chemical reactions are slow in comparison with adsorption–desorption. However, when adsorption–desorption is slow in comparison with surface chemical reactions (40, 41), the gas-phase analog becomes

$$
\begin{array}{c}
B \\
k_{v_1} \nearrow \quad \searrow k_{v_2} \\
A \xrightarrow{\;\;k_{v_3}\;\;} C
\end{array}
\tag{III}
$$

In the most general case, all reactions in Scheme I would be represented as reversible; and reaction Schemes I, II, and III become special cases of this general case. In Scheme II, gas-phase B is a well-defined intermediate, but Scheme III shows that under other conditions C may be formed by the two reactions

$$
\begin{aligned}
A \to A_{ads} \to B_{ads} \to C_{ads} \to C, &\qquad A \xrightarrow{k_{v_3}} C, \\
B \to B_{ads} \to C_{ads} \to C, &\qquad B \xrightarrow{k_{v_2}} C.
\end{aligned}
\tag{90}
$$

In this case, gas phase B is not necessarily an intermediate.

When the conversion steps A → B and B → C are catalyzed by two distinct kinds of sites, X and Y, the reaction scheme becomes in the most general case,

$$
\begin{array}{ccccccc}
A & & B & & & C \\
\updownarrow & & \nearrow \searrow & & & \updownarrow \\
A\!-\!X & \rightleftharpoons & B\!-\!X & & B\!-\!Y \rightleftharpoons C\!-\!Y
\end{array}
\tag{IV}
$$

B must have at least a transitory existence in the gas phase. Thus gas-phase B is a well-defined intermediate in all special cases of reaction Scheme IV which illustrates a *bifunctional catalyst system*. Polyfunctional catalysis allows a wide variety of syntheses to be carried out in a single step. Weisz and Swegler (42) have shown that a catalyst composed of platinum and silica–alumina behaves bifunctionally with respect to the conversion of hexadecane into a mixture of octanes. The supported platinum dehydrogenates the hexadecane to hexadecene which is cracked to octenes on the silica–alumina. In the presence of hydrogen, the octenes are hydrogenated to octanes on the platinum. Heinemann et al. (43) have given an example of a bifunctional catalyst which converts heptane into butane in one operation.

Not all polyfunctional catalyst systems show significant advantages over systems in which the chemical processes are carried out separately. In the irreversible reaction Scheme II, for example, let reaction A → B be catalyzed by catalyst X and the reaction B → C by catalyst Y. Whether catalyst X and catalyst Y are intimately mixed in a single reactor or placed in separate reactors, similar overall conversion of A into C is expected. Such systems are classed as trivial polyfunctional reactions.

On the other hand, if the conversion of A into B is limited by chemical equilibrium according to the scheme

$$A \underset{k_{v'_1}}{\overset{k_{v_1}}{\rightleftarrows}} B \overset{k_{v_2}}{\longrightarrow} C, \qquad (V)$$
$$ X Y$$

considerably higher conversions of A into C may be obtained in a single reactor with a bifunctional catalyst consisting of X and Y. When catalysts X and Y are in separate reactors, the conversion to B in the first reactor cannot exceed equilibrium. If the equilibrium conversion of B is small, then the yield of C in the second reactor will also be small. However, if X and Y are intimately mixed in the same reactor, the small concentration of B limited by equilibrium may be continuously converted to C, so that the reaction sequence proceeds. These statements can be put in quantitative form. When X and Y are in separate reactors, the maximum possible fractional conversion of A into C at infinite time (t) is easily shown to be

$$y_{\max} = 1/[1 + (1/K)], \qquad t = \infty, \qquad (91)$$

where K is the equilibrium constant for A ⇆ B. If K is small, $y_{\max}$ will be small. However, if catalysts X and Y are intimately mixed in the same reactor, reaction Scheme V takes place in a continuous manner. Under these conditions, $y_{\max}$ is unity, even though the concentration of intermediate B may be limited to a very low value by equilibrium. In the above considerations, diffusion of B from site X to site Y has been considered to be fast in comparison with surface reactions. If the rate of diffusion of B through the gas phase is comparable to the rate of the surface reactions, additional equations involving the transport of B are needed.

Gunn and Thomas (*41*) have considered diffusion effects in relation to reaction Schemes II, V, and VI,

$$A \underset{X}{\overset{}{\rightleftarrows}} B \overset{X}{\underset{Y}{\diagdown}} \overset{C}{\underset{D}{}} \qquad (VI)$$

They distinguish two types of catalyst formulation. The first type is a single, mixed catalyst comprising pellets of catalyst X mixed with pellets of catalyst Y. The second type is a compounded catalyst in which X sites and Y sites exist in the same catalyst pellet. The compounded catalyst may be prepared by coimpregnation of the active ingredients of catalyst X and catalyst Y on the same support, or by mechanically compacting particles of the two catalysts into a single pellet. In the first formulation, Gunn and Thomas assume that diffusion effects within the pores of the pellets are appreciable and diffusion effects between pellets of catalyst X and catalyst Y are relatively insignificant. In the second formulation, they assume that diffusion effects from sites X to sites Y are the most significant. The reader is referred to their paper for complete details. In the following, we sketch the case of reaction Scheme V, $A \leftrightarrows B \rightarrow C$, on a compounded catalyst with X sites and Y sites existing in the same catalyst pellet.

We assume a spherical porous pellet of radius R composed of small particles of catalyst X and Y compressed together. Such pellets will consist of a honeycomb of small voids between the ultimate particles of X and Y. It is supposed that the transport of species through these voids is slow and comparable to the rates of surface chemical reactions. We are confronted with the problem of diffusion through the interparticle space. Each pellet is assumed to contain a fraction ξ of the X component. Referring to Eq. (45), we find that the diffusion equation for A may be written,

$$d^2c_A/dr^2 + (2/r)(dc_A/dr) - (\xi k_{v_1}/D_e)c_A + (\xi k'_{v_1}/D_e)c_B = 0, \qquad (92)$$

where the reversibility of the reaction between A and B in reaction Scheme V is taken into account, D_e is the effective diffusivity for the composite catalyst assumed independent of catalyst composition, and $0 \lesssim r \lesssim R$. A similar equation for B is

$$\frac{d^2c_B}{dr^2} + \frac{2}{r}\frac{dc_B}{dr} + \frac{\xi k_{v_1}}{D_e} c_A - \frac{\xi k'_{v_1}}{D_e} c_B - \frac{(1-\xi)k_{v2}}{D_e} c_B = 0. \qquad (93)$$

The k_v's in Eqs. (92) and (93) are expressed per unit of bulk volume of catalyst pellet of pure X or pure Y. The boundary conditions are,

$$c_A = c_{A0}, \qquad c_B = c_{B0} \quad \text{at} \quad r = R,$$

$$dc_A/dr = dc_B/dr = 0 \quad \text{at} \quad r = 0, \qquad (94)$$

$$c_A, c_B \quad \text{finite} \quad \text{at} \quad r = 0.$$

From Eqs. (92) and (93), the concentrations of A and B as functions of penetration into the pellet can be determined. The concentration of C is determined by a simple material balance. As in Section 12.2, we make use of the fact that the rate of reaction inside a pellet is equal to the rate of diffusion (flux) into the pellet. Equation (48) gives the total flux for a spherical pellet of radius R. Dividing this expression by the volume of the pellet and multiplying by

$(1 - \psi_b)$, where ψ_b is the porosity of the catalyst bed in the reactor, we find the flux per unit reactor volume for component A

$$N_A(X) = [3(1 - \psi_b)/R]D_e(dc_A/dr)_{r=R} \qquad (95)$$

Similar equations can be obtained for components B and C. Explicit expressions for the concentration gradients, dc/dr, are obtained as previously by differentiation of the solutions of Eqs. (92) and (93), and the results are substituted into the flux equations for N_A, N_B, and N_C. In order to obtain profiles of A, B, and C as functions of the length of the reactor, we make use of material balances over an infinitesimal element of length dz of the reactor. The balance for component A is,

$$v_1(dc_{A0}/dz)\,dz = -N_A(X)\,dz, \qquad (96)$$

where v_1 is the superficial velocity of reactant through the reactor. Similar equations for N_B and N_C are

$$v_1(dc_{B0}/dz)\,dz = [-N_B(X) + N_B(Y)]\,dz, \qquad (97)$$

$$v_1(dc_{C0}/dz)\,dz = N_B(Y)\,dz. \qquad (98)$$

Equations (96)–(98) are simultaneous, first-order linear differential equations which can be solved for the concentration profiles of A, B, and C as functions of reactor length. Figure 12.5 compares the concentration profiles of reaction

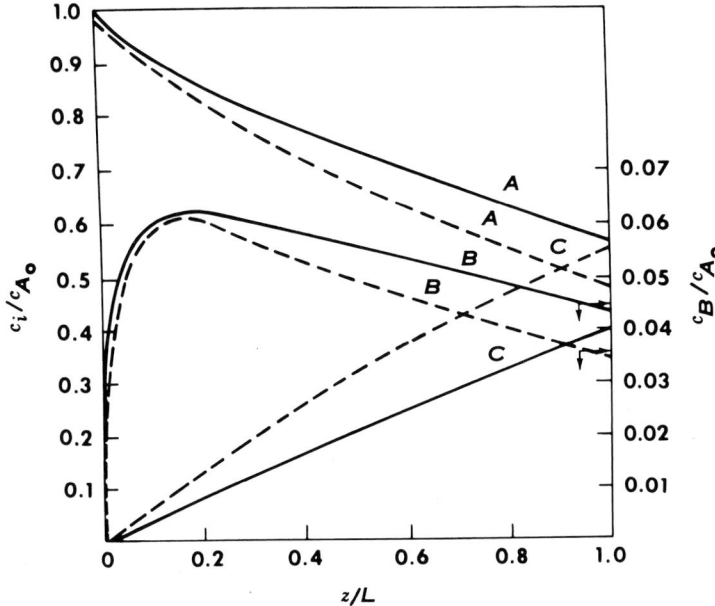

FIG. 12.5. Comparison of compounded (—) and mixed (---) catalysts. [After Gunn and Thomas, *Chem. Eng. Sci.* **20**, 89 (1965). By permission of Pergamon Publishing Company.]

Scheme V, A ⇆ B → C, for compounded catalysts with X and Y sites in the same catalyst pellet and for catalysts consisting of a mixture of discrete pellets of catalyst X and catalyst Y. In both cases, the same values for reaction velocity constants, pellet radius, and diffusivities were assumed. The proper amount of X has been used in each case to give the maximum yield of C. For the discrete mixture, the amount of X was calculated to be 30 %, and for the compounded catalyst, 20 %. The lower proportion of X required for maximum yield of C in the case of the compounded catalyst reveals that compounding brings Y sites to the region where B is formed, which eliminates decrease in reaction rate caused by interpellet diffusion, and allows B to form C more rapidly. Figure 12.5 shows that the rate of disappearance of B, after it has reached a maximum concentration, is faster for the compounded catalyst than for the mixed catalyst. The rate of disappearance of A is also greater, while the rate of formation of C increases. Figure 12.5 also shows that the yield of C is greater for compounded catalyst than for mixed catalyst, illustrating the marked effect which interpellet diffusion has on the formation of desired product.

REFERENCES

1. Satterfield, C. N., and Sherwood, T. K., "The Role of Diffusion in Catalysis." Addison-Wesley, Reading, Massachusetts 1963.
2. Peterson, E. E., "Chemical Reaction Analysis." Prentice-Hall, Englewood Cliffs, New Jersey, 1965.
3. Carberry, J. J., *Am. Inst. Chem. Eng. J.* 6, 460 (1960).
4. Yek, G. C., *J. Chem. Eng. Data* 6, 526 (1961).
5. Bradshaw, R. D., and Bennett, C. O., *Am. Inst. Chem. Eng. J.* 7, 48 (1961).
6. de Boer, J. H., *Colston Res. Symp.* Bristol (1958).
7. McBain, J. W., *J. Am. Chem. Soc.* 57, 699 (1935).
8. Thomas, J. M., and Thomas, W. J., "Introduction to Heterogeneous Catalysis," Chapter 4. Academic Press, New York, 1967.
9. Anderson, R. B., et al., *Ind. Eng. Chem.* 39, 1618 (1947); *J. Am. Chem. Soc.* 69, 3115 (1947).
10. Emmett, P. H., *Advan. Catalysis* 1, 65 (1948).
11. Everett, D. H., *Colston Res. Symp.*, Bristol (1954).
12. Ritter, H. L., and Drake, L. C., *Ind. Eng. Chem. Anal. Ed.* 17, 787 (1945).
13. Shull, C. G., *J. Am. Chem. Soc.* 70, 1405 (1948).
14. Wakao, N., and Smith, J. M., *Chem. Eng. Sci.* 17, 825 (1963).
15. Wheeler, A., *Advan. Catalysis* 3, 258 (1951).
16. Knudsen, M., *Ann. Physik* 28, 75, 999 (1909).
17. Loeb, L. B., "Kinetic Theory of Gases," p. 252. McGraw-Hill, New York, 1927.
18. Taylor, H. S., and Glasstone, S., "A Treatise on Physical Chemistry," p. 100. van Nostrand, Princeton, New Jersey, 1951.
19. Moelwyn-Hughes, E. A., "Physical Chemistry," p. 58. Pergamon Press, New York, 1961.

20. Bradshaw, R. D., and Bennett, C. O., *Am. Inst. Chem. Eng. J.* **7**, 48 (1961).
21. Kammermeyer, K. A., and Rutz, A., *Chem. Eng. Prog. Symp. Ser.* **55**, No. 24, 163 (1959).
22. Whang, H. Y., ScD Thesis in Chem. Eng., Massachusetts Inst. Technology (1961).
23. de Boer, J. H., *Pittsburgh Catalysis Soc., 7th Ann. Symp.*, Pittsburgh, April 24–26 (1968).
24. Poiseuille, J., *Inst. France Acad. Sci. Mem.* **9**, 433 (1946).
25. Hagen, G., *Ann. Physik* **46**, 425 (1939).
26. Thiele, E. W., *Ind. Eng. Chem.* **31**, 916 (1939).
27. Bischoff, K. B., *Ind. Eng. Chem. Fundamentals* **5** (1), 135 (1966).
28. Mingle, J. O., and Smith, J. M., *Am. Inst. Chem. Eng. J.* **7**, 243 (1961).
29. Carberry, J. J., *Am. Inst. Chem. Eng. J.* **8**, 557 (1962).
30. Aris, R., *Chem. Eng. Sci.*, **6**, 262 (1957).
31. Weisz, P. B., and Hicks, J. S., *Chem. Eng. Sci.* **17**, 265 (1962).
32. Damköhler, G., *Z. Phys. Chem.* **A193**, 16 (1943).
33. Weisz, P. B., *Z. Phys. Chem. Neue Folge* **11**, 1 (1957).
34. Carberry, J. J., and Gorring, R. L., *J. Catalysis* **5**, 529 (1966).
35. Wheeler, A., *Catalysis* **2**, 153 (1955).
36. Thomas, J. M., and Thomas, W. J., "Introduction to Heterogeneous Catalysis," Chapter 7. Academic Press, New York, 1967.
37. Van de Vusse, J. G., *Chem. Eng. Sci.* **21**, 631 (1966).
38. Carberry, J., *J. Chem. Eng. Sci.* **17**, 675 (1962).
39. Ostergaard, K., *Proc. 3rd Intern. Congr. Catalysis*, Paper II, p. 12, North-Holland, Amsterdam (1964).
40. Weisz, P. B., *Advan. Catalysis* **13**, 137 (1962).
41. Gunn, D. J., and Thomas, W. J., *Chem. Eng. Sci.* **20**, 89 (1965).
42. Weisz, P. B., and Swegler, E. W., *Science* **126**, 31 (1957).
43. Heinemann, H., Mills, G. A., Hattman, J. B., and Kirsch, F. W., *Ind. Eng. Chem.* **45**, 130 (1953).
44. Roberts, G. W., and Satterfield, C. N., *Ind. Eng. Chem. Fundamentals* **5**, 317 (1966).
45. Knudsen, C. W., Roberts, G. W., and Satterfield, C. N., *Ind. Eng. Chem. Fundamentals* **5**, 325 (1966).
46. Schmalzer, D. K., *Chem. Eng. Sci.* **24**, 615 (1969).

XIII

GEOMETRIC FACTORS AND CATALYSIS

13.1 Introduction

The separation of factors affecting catalysis into geometric factors and electronic factors is arbitrary. It is an *interim* expedient until more basic approaches become possible. Obviously, geometric factors in the last analysis may be the result of the action of electronic factors. Historically, the geometric factor has referred exclusively to the effect on catalysis of lattice spacing. But, for convenience, we shall include other effects such as those of (i) arrangements of atoms in the face exposed at the surface, (ii) lattice distortions, and (iii) the state of dispersion of supported catalysts. The effects on catalysis of lattice imperfections more conventionally associated with electronic properties will be considered in the two following Chapters XIV and XV.

13.2 Lattice Spacings and Catalysis

EARLY WORK

In 1932, Sherman and Eyring (*1*) calculated the energy of activation of the chemisorption of hydrogen on carbon, and associated this energy with the ortho–para hydrogen conversion. They employed the semiempirical methods developed initially by Eyring and Polanyi (*2*) and Eyring (*3, 4*) which involve the construction of potential energy surfaces. As a start, they calculated activation energies for the analogous processes of the hydrogenation of ethylene and acetylene. Values used for equilibrium distances and strengths of bonds are given in Table 13.1. For the purposes of these approximate calcu-

TABLE 13.1

BOND DISTANCES AND STRENGTHS

Bond	Equilibrium distance, Å	Strength, kcal/mole
C—H	1.13	96.0
H—H	0.76	101.5
C—C	1.54	73.5
C=C	1.54	123.0
C≡C	1.54	162.0

lations, the carbon–carbon distances for the single, double, and triple bond were all taken equal to the value for ethane. Two cases were calculated: one assuming that the effective carbon–carbon bond broken was the difference between the strengths of a double and a single bond (49.5 kcal); and the other, the difference between a triple and a double bond (39 kcal). The Morse functions for H—C, C—C, and C—H are known, and it was assumed that 10 % of the total bond energy was coulombic. Calculations of the potential energy surfaces are simplified by assuming that the hydrogen molecule approaches the two carbon atoms in such a manner that the system is always symmetrical about the perpendicular bisector of the line joining these atoms. When a definite value for the carbon–carbon distance is taken, the potential energy can be expressed as a function of two parameters: the distance between the two hydrogen atoms and that between the centers of the two carbon and the two hydrogen atoms. Activation energies for the hydrogenation of ethylene and acetylene for the assumed equilibrium carbon–carbon distance (1.54 Å) are calculated and found to be 51.5 and 46.4 kcal/mole, respectively at $0°K$.

The process for the adsorption of hydrogen on two surface carbon atoms is considered analogous,

$$\begin{array}{ccc} \text{H—H} & & \text{H} \quad \text{H} \\ & \xrightarrow{} & | \quad | \\ \text{C—C} & & \text{C—C} \end{array}$$

Using the same value for C—H bond strength (96 kcal) as above, potential energy surfaces are constructed for a series of carbon–carbon distances between 1.5 and 4.5 Å. The carbon–carbon bond broken was taken as the difference between a triple and double bond at the appropriate distance. The

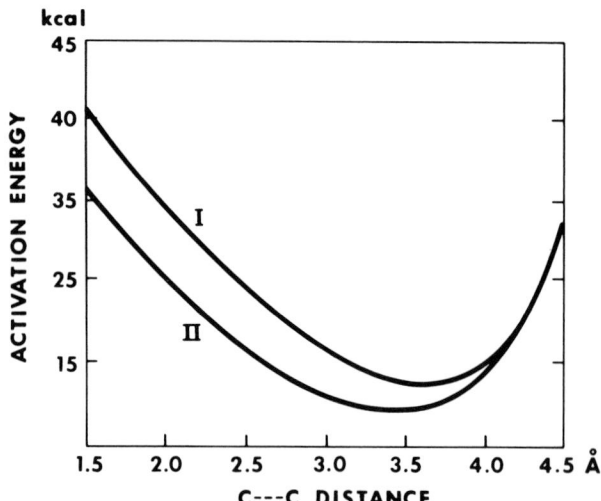

Fig. 13.1. Dependence of activation energy for adsorption of a hydrogen molecule on the C—C distance (*1*). [After Sherman and Eyring, *J. Am. Chem. Soc.* **54**, 2661 (1932). By permission of The American Chemical Society.]

result is shown in Fig. 13.1, curve I. The minimum in the activation energy curve is seen to occur at a carbon–carbon distance of 3.6 Å. An important result of the calculations of Sherman and Eyring is that the minimum in the activation energy occurs at a carbon–carbon distance of approximately 3.6 Å, regardless of the assumed nature of the carbon–carbon bond broken.

We note that the heat of adsorption is the algebraic sum of the strengths of the H—H and carbon–carbon bonds broken and the two C—H bonds formed. If a single carbon–carbon bond is broken, the data of Table 13.1 give a heat of adsorption of 17 kcal. If no carbon–carbon bond is broken, the heat is 90.5 kcal. It is presumed that carbon–carbon bonds in the surface are not broken, and, therefore, the calculated heat of adsorption is much greater than the experimental value which is approximately 2 kcal. On the assumption of no breaking of carbon–carbon bonds and using the experimental value for the heat of adsorption, the strength of the C—H adsorption bond is calculated to be 51.8 kcal. Activation energies as a function of carbon–carbon distance were recalculated on this new basis and the result is shown in curve II of Fig. 13.1. The minimum activation energy occurs once again at 3.6 Å, so the carbon–carbon distance for the minimum appears to be insensitive both to variations in the nature of the carbon–carbon bond broken and in the strength of the C—H bond. The activation energy at the minimum is 8.8 kcal, and is higher than the experimental value which is probably no higher than 2 kcal. But the agreement is as good as can be expected in view of the approximations made. By taking coulombic energies of the C—H and C—C bonds as 14% of the total binding energy instead of 10%, the calculated activation energy is reduced to 5.6 kcal. Detailed potential energy surfaces show that the distance between the hydrogen atoms remains unchanged until the molecule is very close to the line joining the carbon atoms and then expands to the distance, 3.6 Å. The fact that the optimum distance (3.6 Å) is considerably larger than the normal H—H distance may be explained qualitatively as follows: When the carbon–carbon distance is very large, the reaction would virtually involve the dissociation of the hydrogen molecule and the activation energy would be very large. When the carbon–carbon distance is equal to the normal H—H distance, hydrogen on one carbon atom will also interact strongly with the other carbon atom. The London equation shows that such cross-interaction leads to a high activation energy. At an intermediate carbon–carbon distance, there must be a minimum, and the distance 3.6 Å turns out to be that point.

For the ortho-para hydrogen conversion to take place, the Langmuir–Hinshelwood model requires that a second hydrogen molecule be adsorbed on a pair of carbon atoms adjacent to the pair on which the first molecule is adsorbed. Because of repulsion by the first two hydrogen atoms already adsorbed, it might be expected that a higher activation energy would be required for the second molecule of hydrogen. This is not true, however, because of the large carbon–carbon distance, so that the activation energy for the adsorption of two successive molecules is just that for the individual processes. Sherman and Eyring (1) also consider the case of the Rideal model in which one hydrogen molecule is adsorbed and the other approaches it from the gas phase during interaction.

In 1936, Horiuti et al. (5) calculated activation energies for adsorption of hydrogen on nickel, using the two possible Ni–Ni distances, 2.49 and 3.52 Å, as represented in Fig. 13.2. Coulombic energies were taken as 11% of the total binding energy for H—H, 37% for Ni–Ni, and 24%, the mean of the other two, for the Ni—H bond. For the 2.49 Å separation, the activation energy was found to be 75 kcal, and that for 3.52 Å was 57 kcal. Thus, adsorp-

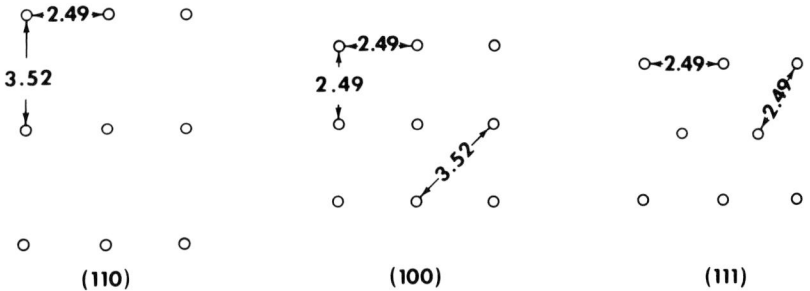

FIG. 13.2. Spacings in three lattice planes in a crystal of nickel.

tion should occur preferentially on the (100) and (110) planes, where the activation energy is lower.

Lattice spacing may also affect the magnitude of the heat of adsorption. For example, we assume adsorption of ethylene in the form

$$CH_2\text{---}CH_2$$
$$Ni\text{------}Ni$$

We take the Ni–C bond distance at 1.82 Å, the same as it is in $Ni(CO)_4$; the carbon–carbon bond at its normal length 1.54 Å, and the Ni–Ni bond at the two lengths 2.49 and 3.52 Å. The Ni–C–C angle may now be calculated. For the Ni–Ni length 2.49 Å, the Ni–C–C angle is close to 105° and is 123° for the 3.52 Å spacing. The former angle is close to the strain-free condition 109° 28', and the small amount of strain may be alleviated by twisting, so that the Ni–Ni and C—C axes are slightly inclined toward one another. Considerably more strain exists when the spacing is 3.52 Å. A higher heat of adsorption would be expected on the least strained complex at 2.49 Å spacing. Now, a higher catalytic reactivity of ethylene would be expected on the spacing with a lower heat of adsorption, that is, at a spacing of 3.52 Å.

In 1939, Beeck et al. (6) produced oriented evaporated metal films and discovered large differences in catalytic activity between oriented and unoriented films. For example, nickel films with the (110) face preferentially exposed were

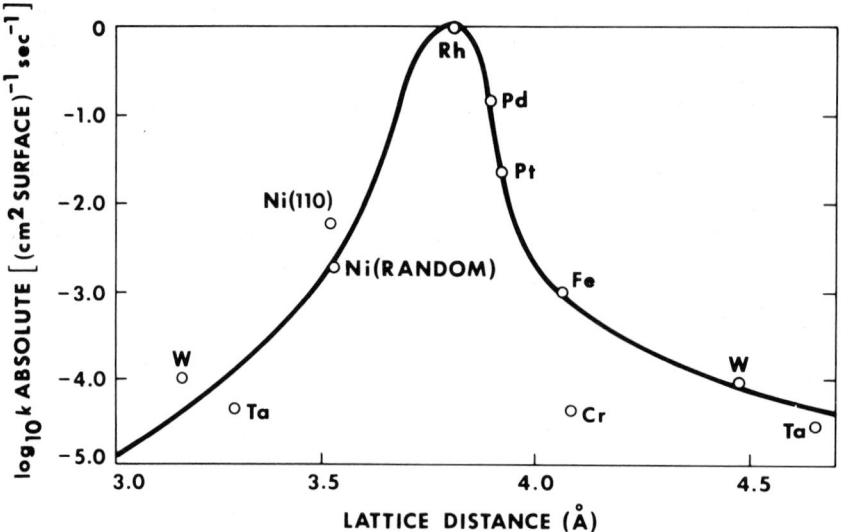

FIG. 13.3. Activity at 0°C of a series of thin metallic films (per unit area) as a function of the distance between particular pairs of atoms in the lattice (7). [Beeck, *Discussions Faraday Soc.* **8**, 118 (1950). By permission of The Faraday Society.]

five times more active for ethylene hydrogenation than unoriented films which presumably had equal amounts of (100), (110), and (111) faces exposed. We note that the (110) face contains the spacing 3.52 Å which, as deduced above, had the possibility of highest catalytic activity, thus emphasizing the importance of a relationship between catalytic activity and lattice spacing. Beeck (7) further discovered that when the logarithm of the activity of metal films in catalyzing ethylene hydrogenation was plotted against lattice spacing, a relatively smooth curve was obtained (Fig. 13.3). Rhodium, with a spacing of 3.75 Å is evidently the most active. We shall find in the next chapter that rhodium again has the highest activity when a similar plot is made against d-character (see Chapter VII) instead of lattice spacing. More recent work by Gwathmey and Cunningham (8) again stresses the importance of lattice spacing in catalysis.

MULTIPLET THEORY OF BALANDIN

Although the inception of the multiplet theory was contemporaneous with the work described above, active interest in it has persisted to the present (9–13), particularly in the Russian school. In this theory, catalytic power is presumed to be determined solely by the suitability of surface geometry, involving both lattice spacing and arrangement of atoms in the surface. The effective catalyst is considered to be a small group or multiplet of neighboring atoms whose configuration is such that they orient the reactant on the surface. The theory has had its greatest application to the reactions of six-membered rings. In the reaction, benzene ⇆ cyclohexane, for example, a sextet of sites is presumed to be required, the benzene skeleton lying flat on the surface, adsorbed by six metal-to-carbon bonds. Only the (111) type planes of face-centered cubic and the habit faces of hexagonal, close-packed crystals exhibit the correct symmetry. Balandin computed the distance between hydrogens attached to the carbon atoms of benzene and the metal atoms from which additional hydrogen atoms could be transferred to the benzene nucleus. When this distance was plotted against the atomic radii of metal atoms, he found that the only metals which exhibited good catalytic activity were those capable of exposing octahedral faces with internuclear distances of nearest-neighbor atoms lying between 2.48 (Ni) and 2.77 Å (Pt). Such metals as calcium, thorium, and iron do not possess the requisite spacing, while nickel, copper, iridium, platinum, palladium, zinc, osmium, and cobalt do. In Fig. 13.4, the adsorption of cyclohexane according to Balandin's theory is illustrated.

There is considerable evidence in support of Balandin's concept, but there is also much which casts doubt on the validity of correlations between bulk lattice parameters and catalysis. For example, an evaporated film of iron, which crystallizes in a body-centered cubic structure, has been shown to be an active

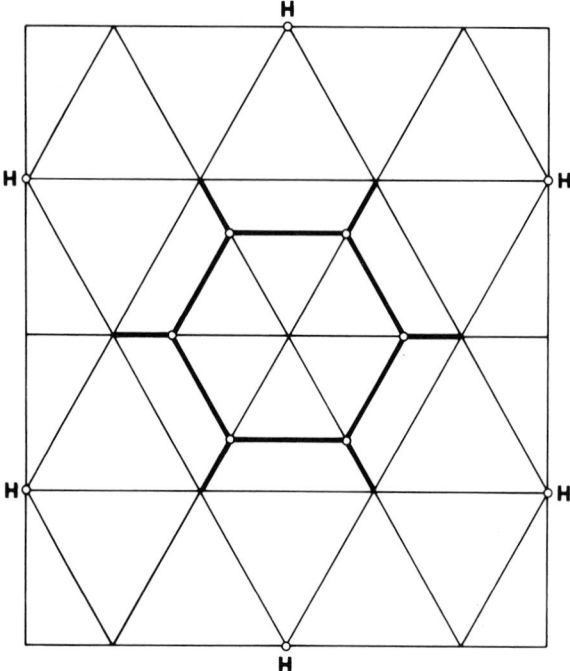

FIG. 13.4. Adsorption of cyclohexane.

catalyst for the above-mentioned hydrogenation–dehydrogenation reaction *(14, 15)*. More serious are the objections raised by the results of low-energy electron diffraction (Section 7.1), electron microscopy, and the investigations of Gwathmey and Cunningham *(8)*, which in some cases reveal considerable disruption of the surface, so that bulk lattice parameters may have little meaning with respect to surface geometry. Even "perfect" crystals show small differences in lattice spacing between the surface layer and the next layer below it. However, there are many cases where no serious disruptions of the surface occur, yet marked variations in catalytic activity among crystal faces are observed. Thus, despite the suspicion surrounding many observations, the importance of the role of lattice spacing and surface geometry in catalysis cannot be denied.

13.3   Dislocations and Catalysis

There is much evidence that lattice imperfections are often the seat of catalytic activity. In this section, we shall present some of the evidence which

correlates one type of lattice imperfection with catalysis, namely the dislocation. The other major category of imperfections, point defects, will be considered in the following chapters.

Dislocations in a crystal are caused by slip. In the ideal case, the simplest
type of slip occurs when planes of atoms slide over one another like cards in
a deck, deforming the crystal. If every atom on one side of a slipped plane has
moved into a position originally occupied by its nearest neighbor in the
direction of slip, we say that the plane has moved one unit of slip. In this ideal
case, crystal perfection viewed atomically has not been disturbed; essentially
all atoms of the crystal possess their original symmetrical environment. In
most real crystals, however, slip is not uniform over a slip plane. Unit slip
may occur over only a *part* of a slip plane, while the remainder of the plane
slips a relatively insignificant amount. Figure 13.5 illustrates a block of crystal

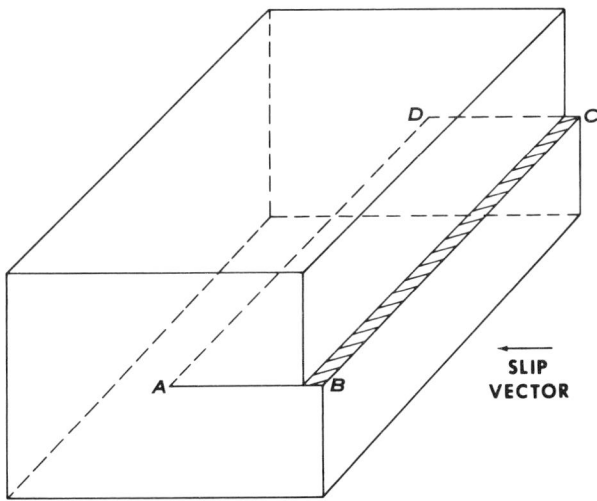

FIG. 13.5. The slip plane producing an edge dislocation. Unit slip has occurred over
the area *ABCD*. The boundary *AD* of the slipped area within the crystal is a dislocation
and is normal to the slip vector (*16*). [After Read, Jr., "Dislocations in Crystals."
Copyright 1953 by McGraw-Hill. Used with permission of McGraw-Hill Book Company.]

in which the lower half has slipped with respect to the upper half. The line *AD*
marks the boundary between the two regions which have slipped by different
amounts. The area of significant slip is *ABCD*. Figure 13.6 shows a possible
arrangement of a plane of atoms viewed along *AD*, i.e., a plane perpendicular
to *AD*. The imperfection along the line *AD*, which is the boundary of the
slipped area *ABCD*, is a type of dislocation known as an *edge dislocation*.
Since the dislocation occurs along a line (*AD*), it is sometimes called a *line*

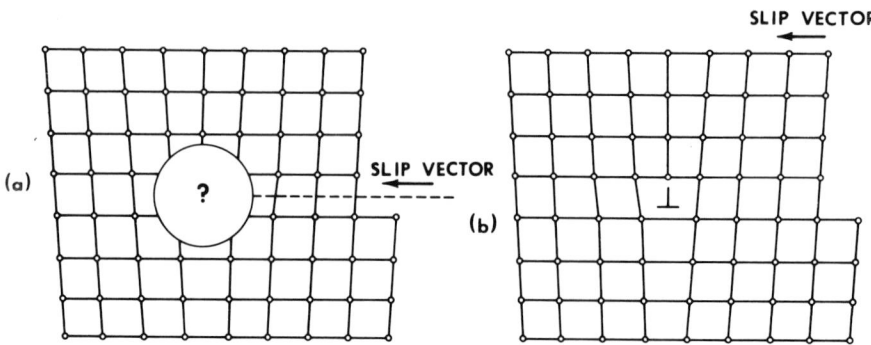

FIG. 13.6. An approximate arrangement of a plane of atoms normal to the line AD of Fig. 13.5. The dislocation represents a region of severe misfit where atoms are not properly surrounded by their neighbors. The symbol $\perp$ denotes the dislocation. The dislocation is the edge of an incomplete atomic plane, hence the name edge dislocation (*16*). [After Read, Jr., "Dislocations in Crystals." Copyright 1953 by McGraw-Hill. Used with permission of McGraw-Hill Book Company.]

dislocation. Note that the slip vector is perpendicular to the dislocation line. Note also that one atom in the plane does not have the proper number of nearest neighbors. The exact arrangement of atoms along AD is not known; a difficult quantum-mechanical calculation would be necessary to find it. However, the approximation in Fig. 13.6 is sufficiently good to show some of the properties of an edge dislocation. The figure reveals that $n + 1$ atomic planes above the slip plane attempt to join n planes below, thus forcing one vertical plane of atoms to terminate on the slip plane. The dislocation is the edge of this extra plane.

Another type of dislocation is the screw dislocation introduced by Burgers

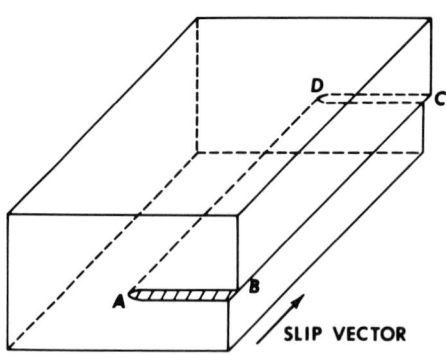

FIG. 13.7. Slip producing a screw dislocation. Unit slip has occurred over $ABCD$. By definition, the screw dislocation AD is parallel to the slip vector (*16*). [After Read, Jr., "Dislocations in Crystals." Copyright 1953 by McGraw-Hill. Used with permission of McGraw-Hill Book Company.]

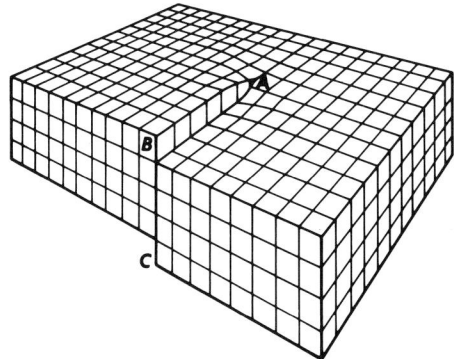

FIG. 13.8. Another view of a screw dislocation. The dislocation AD (of which only end A is visible) is parallel to BC and to the slip vector (*16*). [After Read, Jr., "Dislocations in Crystals." Copyright 1953 by McGraw-Hill. Used with permission of McGraw-Hill Book Company.]

(*17*) in 1939. Figure 13.7 shows the creation of a screw dislocation by slip; the dislocation AD is the boundary within the crystal of the slipped area $ABCD$. The screw dislocation is parallel to the slip vector. In Fig. 13.8, a screw dislocation parallel to a cube edge in a simple cubic crystal is shown and the unit cells are shown as distorted cubes. We can see from this figure why the dislocation is called screw: The crystal consists of a single atomic plane in the form of a helicoid or spiral ramp rather than a set of parallel atomic planes. We note that the same screw dislocation can be produced by slip on any plane containing AD, or on any slip surface ending on AD. For when the dislocation and slip vector are parallel, the slip plane is not uniquely defined. The reader may refer to a more complete and excellent discussion of dislocations given by Read (*16*).

There are many ways of visualizing on an atomic scale the formation of dislocations. For example, instead of imagining an edge dislocation formed by a slip of the lower half of a crystal with respect to the upper half, we may consider the removal of half a plane of atoms below the slip plane or the insertion of half a plane above. A dislocation results similar to that shown in Fig. 13.6. The process requires mass transport and might occur in real crystals by diffusion of vacancies or interstitial atoms.

Duell and Robertson (*18*) produced "superactive" nickel wire by flashing at high temperature, and proposed that lattice vacancies and dislocations were the catalytically active sites. For example, a nickel wire heated to 1330°C in vacuum for 2 min was superactive for the decomposition of formic acid at lower temperatures. At 330 and 430°C, the lowest temperatures investigated, the probability P of decomposition when a formic acid molecule strikes the catalyst surface remained at $P = 1$ for only 20 sec. At temperatures of 530°C and higher, reaction with $P = 1$ lasted for at least 10 min, using formic acid pressures of about 2×10^{-4} mm Hg. Catalytic activities on flashed wires can be increased 10^5 times the normal activity on unflashed wires. The super-activity appears to be frozen into the wire and is not removed by annealing, cooling to room temperature, or by slight oxidation. As mentioned above, superactivity disappears rather quickly on exposure to formic acid. However, the activity reappears without another flashing when formic acid is removed for a time and then re-introduced.

Duell and Robertson quote evidence from other work (*18*) that high, non-equilibrium concentrations of quenched-in vacancies can be produced with their procedures. Furthermore, the formation of dislocations by diffusion and aggregation of vacancies has been observed with the electron microscope (*20, 21*). It is probable that the time required for the aggregation of vacancies at the temperatures employed is less than the duration of superactivity. Thus, according to the authors, both vacancies and dislocations may be the catalytic sites. The authors explain the rapid disappearance of superactivity in the presence of decomposing formic acid by assuming an induced surface mobility of the metal atoms which arises when catalytic action occurs. Thus, reaction may bring about the annihilation of active sites. High activity may be restored by removing formic acid and then allowing sufficient time for more vacancies to diffuse from the bulk to the surface before re-introducing the acid.

We have mentioned already (Section 11.8) the work of Sosnovsky (*22*) in which changes in catalytic activity of silver surfaces for the decomposition of formic acid were found after bombardment with positive argon ions. Drastic changes in the activation energy E and the pre-exponential factor A were observed with a strong compensation effect. Changes in catalytic activity were attributed to an increase in the number of dislocation lines intersecting the surfaces.

Actually, the investigations described above are mostly phenomenological in nature. They merely correlate the effect of various kinds of disturbances of metal structure on catalytic activity. No quantitative correlations between dislocations or vacancies and catalytic activity were made. Indeed, imperfections other than vacancies and dislocations could be responsible for the observed increases in catalytic activity. Hall and Rasé (*23*) have given a more quantitative correlation. Working with freshly cleaned lithium fluoride crystals, they produced different dislocation densities by annealing, cooling, and stressing. Dislocation densities were determined by the etching procedures of Gilman and Johnston (*24*). Dehydrogenation of ethyl alcohol was used as a test reaction. Figure 13.9 shows the correlation which they obtained between dislocation density and catalytic activity. Although this work represents a serious attempt at quantitative correlation of dislocation density and catalyst activity, considerably more systematic work is necessary before it is settled unequivocally that dislocations contribute to catalysis in some catalytic reactions. In fact, Bagg and co-workers (*25*) claim that lines of emergence of dislocations at the surface are unimportant for catalysis.

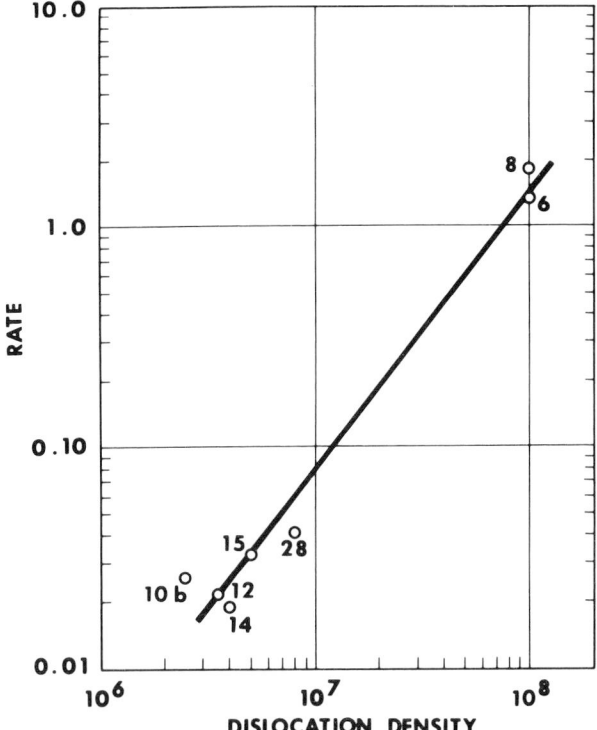

FIG. 13.9. Effect of dislocation density on catalytic activity of lithium fluoride crystals. Rate = [(moles converted/moles charged)(cm³ ethanol/hr)]/(cm² of surface). Dislocation density = No. of dislocations per cm². [After Hall and Rasé, *Nature* **199**, 585 (1963).]

13.4   Determination of Particle Size

As atoms aggregate to form nuclei and crystallites, how does catalytic activity vary with particle size? Is it necessary to have a certain minimum of aggregation exhibiting bulk properties before catalytic activity is observed? Is there a maximum dimension beyond which catalytic activity declines? Do supported catalytic substances disperse uniformly or form clumps? These are some of the fundamental questions which have been asked many times. For valid comparisons, catalytic activity must be referred to the number of *exposed* surface atoms which are responsible for activity. The fundamental measurements are surface area and particle size. We shall discuss first methods for measuring these quantities, and in the next section the relations to catalytic activity. Among the common techniques are: the electron microscope, X ray, adsorption, and magnetic techniques.

ELECTRON MICROSCOPE

With the electron microscope, particle-size measurements can be made, for example, by superimposing on a photographic print of the catalyst surface a lantern slide on which circles of varying diameter have been photographed. Circles are found which most closely match the sizes of the particles to be measured. This technique has been successfully employed for determining the particle size of platinum supported on alumina (*26*). Designating the particle diameter by d_i and the number of particles of a given diameter by n_i, we obtain the number average diameter for a size distribution from the expression,

$$\bar{d}_n = \sum n_i d_i / \sum n_i, \tag{1}$$

the surface average diameter from

$$\bar{d}_s = \sum n_i d_i^3 / \sum n_i d_i^2, \tag{2}$$

and the volume average diameter from

$$\bar{d}_v = \sum n_i d_i^4 / \sum n_i d_i^3. \tag{3}$$

X RAY

By the X-ray technique (*26*), a quantity $\bar{\delta}$, defined as the average of the cube root of the crystallite volume, is determined from X-ray line broadening. This is a volume average. For cubic crystals, $\bar{d}_v = \bar{\delta}$, where $\bar{d}_v$ in this case is the length of a cube edge; for spherical crystallites, $\bar{d}_v = (6/\pi)^{1/3}\bar{\delta}$.

ADSORPTION TECHNIQUE

By the adsorption technique, a surface average diameter is calculated. It is easy to show that $\bar{d}_s$ [Eq. (2)] becomes

$$\bar{d}_s = 6V/S, \tag{4}$$

where V is the volume of the catalytic substance on the support and S is its surface area, for a sphere or a regular polyhedron other than a tetrahedron. The volume V is determined by assuming that the particle density of the supported material is the same as the bulk density. In the past, chemisorption has often been used to determine S, the surface area of a supported material. For instance, Emmett and Brunauer (*27*) determined the free-iron part of the surface of ammonia synthesis catalysts by chemisorption of carbon monoxide, Boreskov and Karnaukhov (*28*) perfected a technique of measuring the surface area of various platinum-containing catalysts by means of selective chemisorption of hydrogen. Schuit and van Reijen (*29*) measured the surface area of nickel-on-carrier catalysts by hydrogen chemisorption. Boudart and co-workers (*30–32*) used a modified technique for determining the surface area of unsupported platinum and of platinum dispersed on alumina which involves reaction of hydrogen at room temperature with oxygen adsorbed on the metal. Scholten and van Montfoort (*33*) found the surface area of palladium on supports using chemisorption of carbon monoxide—a case where hydrogen could not be used because of the high solubility of hydrogen in palladium. Hydrogen adsorption on nickel, platinum, and palladium is dissociative. Guyot and co-workers (*34, 35*) determined the area covered by chromium oxide on a silica alumina support by oxygen chemisorption. In all the above cases, the metal or metal-oxide surface area associated with each molecule of adsorbate is determined by comparison with adsorption on the unsupported substances whose surface areas have been determined by the BET method or other methods. The area per adsorbed molecule is assumed to be the same for supported and unsupported substances, an assumption which may not always be true.

In Table 13.2, electron microscope, X ray, and hydrogen adsorption results obtained by Adams *et al.* (*26*) are given.

TABLE 13.2

COMPARISON OF ELECTRON MICROSCOPE, X RAY, AND ADSORPTION
RESULTS[a]

	Average particle diameter, Å		
Type of diameter	Electron microscope	X ray	Adsorption
$\dfrac{\sum n_i d_i}{\sum n_i}$	28.5		
$\dfrac{\sum n_i d_i^3}{\sum n_i d_i^2}$	30.5		34.4
$\dfrac{\sum n_i d_i^4}{\sum n_i d_i^3}$	31.5	37.9[b]	

[a] See Adams *et al.* (*26*).
[b] Calculated assuming that the crystallites are cubes.

In Table 13.3, results obtained by Scholten and Montfoort (*33*) comparing the electron microscope, X-ray broadening, and carbon monoxide adsorption techniques for determining the particle size of palladium on various supports are summarized. In the case of 0.47% Pd on alumina pretreated at 100°C, the metal was apparently so highly dispersed that the ratio of adsorbed carbon monoxide molecules to palladium atoms was essentially equal to unity.

TABLE 13.3

SURFACE AREAS AND Pd CRYSTALLITE SIZES OF SOME COMMERCIAL Pd-ON-CARRIER CATALYSTS

Sample	Pretreatment evacuation		Total surface area, m^2/gm Cat	Pd surface area, m^2/gm Pd	Mean crystallite size, Å		
	Temp., °C	Time, hr			X-ray	Electron microscope	CO adsorption
γ-Al$_2$O$_3$ + 0.47 % Pd	100	20	122	559		14	8.4
	300	20	129	454			11.3
	500	20	129	150		47	33.0
γ-Al$_2$O$_3$ + 1.1 % Pd	100	20	119	337			15.0
	300	20	119	311			16.0
γ-Al$_2$O$_3$ + 5.5 % Pd	300	20	234	364			13.7
	500	48	233	107	52[a](±10)		47.0
Carbon + 4.9 % Pd	100	20	1000	131			47.0
	300	20	1000	121	57[a](±10)		41.5
Carbon + 10 % Pd	300	20	1250	151	46[a](±10)		33.0

[a] Volume average diameters, other values surface average diameters.

Magnetic Techniques

In certain particle-size ranges, some metals may show a particular magnetic behavior called *super-paramagnetism*. For example, nickel is superparamagnetic in the range 50–100 Å (*36*). This property has permitted independent measurements of particle size and size distribution of supported nickel catalysts to be made (*29, 37*). Basic to this determination is the Langevin function,

$$\sigma/\sigma_s = L(\mu H/kT) = \coth(\mu H/kT) - (kT/\mu H)$$

$$\mu = n\mu_{Ni}$$

(5)

where σ is the magnetization of the sample, σ_s the saturation magnetization, H the magnetic field strength, μ the magnetic moment of a nickel particle, μ_{Ni} the magnetic moment of a nickel atom, and n the number of nickel atoms per particle. Plots of σ/σ_s versus H deviate from the simple Langevin function when there is a spread of particle sizes. Making use of this deviation, an estimate of particle-size distribution can be made. Applying this method to nickel–silica catalysts (Ni : SiO$_2$ = 15 : 100), Schuit and van Reijen (*29*) found a distribution with 75% of the weight of the catalyst sample as particles of 1000 atoms, 17% as particles of 10,000 atoms, and 5% as particles of 100,000 atoms. From this distribution, the number of surface atoms was calculated. It was then assumed that, at saturation, every surface atom could adsorb one hydrogen atom. Results were compared with adsorption experiments and are shown in Table 13.4. Note that experimental adsorptions of hydrogen are considerably lower than adsorptions calculated from magnetic data. Schuit and van Reijen explain their data by postulating that some nickel of very small particle-size may be inaccessible to hydrogen, though magnetically detectable, because the particles become surrounded by the support during reduction of nickel oxide to nickel.

TABLE 13.4

Comparison of Nickel Surface Area as Determined from H$_2$-Chemisorption (V_H) and Estimated from Magnetic Measurements (V_m)[a]

V_H, cm^3 H$_2$/gm Ni	V_m, cm^3 H$_2$/gm Ni
31.0	55
54.0	77
39.5	52

[a] See Schuit and van Reijen (*29*).

Magnetic methods have also been applied to the study of supported transition metal oxides, especially by Selwood (*38, 39*). The basic equations are

$$\mu = 2.84[\chi(T + \Delta)]^{1/2} \tag{6}$$

where μ is the magnetic moment in Bohr magnetons of the transition metal cation, χ the magnetic susceptibility, and Δ the Weiss constant; and the Curie–Weiss Law

$$\chi(T + \Delta) = C, \qquad \mu = 2.84\sqrt{C}. \tag{7}$$

Measuring the magnetic susceptibility over a range of temperatures is sufficient to determine the number of unpaired electrons and thus the oxidation state of the metal ion. Provided independent information is available concerning the state of oxidation, then determinations of magnetic susceptibilities can yield information on exchange demagnetization which is the result of electronic interactions between adjacent cations. The more dilute the supported oxide, the less pronounced is this effect. Selwood has shown that the Weiss constant Δ is a measure of the exchange demagnetization.

In order to illustrate the application of magnetic data, we discuss the case of chromium oxide on γ-alumina. The Curie–Weiss law holds for this system over broad ranges of temperature and chromia concentrations. The magnetic susceptibility rises sharply at a critical point referred to as point "*I*" in Fig. 13.10. The magnetic moment μ was found to be independent of chromium concentration except at quite low concentrations, and its value was close to three Bohr magnetons. On the other hand, the Weiss constant Δ changes appreciably with chromium concentration and clearly reflects point "*I*" on

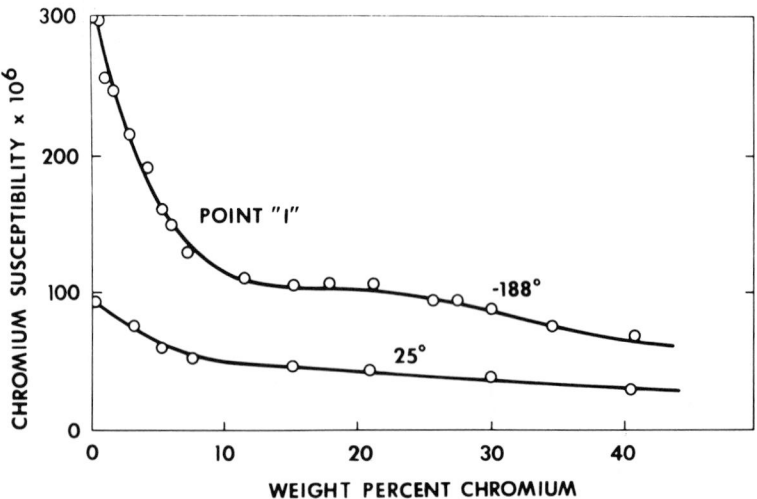

FIG. 13.10. Susceptibility isotherms for chromia supported on γ-alumina (*38*).

the plot of the susceptibility isotherm, Fig. 13.10. As previously stated, changes in Δ are associated with changes in the interactions between neighboring cations. Selwood relates point "*l*" to the concentration at which the chromia is assembled into layers approximately three atoms deep, and it occurs at about 6.0 wt% chromium. Below this concentration, Δ decreases sharply as cation interactions decrease. However, calculations based on the surface area of the support show that, at a concentration of 6.0%, there is only about one-quarter enough chromia to cover the alumina with a monolayer, and, thus, far from sufficient to cover the alumina with three atom-layers. The conclusion is that the surface is covered by widely scattered aggregates of chromia.

STATISTICAL CALCULATIONS OF CLUSTER SIZE

Poole (*40*) has attempted to calculate the number of small clusters of transition-metal ions formed on an alumina surface by adsorption from solution using a statistical model. The calculations are compared with magnetic spin resonance and susceptibility data. He assumes that the substrate contains N adsorption sites and that PN metal ions are randomly adsorbed on PN of these sites where P is the fraction of the total sites covered. Multilayer adsorption will not be considered. The probability of a given site being unoccupied is $(1 - P)$. Consider a square-plane lattice with S isolated metal ions. The probability S/N of finding a given site occupied by an isolated adsorbed ion is the probability of the site being occupied (P) times the probability that the four nearest-neighbor sites are unoccupied $(1 - P)^4$. Thus, the number of isolated metal ions is

$$S = NP(1 - P)^4. \tag{8}$$

Similar expressions may be obtained for the number of ions in isolated pairs D, triplets T, and quadruplets Q. Table 13.5 lists these expressions as determined by Poole for the plane square lattice as well for the plane triangular lattice. It is evident from the table that all formulas are of the form $KP^m(1 - P)^n$ where K is a constant. The value of P_{mn} which corresponds to a maximum in the number of sites occupied by a cluster of m ions is given by

$$d[KP^m(1 - P)^n]/dP = 0, \tag{9}$$

whose solution $P = P_{mn}$ is

$$P_{mn} = m/(m + n). \tag{10}$$

The maximum number of sites occupied by clusters $S, D, \ldots,$ is denoted by the symbols $S_0, D_0, \ldots.$ Thus, for isolated atoms

$$S_0 = NP_{1n}(1 - P_{1n})^n, \tag{11}$$

which for a square lattice where $n = 4$,

$$S_0 = 0.082N \quad \text{and} \quad P_{14} = 1/5. \tag{12}$$

TABLE 13.5

NUMBER OF ADSORPED ATOMS IN SINGLE S, DOUBLE D, TRIPLE T, AND QUADRUPLE Q
ISOLATED GROUPS

Plane triangular lattice	Plane square lattice

Plane triangular lattice:

$\cdot S = NP(1 - P)^6$

$\cdot\cdot D = 6NP^2(1 - P)^8$

$\therefore T_0 = 6NP^3(1 - P)^9 = P(1 - P)D$

$\cdots T_1 = 9NP^3(1 - P)^{10} = \frac{3}{2}(1 - P)T_0$

$\cdot\cdot T_< = 18NP^3(1 - P)^{10} = 2T_1$

$T = T_0 + T_1 + T_<$

$\because Q_0 = 12NP^4(1 - P)^{10} = 4PT_1/3$

$\cdots Q_1 = 12NP^4(1 - P)^{12} = (1 - P)^2 Q_0$

$\therefore Q_\Lambda = 48NP^4(1 - P)^{11} = 4(1 - P)Q_0$

$\because Q_\cap = 24NP^4(1 - P)^{12} = 2Q_1$

$\cdots Q_N = 24NP^4(1 - P)^{12} = 2Q_1$

$\cdots Q_< = 48NP^4(1 - P)^{12} = 4Q_1$

$\therefore Q_\lambda = 8NP^4(1 - P)^{12} = \frac{2}{3}Q_1$

$Q = Q_0 + Q_1 + Q_\Lambda + Q_\cap + Q_N + Q_< + Q_\lambda$

Plane square lattice:

$\cdot S = NP(1 - P)^4$

$\cdot\cdot D = 4NP^2(1 - P)^6 = 4P(1 - P)^2 S$

$\because T_L = 12NP^3(1 - P)^7 = 3P(1 - P)D$

$\cdots T_1 = 6NP^3(1 - P)^8 = \frac{1}{2}(1 - P)T_L$

$T = T_1 + T_L$

$\because Q_0 = 4NP^4(1 - P)^8 = \frac{2}{3}PT_1$

$\because Q_T = 16NP^4(1 - P)^8 = 4Q_0$

$\because Q_L = 32NP^4(1 - P)^9 = 8(1 - P)Q_0$

$\cdots Q_1 = 8NP^4(1 - P)^{10} = 2(1 - P)^2 Q_0$

$\because Q_N = 16NP^4(1 - P)^8 = Q_T$

$Q = Q_0 + Q_T + Q_L + Q_1 + Q_N$

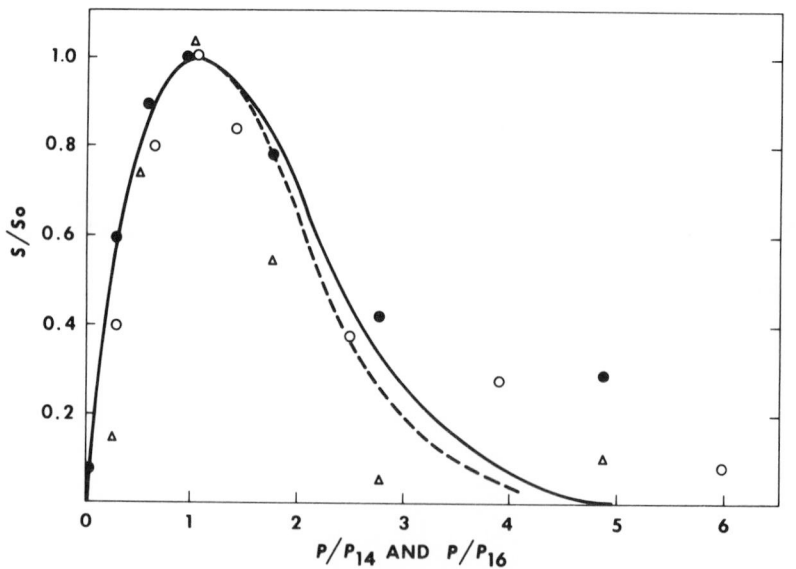

FIG. 13.11. Comparison of plane square lattice (- - - -) and plane triangular lattice (——) theoretical curves with experimental data from isolated Cr^{+3} ions detected by ESR of reduced Cr/Al_2O_3 (●), from the ^{27}Al NMR of reduced Cr/Al_2O_3 (△), and from the Co^{+2} isolated ions detected by the magnetic susceptibility of Co/Al_2O_3 (○) *(40)*. [After Poole, Jr., *J. Phys. Chem.* **67**, 1297 (1963). By permission of The American Chemical Society.]

In Fig. 13.11, S/S_0 is plotted against P/P_{14} for a square lattice and P/P_{16} for a triangular lattice. Maxima occur at $(S/S_0) = (P/P_{14}) = (P/P_{16}) = 1$. These curves are now compared with experimental data from ESR, NMR, and magnetic susceptibility. The ESR spectrum from alumina impregnated with a low concentration of chromia originates from isolated Cr^{+3} ions (*41, 42*). Thus the comparison is made by plotting the concentration of chromia versus the ESR signal, both normalized in such a way that abscissa and ordinate values in the graph are equal to unity at the maximum point. Magnetic susceptibility data for low concentrations of cobalt oxide on alumina were normalized in a similar way. In the case of NMR data, use was made of the fact that when alumina is impregnated with a transition metal, the amplitude of the ^{27}Al NMR absorption envelope at first decreases with increasing metal content, reaches a minimum, and then increases for higher concentration. The decrease in ^{27}Al amplitude was attributed to the removal from detection of ^{27}Al nuclei near small clusters of transition metal ions. The number of aluminum nuclei removed from detection by chromium ions was normalized to the maximum number and plotted against normalized concentration. The results are shown in Fig. 13.11 for ESR, NMR, and magnetic susceptibility data in comparison with the statistical curves. Qualitative agreement is evident except at high metal-concentration.

13.5 Particle Size and Catalytic Activity

STRUCTURE-INSENSITIVE REACTIONS

Systematic investigations of the problem of specific catalytic activity (activity per unit area of catalytic substance) were first carried out by Boreskov and co-workers based on their development of a technique for measuring the surface area of platinum catalysts by means of the selective chemisorption of hydrogen (*28*). They showed that the specific activity of platinum in the oxidation of sulfur dioxide (*43*) and of hydrogen (*44*) varied less than an order of magnitude for catalyst samples differing in platinum surface area by as much as four orders of magnitude. Catalysts included platinum supported on silica gel, platinum sponge, wire, and gauze. An exception to this insensitivity of specific activity to platinum surface area was found in the case of hydrogen–deuterium exchange at low temperature (*45*).

The investigations of the Boreskov school have been extended by Boudart and co-workers (*30, 31, 46*). They showed that the hydrogenation of cyclopropane at 0°C was remarkably insensitive to platinum surface area over a range of catalysts including platinum foil, sintered and unsintered dispersions of platinum on η- and γ-aluminas. The results are shown in Table 13.6 where it is evident that the difference between highly dispersed samples and others (sintered, less dispersed, foil) was only two-fold in specific activity, while the

TABLE 13.6

<small>ACTIVITY AND PLATINUM SURFACE AREA OF CATALYSTS FOR THE 0°C HYDROGENATION
OF CYCLOPROPANE[a]</small>

Wt % Pt on catalyst	Specific surface area of Pt, moles H_2/gm sample	Activity, moles cyclopropane/ min/gm sample	Dispersion D, %
0.3 % on η-alumina	10.2	82.4	44
0.3 % on γ-alumina	13.5	100.0	59
0.6 % on γ-alumina	33.5	186.0	73
1.96 % on η-alumina	97.0	637.0	64
Same as above, sintered	11.5	44.0	7.6
4.3 % on silica gel	57.3	160.0	17.0
100.0 % (foil)	0.55	1.4	0.0039

[a] See Boudart *et al.* (*31*).

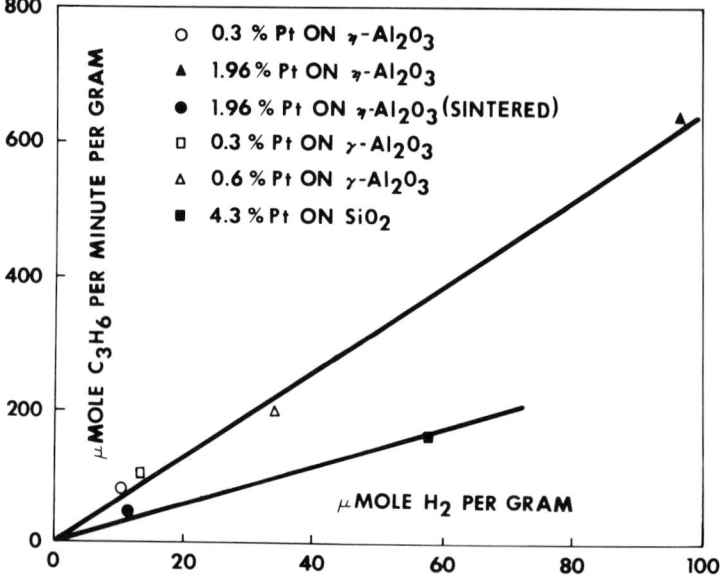

FIG. 13.12. Activity as a function of platinum surface area.

platinum specific surface area (area/gm Pt) varied by more than four orders of magnitude.

The data of Table 13.6 are plotted in Fig. 13.12 and show that the catalysts may be divided into two groups; those with high dispersion and those with much lower dispersion. In each group, the ratio of activity to platinum surface area is approximately constant. The difference in specific activity between the two groups as shown by the slopes of the two lines is only a factor of two which is insignificant relative to the changes in platinum surface area. One sharp difference between highly dispersed catalysts and other catalysts is the marked susceptibility of the former to oxygen poisoning. However, it must be concluded that certain reactants are immune to whatever property causes this difference as indicated by their constant specific activity as a function of particle size. Boudart suggests that such reactions be called *facile reactions.* For these reactions, the majority of sites possess ample activity, and the reactants fail to sense the nonuniformity of the surface.

Poltorok and Boronin (*47*) have obtained further evidence for the insensitivity of specific activity to particle size for platinum on silica gel in the reactions: hydrogenation of cyclohexene, dehydrogenation of cyclohexane, hydrogenation of hexene-1, dehydrogenation of isopropanol, and the hydrogenolysis of cyclopentane. They pointed out that reactions have the greatest possibility of being sensitive to changes in structure for particles smaller than about 50 Å, because in that range occur the greatest and the most rapidly changing deviations from bulk metal. Table 13.7 summarizes their calculations on the properties of regularly faceted platinum crystals of various sizes. The fraction of surface atoms and mean number of bonds per atom in the

TABLE 13.7

PROPERTIES OF REGULARLY FACETED PLATINUM CRYSTALS OF VARIOUS SIZES[a]

No. of atoms	Length of crystal edge Å	Fraction of atoms on surface	Total atoms in crystal	Mean no. of bonds per atom in crystal
2	5.50	1.00	6	2.00
3	8.95	0.95	19	3.16
4	11.00	0.87	44	3.81
5	13.75	0.78	85	4.23
6	16.50	0.70	146	4.51
7	19.25	0.63	231	4.72
8	22.00	0.575	344	4.89
9	24.75	0.53	489	5.00
10	27.50	0.49	670	5.10
11	30.25	0.45	891	5.18
12	33.00	0.42	1156	5.25
13	35.75	0.39	1469	5.31
14	38.50	0.37	1834	5.35
15	41.25	0.35	2255	5.40
16	44.00	0.33	2736	5.44
17	46.75	0.31	3281	5.48
18	49.50	0.30	3894	5.50

[a] See Poltorok and Boronin (*47*).

crystal have reached nearly a constant value when the crystal edge has attained a length of 50 Å. Some of their catalyst preparations reached the limit of dispersibility with the fraction of surface platinum atoms equal to unity. Yet the hydrogenation–dehydrogenation reactions which they studied showed very little sensitivity to particle size. Further evidence for insensitivity of specific activity to particle size was found by Dorling and Moss (*48*) who studied the hydrogenation of benzene over platinum–silica catalysts. They found that specific activity remained constant for catalysts fired below 400°C. Catalysts fired at 400 and 500°C contained larger crystallites, and had smaller specific activities, while those fired at 600 and 800°C were inactive. No explanation is given for the sensitivity to structure which developed in catalysts fired in excess of 400°C. Similar studies have been made of the hydrogenation of ethylene (*54*).

STRUCTURE-SENSITIVE REACTIONS

The question arises as to whether there are reactions in addition to H–D exchange found by Boreskov and co-workers (*45*) which show sensitivity to structure in well-dispersed catalyst systems. Boudart et al. (*46*) have found such a reaction. They studied the initial rates of the reaction of neopentane on platinum powder catalysts and supported platinum catalysts at 300°C, 1 atm total pressure, and a hydrogen-to-neopentane ratio of 10. The following supports were used: silica gel, carbon, and η- and γ-alumina. The number of exposed platinum atoms was measured by the hydrogen–oxygen titration method previously mentioned in Section 13.4 (*32*). Values of dispersion, defined as the percentage of platinum atoms which are surface atoms, ranged from 3×10^{-2} for the powder to about 70 for the supported catalysts. Under the conditions of the experiments, neopentane undergoes two parallel reactions, isomerization to isopentane and hydrogenolysis to isobutane and methane. Boudart et al. define a selectivity factor, S, as the ratio of the isomerization rate, N_I, to the hydrogenolysis rate, N_H. This ratio varied by as much as a factor of 100. For supported catalysts, selectivity increased with increasing catalyst pretreatment temperature (425–900°C). The hydrogenolysis reaction was strongly structure sensitive, the specific activity decreasing markedly with increasing temperature of pretreatment. Boudart proposes to call such structure-sensitive reactions *demanding reactions*. The results are shown in Table 13.8.

To explain their results, Boudart et al. followed the considerations of Anderson and Avery (*49*) who studied the hydrogenolysis of ethane and the isomerization and hydrogenolysis of *n*-butane, isobutane, neopentane, and isopentane on platinum films. They postulated that neopentane can react through two different types of adsorbed intermediates, one diadsorbed and the other triadsorbed. The formation of the triadsorbed intermediate is more

TABLE 13.8

SPECIFIC ACTIVITIES AND SELECTIVITIES FOR THE REACTION OF
NEOPENTANE UNDER STANDARD CONDITIONS

Catalyst		Disper- sion *D*, %	Specific activities, 10^3 molecules converted/Pt atom/sec		Isomerization selectivity $S = N_I/N_H$	Max. temp. pretreatment in reduced state, °C
Wt% Pt	Support		Isomeri- zation N_I	Hydro- genolysis N_H		
0.60	$Al_2O_3(\gamma)^a$	73	2.6	4.70	0.55	425
1.96	$Al_2O_3(\eta)^a$	64	5.3	13.00	0.41	425
0.60	$Al_2O_3(\gamma)^b$	73	2.8	1.90	1.50	500
1.96	$Al_2O_3(\eta)^b$	64	3.4	2.30	1.50	500
4.30	$SiO_2{}^b$	17	3.8	13.00	0.29	500
1.0	Spheron^c	12	1.1	0.04	27.00	900
1.96	$Al_2O_3(\eta)^d$	7.6	11.0	4.50	2.40	650
100.00^e		0.028	0.73	0.08	9.10	500

^{a} Reduced at 425°C in hydrogen for 10 hr.
^{b} Same, but at 500°C.
^{c} Reduced at 500°C, evacuated at 900°C for 16 hr, stored in air before reduction as in *b*.
^{d} Sintered in flowing hydrogen at 650°C before reduction as in *b*.
^{e} Treated alternately and repeatedly in hydrogen and oxygen before reduction as in *b*.

critical than the diadsorbed and requires a particular geometric configuration that is not found to the same degree on all types of statistical platinum surfaces. Because of this limitation, the specific activity for hydrogenolysis of neopentane was found to depend strongly on catalyst pretreatment.

Another example of a structure-sensitive reaction is propylene polymerization over supported chromium oxide catalysts (*34*). Surface area of chromium oxide on silica–alumina was measured by oxygen chemisorption and was found always to be a low fraction of the total surface area. It goes through a maximum with chromium content and decreases with the average oxidation number. Specific activity remained constant only in a very limited range of chromium concentration. Chromium oxide appears to interact with the support, causing changes in texture as a function of concentration. The quality of the coverage seems to be as important as the value of the area covered.

KOBOSEV ENSEMBLES

Pertinent to the above discussion is the theory of ensembles first proposed by Kobosev in 1938 (*50–52*). According to his theory, the seat of catalytic activity is an amorphous phase. The noncatalytic crystalline phase is pictured as having a mosaic structure composed of an aggregate of closed cells (regions of migration) surrounded by energy and geometrical barriers (e.g., micro-fissures) which cannot be crossed by surface atoms. These regions may be visualized as deep potential wells and the amorphous phase is assumed to accumulate in these wells forming ensembles. The distribution of amorphous phase in these potential wells is supposed to vary according to the rules of random statistics. On the basis of experimental data, Kobozev deduces that the number of atoms of amorphous phase in a single potential well forming an ensemble corresponding to maximum specific activity is small. For example, the maximum in specific activity for the ammonia synthesis over an iron–charcoal catalyst occurs at the very dilute fractional coverage of the charcoal surface by iron of 0.0006 (53), corresponding by Kobozev's calculations to 2–3 atoms of iron per ensemble. With increasing dilution, the number of atoms per ensemble decreases, and so does the specific activity. Single-atom ensembles are absolutely inactive. Contrary to Balandin's sextet theory for the hydrogenation of six-membered aromatics, Kobozev claims that hydrogenation of both ethylene and aromatic bonds occurs on the same two-atom ensembles. Kobozev gives many examples of maxima in specific activity at extremely high dilution of supported substance. His results indicate structure-sensitivity for many reactions, thus apparently contradicting in some cases recent work. However, in recent studies discussed earlier in this section, the extreme dilutions of supported substances employed by Kobozev were not investigated. More work is needed to settle the question of structure sensitivity at very high dilutions. It should be pointed out that experimental difficulties in determining specific activities accurately increase enormously at such high dilutions.

REFERENCES

1. Sherman, A., and Eyring, H., *J. Am. Chem. Soc.* **54**, 2661 (1932).
2. Eyring, H., and Polanyi, M., *Z. Phys. Chem.* **B12**, 279 (1931).
3. Eyring, H., *J. Am. Chem. Soc.* **53**, 2537 (1931); **54**, 3191 (1932).
4. Eyring, H., *Chem. Rev.* **10**, 103 (1932).
5. Horiuti, J., Okamoto, G., and Hirota, K., *Sci. Papers Inst. Phys. Chem. Res. (Tokyo)* **29**, 223 (1936).
6. Beeck, O., Smith, A. E., and Wheeler, A., *Proc. Roy. Soc.* **A177**, 62 (1940).
7. Beeck, O., *Discussions Faraday Soc.* **8**, 118 (1950).
8. Gwathmey, A. T., and Cunningham, R. E., *Advan. Catalysis* **10**, 57 (1958).
9. Balandin, A. A., *Z. Phys. Chem.* **132**, 289 (1929).
10. Balandin, A. A., *Advan. Catalysis* **10**, 96 (1958).
11. Balandin, A. A., *Russ. Chem. Rev. (English Transl.)* **11**, 589 (1962).
12. Balandin, A. A., Bielanski, A., et al., "Catalysis and Chemical Kinetics," Chapter 1. Academic Press, New York and Wydawnictura Nankowo-Techniczne, Warszawa (1964).

13. Trapnell, B. M. W., *Advan. Catalysis* **3**, 1 (1951).
14. Emmett, P. H., New Approaches to the Study of Catalysis (Priestly Lecture), Chapter 3. The Pennsylvania State Univ., University Park (1962).
15. Beeck, O., and Ritchie, A. W., *Discussions Faraday Soc.* **8**, 159 (1950).
16. Read, W. T., Jr., "Dislocations in Crystals." McGraw-Hill, New York, 1953.
17. Burgers, J. M., *Proc. Koninkl. Ned. Akad. Wetenschap* **42**, 293, 378 (1939).
18. Duell, M. J., and Robertson, A. J. B., *Trans. Faraday Soc.* **57**, 1416 (1961).
19. Cottrell, A. H., "Vacancies and Other Point Defects in Metals and Alloys," p. 1. Institute of Metals, London, 1958.
20. Hirsch, P. B., Silcox, J., Smallman, R. E., and Westmacott, K. H., *Phil. Mag.* **3**, 897 (1958).
21. Silcox, J., and Hirsch, P. B., *Phil. Mag.* **4**, 72 (1959).
22. Sosnovsky, H. M. C., *J. Phys. Chem. Solids* **10**, 304 (1959).
23. Hall, J. W., and Rasé, H. F., *Nature* **199**, 585 (1963).
24. Gilman, J. J., and Johnston, W. G., *J. Appl. Phys.* **27**, 1018 (1956).
25. Bagg, J., Jaeger, H., and Sanders, J. V., *J. Catalysis* **2**, 449 (1963).
26. Adams, C. R., Benesi, A. A., Curtis, R. M., and Meisenheimer, R. G., *J. Catalysis* **1**, 336 (1962).
27. Emmett, P. H., and Brunauer, J. J., *J. Am. Chem. Soc.* **59**, 310 (1937).
28. Boreskov, G. K., and Karnaukhov, A. P., *Zh. Fiz. Khim.* **26**, 1814 (1952).
29. Schuit, G. C. A., and van Reijen, L. L., *Advan. Catalysis* **10**, 243 (1958).
30. Spenadel, L., and Boudart, M., *J. Phys. Chem.* **64**, 204 (1960).
31. Boudart, M., Aldag, A., Benson, J. E., Dougharty, N. A., and Harkins, C. G., *J. Catalysis* **6**, 92 (1966).
32. Benson, J. E., and Boudart, M., *J. Catalysis* **4**, 704 (1965).
33. Scholten, J. J. F. and van Montfoort, A., *J. Catalysis* **1**, 85 (1962).
34. Guyot, A., Charcosset, H., and Revillon, A., *J. Catalysis* **8**, 326 (1967).
35. Guyot, A., Charcosset, H., and Revillon, A., *J. Catalysis* **8**, 334 (1967).
36. Neel, L., *Compt. Rend.* **228**, 664 (1949).
37. Reinen, D., and Selwood, P. W., *J. Catalysis* **2**, 109 (1963).
38. Selwood, P. W., *Advan. Catalysis* **3**, 27 (1951).
39. Selwood, P. W., "Magnetochemistry." Wiley (Interscience), New York, 1956.
40. Poole, C. P., Jr., *J. Phys. Chem.* **67**, 1297 (1963).
41. O'Reilly, D. E., *Advan. Catalysis* **12**, 31 (1960).
42. O'Reilly, D. E., and MacIver, D. S., *J. Phys. Chem.* **66**, 276 (1962).
43. Boreskov, G. K., and Chesalova, V. S., *Zh. Fiz. Khim.* **30**, 2560 (1956).
44. Boreskov, G. K., Slin'ko, M. G., and Chesalova, V. S., *Zh. Fiz. Khim.* **30**, 2787 (1956).
45. Abdeenko, M. A., Boreskov, G. K., and Slin'ko, M. G., *Probl. Kinetiki i Kataliza Akad. Nauk. SSSR* **9**, 61 (1957).
46. Boudart, M., Aldag, A. W., Ptak, L. D., and Benson, J. E., *J. Catalysis* **11**, 35 (1968).
47. Poltorok, O. M., and Boronin, V. S., *Intern. Chem. Eng.* **7**, 452 (1967).
48. Dorling, T. A., and Moss, R. L., *J. Catalysis* **5**, 111 (1966).
49. Anderson, J. R., and Avery, N. R., *J. Catalysis* **5**, 446 (1966).
50. Kobozev, N. I., *Acta Physicochim. URSS* **9**, 1 (1938).
51. Kobozev, N. I., *Acta Physicochim. URSS* **21**, 943 (1946).
52. Kobozev, N. I., *Acta Physicochim. URSS.* **21**, 7 (1946).
53. Kobozev, N. I., and Klichko-Gurvich, L., *Acta Physicochim.* **10**, 1 (1939).
54. Dorling, Y. A., Eastlake, M. J., and Moss, R. L., *J. Catalysis* **14**, 23 (1969).

XIV

ELECTRONIC FACTORS IN CATALYSIS—METALS

14.1 The Electronic Factors

From Section 7.5, it is evident that information on surface states of metals is not sufficiently developed to serve as a basis for theories of catalytic activity. Furthermore, in spite of the great amount of work relating catalytic activity to point defects in semiconductors (see Chapter XV), there has been scarcely any corresponding work on metals. Most of the attempts to correlate catalytic activity and the electronic structure of bulk metal involve the d electrons of transition metals. This is reasonable in view of similar attempts (see Section 7.2) to correlate the d-electron pattern and chemisorption which bears a close, but complex, relationship to catalytic activity.

In Chapters VII and VIII we described both the electron band theory and the valence bond theory of metals, either of which might be employed in attempts to correlate catalytic activity with electronic structure. According to the band theory, electrons retain much of the character which they possess in the isolated atom except that discrete electron levels such as the s-, p- or d-levels become bands of permitted values in the metal crystal, and the valency electrons are considered to move quite freely through the assembly of nuclei with their tightly held inner electrons. These bands may overlap. For example, the electron configuration of an isolated iron atom is $3d^6 4s^2$, whereas in the metallic state, as a result of d- and s-band overlapping, the average structure is $3d^{7.8}4s^{0.2}$. In either terminology, the d-levels are incompletely filled, which is characteristic of transition metals. Bonding of iron atoms together is attributed to electrons in overlapping d- and s-bands, and magnetic and conductive properties are associated with holes in the d-band. In the valence bond theory, bonding electrons are postulated to occupy dsp hybrid atomic orbitals, and electrons governing magnetic and conductive properties are assigned to atomic d-orbitals. The electron band theory is more fundamental, but the valence bond theory, though empirical, offers more chance of obtaining numerical relationships, at least in the case of single metals. The band theory has found its greatest application in the studies of metal alloys.

The numbers which may be calculated and used in the valence bond theory are δ, the d-character, and λ, the heat of sublimation. In Section 7.2, the relationships of these numbers to heats of chemisorption were discussed. Attempts to correlate δ with heats of chemisorption are based on the assumption that the adsorption bond is a covalence between electrons from the adsorbate and those unpaired electrons in atomic d-orbitals of the metal not used according to Pauling in dsp metal–metal bonding. Correlations of $\lambda/6 = D_{MM}$ with heats of chemisorption are based on the assumption that the adsorption bond is similar to the metal–metal bond and that it uses an unengaged dsp-orbital. If the adsorption bond possesses significant ionic character, which is unusual except in the case of oxygen, an electronegativity term is added to the sublimation term [Eq. (4), Chapter VII]. As explained in Section 7.2, the latter method helps to explain the adsorption of atoms on the surfaces of nontransition metals, especially those of Group IB. The two methods of attempting correlations reveal strong conceptual differences. Thus, it is assumed that δ measures the extent to which d-electrons participate in metal–metal dsp bonding orbitals, making them available for chemisorption, so that as δ increases, the heat of chemisorption is expected to decrease. On the other hand, Eley's equation [Eq. (4), Chapter VII] shows the heat of chemisorption increasing as λ increases, assuming insignificant ionic effects, where λ is a measure of the strength of a metal–metal dsp bond. It will become evident that catalytic activity, like the heat of chemisorption, frequently does not correlate satisfactorily with the general electronic properties, δ and λ. The reasons for this will be pointed out.

We stated in Chapter XIII that geometric and electronic factors must be interrelated. Pauling has given such a relationship between δ, the d-character and lattice spacing expressed in terms of R, the single bond radius of the atoms in a metal crystal. His equation for the first transition series is

$$R = 1.825 - 0.043z - (1.600 - 0.100z)\delta \tag{1}$$

where z is the number of electrons in the neutral atom outside the inert gas shell. Since the constants of the terms containing z are small, R is largely determined by δ. Thus, δ controls only lattice spacing. It has nothing to do with the particular arrangement of catalytically active points at a given surface. In the light of this, we cannot expect d-character to explain the differences in activity between different faces on the same metal. It can only determine the general level of activity of polycrystalline catalysts. With respect to λ, the quantum mechanical relationships between bond length (lattice spacing in a crystalline solid) and bond strength ($\lambda/6 = D_{MM}$ for a face-centered cubic lattice) are well recognized, but difficult to calculate with much accuracy except for simple molecules.

14.2   Valence-Bond Theories of Catalysis

In 1950, Boudart (*1*) and Beeck (*2*) showed that a smooth curve was obtained when the percentage of d-character δ of certain transition metals was plotted against the logarithms of their activities for the hydrogenation of ethylene. Highest activity was reached with rhodium as was observed in the plot of lattice spacing against catalytic activity for the same reaction, Fig. 13.3. Beeck's curve in Fig. 14.1 shows activity for ethylene hydrogenation increasing with increasing δ. Beeck (*2*) obtained the inverse of this relationship on the same metals for the heat of chemisorption of hydrogen as a

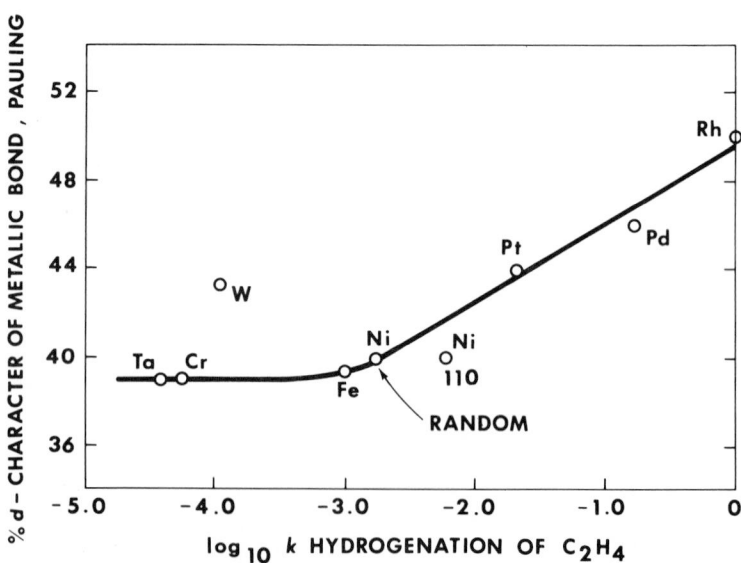

FIG. 14.1. Variation of δ, the percentage of d-character, for certain transition metals against the logarithm of the activity per unit area for the hydrogenation of ethylene (*2*). [After Beeck, *Discussions Faraday Soc.* **8**, 118 (1950). By permission of The Faraday Society.]

function of δ. We note that decreasing heats of chemisorption with increasing δ has been interpreted by Beeck (2) (see also Sections 7.2 and 14.1) to mean the increasing use of d-electrons in metal–metal dsp-bonding and the consequent decrease in unpaired d-electrons available for chemisorption. However, Pauling's treatment of d-character does not give an inverse relationship between δ and unpaired d-electrons (3). The situation is contradictory and confusing. To add to the confusion, we have seen in Chapter VII that the inverse relationship between chemisorption and δ is not at all satisfactory when a broader range of metals is studied. Insufficient data are available on relations between δ and catalytic activity; more reactions would have to be studied before firm conclusions could be reached. It is suspected that many divergences would arise.

Just as confusing are the relationships between λ and catalytic activity. These relationships are postulated on the basis that there is a direct relationship between catalytic activity and chemisorption and between chemisorption and metal–metal bond strength as measured by λ. In the simplest of all reactions, the recombination of hydrogen atoms, there appears to be a satisfactory relationship, with the exception of platinum, between λ and the efficiency of recombination, between λ and initial heat of chemisorption of hydrogen, and, thus, between efficiency of recombination and heat of chemisorption for a range of transition metals (4). The efficiency of recombination decreases with increasing λ. However, there is an approximate tendency for activity to increase with increasing λ for ortho–para hydrogen conversion (5) and ethane exchange (6). Finally, for the hydrogenation of ethylene, Schuit *et al.* (6) find a maximum in activity at rhodium. The results may be interpreted by assuming an optimum heat of chemisorption of ethylene below which the concentration of ethylene with respect to hydrogen is too low and above which it is too high.

The reasons for the confusion are obvious when the complexities of the problem are considered. We summarize a few of the factors involved:

(1) δ and λ can be expected to explain only the general level of activity because, though related to lattice spacing, they do not decide the particular arrangement of active centers on a given surface.

(2) Surface states are probably different from bulk states.

(3) Heats of chemisorption on a real surface usually cover a broad spectrum of site energies, and the particular subrange used in catalysis probably depends on the reaction and may bear no relation to δ and λ. In case of two or more adsorbed reactants, their relative adsorbabilities are important in determining the optimum ratio of concentrations for maximum reaction rate.

(4) It is difficult to know whether to compare reaction velocity constant,

activation energy, or pre-exponential factor with δ and λ. The pre-valence of the compensation effect in heterogeneous reactions im-plies the existence of complex and unknown factors which cannot be explained entirely by δ and λ.

(5)   Different methods of preparing metal catalysts give different patterns of catalytic activity and chemisorption.

(6)   Point defects may have far more important effects on catalytic activities than δ or λ.

14.3   Electron-Band Theories of Catalysis

According to the free-electron theory of metals, we found (Section 7.4) that the distribution function $N(E)$ is proportional to $E^{1/2}$, where $N(E)\,dE$ is the number of electron energy states per unit volume with energies between E and $E + dE$. The corresponding expression derived from the band theory is more complex because of the energy discontinuities which produce peculiar effects as the filling of a band approaches completion. The energy discontinui-ties arise as a result of including the periodic field of the metallic nuclei, which is neglected in the free-electron theory. Figure 14.2 shows a typical plot of $N(E)$ versus E when no overlapping of bands occurs. When overlapping does occur, even more complicated plots result. Since the number of states in the energy band δE_p is related to the volume of the spherical shell, we see that $N(E)$ will

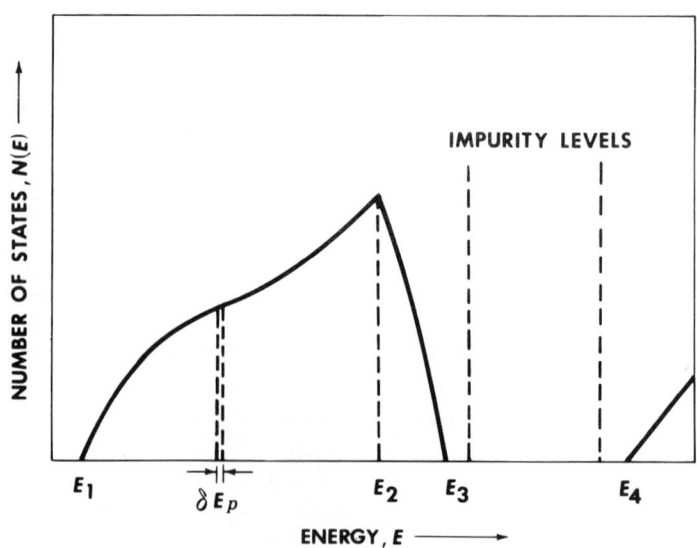

FIG. 14.2. Typical electron bands.

increase in the range δE_p. The maximum at E_2 occurs when the Fermi surface first touches in some direction the Brillouin zone which marks the boundary of an energy discontinuity. Detailed calculations in the band theory show that energy discontinuities are reached at different distances in k-space (wave-number space) depending on direction. A rapid decrease in $N(E)$ occurs after passing E_2 because there are no energy states available for some directions of the wave number k. The higher the $N(E)$ curve, the greater the number of electrons which may be accommodated in states within a given energy range, and, hence, the lower the energy necessary to accommodate a given number of electrons. It is obvious that chemisorption which involves transfer or sharing of electrons between adsorbate and catalyst, will depend, at least in part, on the nature of the $N(E)$ curve, provided that adsorption happens to depend on the collective electronic properties of the catalyst. Also, since chemisorption is the precursor of catalysis, it is reasonable to expect some relationship between catalysis and the nature of the $N(E)$ curve. Then, as discussed in Section 7.2, it might be expected that chemisorption, and hence catalysis, would depend on φ, the work function of the metal.

Dowden (7) has considered the factors favoring positive ion, negative ion, and covalent-bond formation at metal surfaces, and has given formal equations for determining the bond characteristics. Unfortunately, there are insufficient data to test these equations. Furthermore, the approach assumes that chemisorption, and so catalysis, is related to the collective electronic properties of the bulk solid, which may not always be true. We summarize Dowden's rules for bond formation. Positive ion formation is favored by (i) a large exit work function, (ii) a large positive value of the density gradient $[d \ln N(E)/dE]$ at the Fermi surface, $E = \varphi$, and (iii) controlled quantities of electronegative additives which will increase φ. Negative ion formation is favored by (i) a low exit work function, (ii) a large negative value of $[d \ln N(E)/dE]$ at the Fermi suface, and (iii) controlled quantities of electropositive additives. Covalent-bond formation is favored by (i) large values of the exit work function, (ii) large positive values of $[d \ln N(E)/dE]$, and (iii) the presence of unfilled atomic orbitals.

The electron band theory has been applied chiefly in attempts to explain the variation in catalytic activity of alloys with composition. In particular, it has been applied to the alloys of Group IB metals with transition metals. Changes in alloy composition alter the energy density, $N(E)$, of electrons at the Fermi surface; and in the special case of transition metals, alloying with a Group IB metal decreases the number of holes in the d-band. We illustrate with nickel–copper alloys. According to Mott and Jones (8), the $3d$ and $4s$ bands for both nickel and copper in the crystalline state overlap and to about the same extent in each element. In nickel, it was assumed that the Fermi surface lay a little below the top of the $3d$ band and that magnetism is due to the uncompensated spin moments of 0.6 holes per atom in the $3d$-band. The electron density in the $3d$-band is therefore 9.4 per atom, and the remaining 0.6 electron is in the $4s$ band. The magnetic moment of nickel is almost entirely due to the spins of d-electrons, and the intensity of magnetization is approximately equal to the number of holes in the d-band ($0.6\ \mu_B$). The observed magneton number obtained from magnetic measurements is not necessarily equal to the effective number of uncompensated electronic magnetic moments. A small difference due to contribution of orbital magnetic moment to the total moment may exist. In the case of nickel, the correction for orbital moment is believed to reduce the number of holes in the d band to 0.55, but for our purposes this correction will be ignored. The noninteger values for the number of electrons in the d-band may be explained by assuming that they are average values for the crystal made up, for example, of appropriate quantities of $(3d)^{10}$ and $(3d)^8$ configurations. Copper possesses one more electron than nickel, and has a completely filled d-band. It is assumed that the single s-electron of copper goes to the nickel d-band tending to reduce the magnetic moment of the nickel. The density of states in the d-band of nickel is believed to be over ten times greater than that in its s-band, so that the extra s-electron per atom of added copper should go at least 90 % into the nickel d-band. With 0.6 holes per atom in the nickel d-band and a contribution of 1.0 electron per added copper atom, the number of holes per atom of alloy as a function of copper content should be

$$n_{\mathrm{H}} = (0.6 - x)/(1 + x), \qquad x \leqslant 0.6,$$
$$n_{\mathrm{H}} = 0, \qquad x > 0.6, \tag{2}$$

where x is the number of copper atoms per nickel atom; or

$$n_{\mathrm{H}} = 0.6(1 - y) - y, \qquad y \leqslant 0.375,$$
$$n_{\mathrm{H}} = 0, \qquad y > 0.375, \tag{3}$$

where y is the atomic fraction of copper in the alloy and $(1 - y)$ is the atomic fraction of nickel. It is evident that the magnetic moment would be expected to drop to zero at a copper–nickel atomic ratio of 0.6 ($=x$), or at 37.5 atomic percent copper ($=y$). There are discrepancies in the literature concerning the experimental determination of the alloy composition at which the magnetic moment becomes zero. For example, Reynolds (9) examined skeletal nickel–copper alloys prepared by exhaustive extraction of nickel–copper–aluminum alloy powders with sodium hydroxide and found that the magnetic moment had dropped to zero at about 40% copper in the alloy. In nickel–copper catalysts prepared by reduction of the oxides, however, the magnetic moment was found to drop sharply to a low value at about 60% copper (9), and Sucksmith et al. (10) using alloys produced by melting together pure nickel and copper in an induction furnace give an extrapolated value for zero magnetic moment at about 53% copper.

Platinum and palladium are also believed to have about 0.6 holes in their d-bands. It is interesting to note that the susceptibility of palladium drops to zero at a ratio of dissolved hydrogen atoms to palladium of 0.6, suggesting that each hydrogen atom supplies one electron to palladium d-bands (*11*). On the other hand, Couper and Eley (*12*) find saturation of the d-band at about 60% gold in a palladium–gold alloy, and Couper and Metcalfe (*13*) find the same saturation composition with palladium–silver alloys.

In using alloys of this nature for catalytic studies, it is desirable to select two metals whose lattice distances are approximately the same, so that this variable can be eliminated from consideration over the entire range of composition. The lattice constants of copper and nickel are approximately the same and so are those of gold–palladium and silver–palladium. Thus, these alloys have been used extensively in catalytic studies. We shall now summarize the results of catalytic activity studies on alloys.

Couper and Eley (*12*) determined the catalytic activity of a series of palladium–gold alloys for ortho–para hydrogen conversion. Catalyst in the form of wires was used. The wires were cleaned by bombardment with hydrogen atoms from an electric discharge, and the dissolved hydrogen was removed by heating to 400–600°C *in vacuo*. First-order rate constants and activation energies were determined. The activation energy remained low and approximately constant at about 3.5 kcal from zero up to 60% gold (60 Au : 40 Pd) and then rose sharply to about 8.5 kcal. The magnetic susceptibility was stated to fall to zero at the point where the activation energy for ortho–para hydrogen conversion suddenly rose, namely, 60 Au : 40 Pd, which, as we have seen, is at considerable variance with the composition predicted by simple band theory. Experimental activation energies for ortho-para hydrogen conversion on alloys of palladium–silver are qualitatively similar and the magnetic susceptibility drops to zero at the same composition (*13*). The changes, however, are not so abrupt. The activation energy rises from 2 to 4 kcal as the composition goes from 0 to 60% silver and is followed by a more rapid, smooth increase to a value of 11.5 kcal for pure silver. In order to explain the continued low activation energies up to the completion of the d-band at 60% gold or silver, despite its progressive filling, it has been suggested that necessary d-band vacancies are produced by excitation of electrons from d-levels up to the Fermi surface. Both systems indicate that the incomplete d-band of palladium-rich alloys is associated with low values of the activation energy and that the completed d-band is associated with higher energies. The association is not simple.

In a study of the gas-phase hydrogenation of 1,3 butadiene to butenes over palladium–gold alloys, activation energies exhibited a sharp maximum at 60–75% gold (*14*). The simple electronic theory is not completely adequate to explain the results because of the continued change in activity after completion of the *d*-band. The authors suggest that in the compositional range of *d*-band saturation, palladium atoms may be modified by strongly adsorbed butadiene so that they behave as atoms with their *d*-band partially filled.

Reynolds (*9*) studied the hydrogenation of benzene over skeletal nickel–copper catalysts prepared by extraction of nickel–copper–aluminum alloy powders with aqueous sodium hydroxide and the hydrogenation of styrene over catalysts prepared by reduction of mixtures of nickel and copper oxides. Magnetic susceptibility and activity fell continuously to zero over the range 0–40% copper for the skeletal alloy catalyst and thus show agreement with the simple band theory. Reduced oxide mixtures, showed similar declines but did not reach low values of magnetic susceptibility until the composition had reached about 60% copper, while the catalytic activity reached zero at 30–40% copper similar to the skeletal catalyst. Contrary to the work of Reynolds, Russell (*15*) found a maximum in activity for hydrogenation of ethylene at 60% copper over reduced copper–nickel catalysts prepared by reduction of oxides formed from precipitated carbonates. Pure nickel is known to adsorb ethylene so strongly that the surface is poisoned. Coincident with the filling of *d*-band holes, there may be a reduction in heat of adsorption of ethylene and hydrogen and an increase in catalytic activity. Ultimately, as copper content is increased, the number of *d*-band holes will become so few that activity will decrease. For the ortho–para hydrogen conversion at $-196°C$ and at near room temperature, Russell and Shield (*16*) found that there was little change in activity with alloy composition in the range from 5 to over 90% copper. Outside of this range, however, there were sharp changes in activity, increasing with nickel content. Using copper–nickel alloy films, Gharpurey and Emmett (*17*) investigated the hydrogenation of ethylene and found a broad region from nearly pure nickel to about 80% copper where the activity was relatively constant, while in the region above 80% copper the activity increased drastically with increasing nickel content. These results are similar to those of Russell on ortho–para hydrogen conversion but do not agree with Russell's results for the hydrogenation of ethylene, as described above.

According to Dowden's rules (*7*) given above, the velocity of a reaction should depend not only on *d*-band vacancies but also on the density gradient $[d \ln N(E)/dE]$ at the Fermi surface since this factor in the electronic theory contributes to the determination of the nature and the amount of the adsorbed species. For the nickel–copper alloys, low temperature specific heat data indicate a sharp fall in the energy density of electron levels $N(E)$ at the Fermi surface near the point where the *d*-band of nickel becomes filled as determined from experimental measurements of the magnetic susceptibility (*18*). We have seen that catalytic activities for hydrogenation of benzene and styrene decreases sharply in the same region in which $N(E)$ decreases. This is the result that would be expected from Dowden's rules for the formation of chemisorbed positive ions. Similar results (*18*) were obtained for the hydrogenation of styrene over nickel-ion alloys with the critical composition occurring at 80 Ni : 20 Fe, the rapid fall in activity in going from 100% nickel to 80 Ni : 20 Fe being attributed to the sharp decrease in $N(E)$. When no d-band vacancies exist in either constituent of a binary alloy, then catalytic activity would be expected to depend more heavily on the density gradient at the Fermi surface. Schwab and co-workers (*19–21*) have studied such alloys, mainly silver alloyed with Zn, Cd, Hg, Ga, In, Tl, Sn, Pb, Sb, or Bi for the decomposition of formic acid into carbon dioxide and hydrogen. They found that for certain alloys the activation energy increased as the Brillouin zone approached a saturation limit; and they reached the conclusion that empty levels in the zone favored catalytic reaction, thus indicating that the activated formic acid adsorption complex should bear a positive charge. In other words, adsorption is favored by large, positive values of the density gradient $[d \ln N(E)/dE]$ at the Fermi surface.

In view of the rather confused picture of the formation of d-band holes in catalysis, it is not surprising that fresh attempts are being made to introduce more order. In order to explain the results of ortho–para hydrogen conversion on palladium–gold alloys obtained by Couper and Eley (*12*), Wise (*22*) introduces an entropy term. Since there is little variation in activation energy of this reaction over the range 0 to 60 at. % gold, the changes in reaction rate must be associated with the pre-exponential term of the Arrhenius expression which is associated with the entropy of activation. He associates the activation process with the promotion of an adsorbate electron into the d-band of the metal. The entropy change per electron of such a process is influenced by changes in the density gradient at the Fermi surface of the solid as a function of alloy composition. The relationship is

$$\Delta S = \tfrac{1}{3}\pi^2 k^2 T [d \ln N(E)/dE]_{E_F}, \tag{4}$$

where k is the Boltzmann constant. Insufficient data are available to test this equation.

Sachtler and co-workers (*23–28*) have recently shed new light on the problem of alloy catalysis. Using copper–nickel films prepared by successively depositing the two metals from the vapor and subsequently sintering the films under ultra-high vacuum at 200°C, they studied the hydrogenation of benzene (*28*). Photoelectric emission, X-ray diffraction studies, and data on diffusion led them to conclude that over a wide range of overall composition (nearly pure nickel to approx. 80% copper) the surface consists of a copper-rich alloy and an almost pure nickel phase. In equilibrated films, each crystallite contains a kernel of almost pure nickel enveloped in a skin of copper-rich alloy. As the composition of either phase is independent of the overall composition, the surface properties of these alloys are constant, which is confirmed by experimental data on the work function. One would also expect the catalytic properties to remain constant over this broad range. The results of Shield and Russell (16) on ortho–para hydrogen conversion and Gharpurey and Emmett (17) on the hydrogenation of ethylene reported above are in good agreement with the model of Sachtler and co-workers, for they show relatively constant activity in the broad two-phase region and rapidly increasing activity on the high-nickel end outside of this region. The outer-alloy phase in the two-phase region, whose nickel content can be determined by selective chemisorption of hydrogen, contains much more copper than is required to fill the holes in the d-band of nickel. Yet, there is appreciable catalytic activity in this region, though much less than with pure nickel. It appears that the widespread notion that holes in the d-band are a prerequisite for catalysis is not strictly true.

There does seem to be some general relationship between d-band vacancies and catalysis, though there are exceptions, and at best the relationships are complex and never fully in line with the simple d-band theory. Many of the discrepancies may be experimental; homogeneous alloys are not easy to prepare and, in some cases, the state of the alloy may be suspect. Tightly adsorbed residues of reactant on catalyst surfaces definitely introduce kinetic complexities that are difficult to interpret, and such residues are prevalent in catalysis. Furthermore, the use of collective bulk properties to explain localized surface phenomena must always be open to suspicion as Sachtler points out; relationships undoubtedly exist, but they are not simple.

14.4   Catalytic Activity and Relations between Surface and Bulk Energetics

Because of the poor correlations, in many cases, between bulk electronic factors and catalytic activities, and despite the complex relationships that may exist between bulk and surface properties, attempts have persisted to relate other bulk factors to catalysis. In particular, the connections between the energetics of bulk and surface compounds and their relation to catalysis have been explored. Inducement for such studies comes from the satisfactory linear correlation between heats of adsorption of oxygen and heats of formation of the most closely corresponding bulk metal oxides (29). Surface potential measurements in the presence of adsorbed oxygen show that the metal–oxygen bond is substantially ionic, an electron from the metal being donated to the oxygen. Thus, we might expect to find a close relationship between heats of oxygen chemisorption and the heats of formation of the bulk oxides, much closer at least than the corresponding relationships for adsorbates such as hydrogen, ethylene, or carbon monoxide which are essentially covalent. The work of the Dutch school (*30–33*) suggests that the metal-catalyzed decomposition of formic acid proceeds by way of a formate intermediate on metals. Infrared studies show that a formate is formed when formic acid is adsorbed on a catalyst surface. Figure 14.3 shows that the activity of metals for the decomposition of formic acid can be related to known or computed heats of formation or metal formates by a curve described as "volcano-shaped." Activity is measured by the temperature T_r at which a predetermined reaction rate (0.16 molecules site^{-1} sec^{-1}) is attained. The volcano-shaped curve is interpreted as follows: When the heat of formation is low as for silver and gold, the rate is low because the concentration of adsorbed species is low; but when the heat of formation is high as for tungsten, the rate is low because the adsorbed species are held too tightly. Thus, a maximum rate is to be expected at some intermediate value of the heat of formation corresponding to some optimum in coverage and rate of desorption. A full test of the theory awaits the compilation of more experimental data.

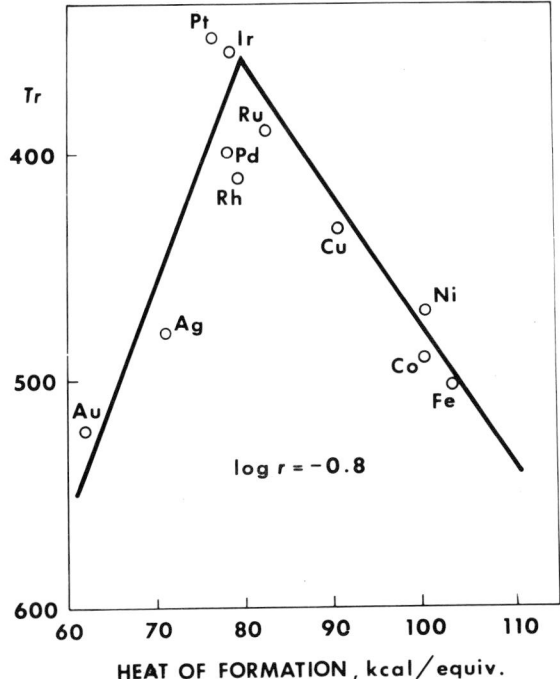

Fig. 14.3. Activity of various metals for the decomposition of formic acid as a function of the heat of formation of their formates. [After Fahrenfort *et al.*, "The Mechanism of Heterogeneous Catalysis" (de Boer *et al.*, eds.). Elsevier, Amsterdam, 1956.]

The concept of surface intermediates attached by localized bonds of one type or another is turning out to be fruitful as we shall see in Chapter XVI. But the application of bulk factors, electronic or other, to surface problems is bound to run into difficulties. The inherent heterogeneities of surfaces, such as those represented by normal surface states and by imperfections, and the heterogeneities induced by strongly adsorbed residues of one sort or another which often clutter the surfaces of catalysts are bound to introduce complexities into the course of a reaction that defy detailed correlation with bulk properties. Each system, reactants and catalyst surface, has its own peculiar combination of controlling factors arising out of the surface properties, collective or localized, which reflect bulk properties with varying degrees of faithfulness. That is what makes generalization so difficult.

REFERENCES

1. Boudart, M., *J. Am. Chem. Soc.* **72**, 1040 (1950).
2. Beeck, O., *Discussions Faraday Soc.* **8**, 118 (1950).
3. Baker, McD., M., and Jenkins, G. I., *Advan. Catalysis* **7**, 1 (1955).
4. Bond, G. C., "Catalysis by Metals," p. 479. Academic Press, New York, 1962.
5. Eley, D. D., and Shooter, D., *Proc. Chem. Soc.* 315 (1959).
6. Schuit, G. C. A., van Reijen, L. L., and Sachtler, W. M. H., *Proc. 2nd Intern. Congr. Catalysis*, Vol. 1, p. 893, Technip, Paris, 1961.
7. Dowden, D. A., *J. Chem. Soc.* 242 (1950).
8. Mott, N. F., and Jones, H., "Properties of Metals and Alloys." Oxford Univ. Press, London, and New York, 1936.
9. Reynolds, P. W., *J. Chem. Soc.* 265 (1950).
10. Ahern, S. A., Martin, M. J. C., and Sucksmith, W., *Proc. Roy. Soc. (London)*, Ser. A **248**, 145 (1958).
11. Kittel, C., "Introduction to Solid State Physics." Wiley, New York, 1956.
12. Couper, A., and Eley, D. D., *Discussions Faraday Soc.* **8**, 105 (1950).
13. Couper, A., and Metcalfe, A., *J. Phys. Chem.* **70**, 1850 (1966).
14. Joice, B. J., Rooney, J. J., Wells, P. B., and Wilson, G. R., *Discussions Faraday Soc.*, **41**, 223 (1966).
15. Best, R. J., and Russell, W. W., *J. Am. Chem. Soc.* **76**, 838 (1954).
16. Shield, L. S., and Russell, W. W., *J. Phys. Chem.* **64**, 1592 (1960).
17. Gharpurey, M. K., and Emmett, P. H., *J. Phys. Chem.* **65**, 1182–1184 (1961).
18. Dowden, D. A., and Reynolds, P. W., *Discussions Faraday Soc.* **8**, 184 (1950).
19. Schwab, G. M., *Trans. Faraday Soc.* **42**, 689 (1946).
20. Schwab, G. M., and Pestamatjoglou, S., *J. Phys. Colloid Chem.* **52**, 1046 (1948).
21. Schwab, G. M., *Discussions Faraday Soc.* **8**, 166 (1950).
22. Wise, H., *J. Catalysis* **10**, 69 (1968).
23. Van der Plank, P., and Sachtler, W. M. H., *J. Catalysis* **7**, 300 (1967).
24. Sachtler, W. M. H., Dorgelo, G. J. H., Jongepier, R., *J. Catalysis* **4**, 100 (1965).
25. Sachtler, W. M. H., and Dorgelo, G. J. H., *J. Catalysis* **4**, 654 (1965).
26. Sachtler, W. M. H., and Jongepier, R., *J. Catalysis* **4**, 665 (1965).

27. Sachtler, W. M. H., Dorgelo, G. J. H., and Jongepier, R., *Proc. Intern. Symp. Basic Probl. Thin Film Phys.* p. 218. Clausthal-Gottingen, 1965.
28. Van der Plank, P., and Sachtler, W. M. H., *J. Catalysis* **12**, 35 (1968).
29. Bond, G. C., "Catalysis by Metals," p. 472. Academic Press, New York, 1962.
30. Fahrenfort, J., van Reijen, L. L., and Sachtler, W. M. H., *Z. Electrochem.* **64**, 216 (1960).
31. Fahrenfort, J., van Reijen, L. L., and Sachtler, W. M. H., "The Mechanism of Heterogeneous Catalysis" (J. H. deBoer *et al.*, eds.), p. 23. Elsevier, Amsterdam, 1960.
32. Mars, P., Scholten, J. J. F., and Zwietering, P., *Advan. Catalysis* **14**, 35 (1963).
33. Sachtler, W. M. H., *Discovery* **26**, 16 (1965).

XV

ELECTRONIC FACTORS IN CATALYSIS—SEMICONDUCTORS

15.1 Introduction

In Chapter VIII, relationships between semiconductivity and chemisorption were developed. In this chapter, we want to extend the relationships to catalysis. For background information on the theory of semiconductors, the reader is referred to Chapter VIII. Practically all of the work has been done with semiconducting oxides, and our remarks are limited chiefly to such substances, although they will apply, in principle, to semiconductors in general. The initial ideas relating electronic properties of semiconductors to catalysis involved the process of charge transfer in which a free electron is transferred completely from the semiconductor to the adsorbate or vice versa. Reaction was then assumed to take place through ionic intermediates. The fundamental idea of charge-transfer catalysis was first put forth by Wagner and Hauffe (*1*) in 1938, and subsequently pursued by Garner *et al.* (*2*), Wolkenstein (*3*), Hauffe (*4*), and Hauffe and Schlosser (*5*). In its simplest form, the charge-transfer theory of catalysis is a qualitative extension of the boundary-layer theory of chemisorption that was reviewed in Chapter VIII. A more refined, quantitative presentation of the theory was given by Garrett (*6*). There is much experimental evidence which can be considered to support the charge-transfer theory, but most of it is qualitative in nature, and no precise, quantitative demonstration of the correctness of the model has appeared. However, the theory has considerable heuristic value. Perhaps, the most valuable experiments have been those in which a single reaction, such as the decomposition of formic acid or of nitrous oxide, has been selected and compared on a series of catalysts of known electronic and ionic structure. Wolkenstein (7) has approached the same problem from a more fundamental and broader point of view. As we saw in Chapter VIII, he prefers to consider transfer of an electron in its quantum mechanical sense as a transfer of an electron from one energy level to another, and not in a "geometrical" sense as a transfer from one particle to another. In this theory, the adsorption bond need not be purely ionic, but may vary from purely covalent to purely ionic depending on the characteristics of the system. However, in either system, we are dealing with the phenomenon of cooperative interaction in which particles virtually interact with the crystal lattice as a whole, since the entire collective of free electrons and holes of the lattice comes into play.

Not all chemisorptions or reactions on semiconductor surfaces can be interpreted in the light of cooperative interactions, that is, interactions with free electrons and holes. Reactions are known where the direct interaction of lattice defects with the reacting molecules represents the decisive step. And here the simple boundary-layer or charge-transfer theory does not apply. Lattice defects, such as ions in interstitial positions or ion vacancies, may have sufficient mobility, especially at elevated temperatures, that interactions with them at the surface may approach the rate of interaction with mobile, free electrons. To distinguish from cooperative interactions, we call such interactions *localized interactions*. We shall discuss both types of interactions in this chapter in relation to catalysis. And, in addition, we shall review some recent work which relates catalysis on semiconductors directly to surface states.

15.2   Cooperative Electronic Interactions and Catalysis

There are several ways to investigate the influence of electronic structure on the catalytic activity of oxides. First, the electronic structure of a catalyst may be modified systematically and the effect on a given reaction determined. Various methods of modifying the electronic structure may be employed: (1) pretreatment of the catalyst at elevated temperatures in oxidizing or reducing atmospheres, (2) preadsorbing donor or acceptor gases, (3) incorporation of altervalent ions. All of these modifications may be followed by measurements of electrical conductivity. A second method of investigating the influence of electronic structure is to determine the activities of a given oxide toward a series of different types of catalytic reactions. Third, a given reaction may be compared over a wide variety of oxides. We shall now review the various experimental studies which have been made that provide the basis for the theories which have been proposed.

SEMICONDUCTIVITY AND CATALYSIS

One of the most complete studies is the decomposition of nitrous oxide on a series of metal oxides including p-type and n-type semiconductors and insulators. There is strong evidence (8) that the reaction proceeds by the mechanism:

$$N_2O_{(g)} + \text{electron (from catalyst)} \leftrightarrows N_{2(g)} + O^-_{(ads)} \tag{1}$$

followed by

$$2O^-_{(ads)} \rightarrow O_{2(g)} + 2 \text{ electrons (to catalyst)} \tag{2}$$

or by

$$O^-_{(ads)} + N_2O_{(g)} \rightarrow N_{2(g)} + O_{2(g)} + \text{electron (to catalyst)}. \tag{3}$$

Reaction (3) is more likely when the coverage with oxygen is high (9, 10). It has been proposed by Engell and Hauffe (11) that the rate-controlling step in the decomposition reaction is the desorption of oxygen. On the other hand, more recent work (12, 13) favors the adsorption reaction as rate controlling.

Combining their own work (9) with that of Schwab and co-workers (14), and Schmid and Keller (15), Dell et al. (9) have shown that the oxides active for decomposing nitrous oxide can be divided into three groups depending on the temperature required to produce measurable decomposition. The results are shown in Fig. 15.1. The first group which is active below 400°C contains only p-type semiconductors (Cu_2O, CoO, Mn_2O_3, and NiO); the last, and least effective group, operates above 550°C and contains only n-type semiconductors (Al_2O_3, ZnO, CdO, TiO_2 Fe_2O_3, and Ga_2O_3). An intermediate group includes certain insulators (e.g., MgO and CaO), though recent work (16) places these oxides less active than the n-type oxides.

When decomposition occurs over one of the n-type oxides, e.g., ZnO, conductivity drops in agreement with the mechanism that free electrons are trapped by the chemisorbed oxygen. It has been suggested by Dell et al. (9)

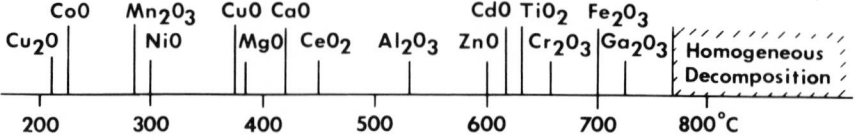

FIG. 15.1. Relative activity of oxide for the decomposition of nitrous oxide showing the temperature at which reaction first becomes appreciable. [After Dell et al., *Trans. Faraday Soc.* **49**, 201 (1953). By permission of The Faraday Society.]

that the high temperature necessary in the case of n-type oxides may be due to the difficulty of these oxides to chemisorb oxygen. The adsorption process may require a deep electron trap which is probably available at only relatively few sites on the surface. When p-type oxides are used, e.g., NiO, conductivity increases, for electrons trapped on chemisorbed oxygen are pulled from the valence band leaving conducting holes. Firm conclusions concerning the insulator oxides cannot be drawn until their position in the series has been established unequivocally. Hauffe *et al.* (*17*) and Winter (*12*) found that lithium addition to nickel oxide, which increases the p-character of the oxide (lowers the Fermi level), also increases the catalytic activity for N_2O decomposition in agreement with the activity pattern discussed above.

Another reaction which has been studied in some detail is the oxidation of carbon monoxide. It is more complex than the decomposition of N_2O because, in addition to oxygen, carbon monoxide and carbon dioxide can also be adsorbed on the catalyst. However, according to Stone (*18*) it may be concluded that the most efficient catalysts are the p-type oxides, which are usually active below 50°C, that the second most active group consists mainly of n-type oxides, and that insulators are the least active.

Investigations of the oxidation of carbon monoxide over lithium-doped nickel oxide catalysts are confusing and contradictory. Parravano (*19*) showed that adding lithium to nickel oxide caused an increase in activation energy, while adding ions such as Cr^{3+} reduced the activation energy. Later, results by Schwab and Block (*20*) revealed the opposite effect. Parravano obtained his results in the range 20–250°C while Schwab and Block worked in the range 250–400°C. Several speculations have been made that there might exist different adsorption mechanisms and rate-determining steps in the two regions. A reinvestigation by Keier *et al.* (*21*) of Parravano's low temperature work confirmed the rise in activation energy with addition of lithium, but they also found a rise in activation energy with addition of chromium contrary to the results of Parravano. In the high-temperature region, Dry and Stone (*22*) obtained an increase in activation energy on addition of chromium in agreement with Schwab and Block, though the increase was less abrupt. Dry and Stone further brought out the important fact that the changes in activation energy are matched by appreciable changes in the frequency factor so that the changes in activity are much smaller than would be anticipated on the basis of activation energy changes alone. A good, speculative discussion of the problem is given by Stone (*23*). Amigues and Teichner (*24*) studied the oxidation of carbon monoxide on pure zinc oxide at 260°C. They found that oxygen appears to be adsorbed both in the ionic and nonionic form, while carbon monoxide and carbon dioxide are adsorbed essentially in nonionic forms. They present evidence to show that the nonionic form of adsorbed oxygen participates in the reaction and that the slowest step of the oxidation does not involve the transfer of electrons. On this basis, no effect of doping of ZnO by Li^+ or Ga^{3+} would be expected. None is observed.

The H_2–D_2 exchange reaction is a good one to study because it proceeds at low enough temperatures that electronic phenomena are not apt to be obscured by ionic phenomena. Molinari and Parravano (25) investigated the H_2–D_2 exchange reaction on ZnO modified by the addition of altervalent ions. Addition of Li_2O to ZnO decreases the n-type character and it was found that activity for H_2–D_2 exchange was also decreased. But the addition of Ga_2O_3 or Al_2O_3, which increases the n-type character, increased the catalytic activity. Voltz and Weller (26) studied the same reaction over chromium oxide, a p-type semiconductor. They measured catalytic activity at -78 or $-195°C$ after pretreatment of the catalyst in an atmosphere of hydrogen or oxygen at 500°C. In the " oxidized " state, the catalyst was found to have a low catalytic activity and a high conductivity, i.e., a large number of positive holes. But in the reduced state, the catalytic activity was high and the conductivity low. Clark (27) shows a similar dependence on pretreatment of the catalyst for a number of n- and p-type oxides. The results of the above studies on the H_2–D_2 exchange reaction are summarized in Table 15.1. Note that in

TABLE 15.1

H_2–D_2 EXCHANGE ON OXIDE CATALYSTS[a]

Catalyst	Author	Pretreatment	Effect on oxide	Effect on exchange activity
ZnO	Molinari and Parravano (25)	Addition of cations with valence greater than 2	Increases number of electrons	Rate increased
		Addition of cations with valence of 1	Decreases number of electrons	Rate decreased
Cr_2O_3	Voltz and Weller (26)	O_2 at 500°C	Increases the number of positive holes	Rate decreased
		H_2 at 500°C	Decreases the number of positive holes	Rate increased
MoO_3, WO_3 UO_3, Cr_2O_3, NiO, FeO, Zno	Clark (27)	O_2 at 500°C	Decreases the number of electrons	Rate decreased
		H_2 at 500°C	Increases the number of electrons	Rate increased

[a] Baker and Jenkins (28).

all cases H_2–D_2 exchange activity increases with increasing electron concentration. In other words, H_2–D_2 exchange is favored by modifications of the catalyst in the direction p-type → n-type. The effect is opposite to that found for the decomposition of N_2O, which is favored by increasing p-character. In order to explain why activity for H_2–D_2 exchange increases with increasing electron concentration in the oxide, Baker and Jenkins (28) relate the phenomenon to the strength of the hydrogen–metal bond. Beeck (29) found that hydrogenation activity on metal films increased with decreasing heat of hydrogen adsorption, which he used as a measure of the hydrogen–metal bond strength. By analogy, Baker and Jenkins suggest that if hydrogen adsorption on metal oxides occurs either with the formation of covalent bonds or positive ions, then an increase in electron concentration should decrease the strength of bonding.

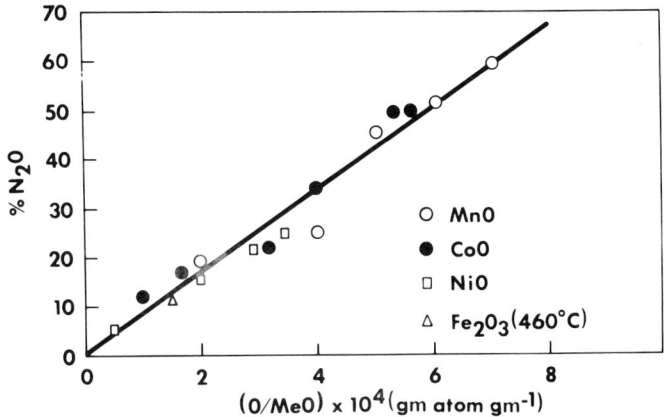

Fig. 15.2. The catalytic activity of a series of metallic oxides MeO for the catalytic oxidation of ammonia to nitrous oxide, as a function of the excess oxygen content of the oxides. [After Krauss, *Z. Elektrochem.* **53**, 320 (1948).]

A good example of the relationship between p-type character and catalytic activity has been given by Krauss (30) who studied the rate of formation of nitrous oxide during the oxidation of ammonia over oxides of manganese, cobalt, nickel, and iron. He determined the degree of p-type character by titrating the amount of excess oxygen in the lattice. Figure 15.2 shows the remarkable linear proportionality which he obtained between the amount of excess oxygen in the lattice and the percentage of N_2O in the exit gases from a flow reactor.

As a final example of the relationships between catalytic activity and electronic structure of semiconducting metal oxides, we mention the work of Hart and Ross (31) who made a thorough study of the decomposition of hydrogen peroxide vapor over various oxides. They found the sequence of activities to be:

$$Mn_2O_3 > PbO > Ag_2O > CoO > Cu_2O > Fe_2O_3 > CdO \gg ZnO = MgO > \alpha\text{-}Al_2O_3.$$

Note that this sequence resembles that for the decomposition of nitrous oxide shown in Fig. 15.1. Again, the *p*-type oxides are found to be the most active. The *n*-type oxides turn out to be more active than the insulator oxides in disagreement with the order given by Stone for nitrous oxide decomposition in Fig. 15.1, but in agreement with the work of Saito *et al.* (*16*) for the same reaction.

Thus, there seems to be a phenomenological relationship between catalysis and semiconductivity, and, therefore, some justification for the development of electronic theories of catalysis. However, the picture is not always clear and there are many examples of confusion and contradiction such as in the oxidation of carbon monoxide. Furthermore, the relationships are qualitative. Probably, quantitative relationships should not be expected because *n*- or *p*-type character in itself is not a sufficiently profound criterion. As Stone (*23*) points out, electron exchange between catalyst and adsorbate is largely determined by "chemical" factors, and changes in the Fermi level produced by dopents, thermal or chemical pretreatments, and the like must be rather specific and not subject to broad generalizations.

CHARGE-TRANSFER THEORIES AND CATALYSIS

We start with a simple example given by Hauffe (*32*) for the decomposition of nitrous oxide and then proceed to more detailed theories. Hauffe *et al.* (*17*) have given strong evidence that the rate-determining step for the nitrous oxide decomposition on *p*-type oxides (e.g., NiO) is the desorption step which may be written by modification of Eq. (3) as

$$\oplus_{(B)} + O^-_{(ads)} + N_2O_{(g)} \rightarrow N_{2(g)} + O_{2(g)}, \tag{4}$$

where $\oplus_{(B)}$ is a positive hole in the boundary layer and $O^-_{(ads)}$ is an adsorbed oxygen anion. If Li_2O is added to NiO the concentration of electron holes will be increased as shown in Fig. 8.4d, and which may be written as

$$\tfrac{1}{2}O_{2(g)} + Li_2O \rightleftharpoons 2Li|Ni|' + 2\oplus + 2NiO, \tag{5}$$

where $Li |Ni|'$ is a lithium ion on a nickel ion lattice site. As the amount of Li_2O added is increased, thus increasing the concentration of holes, the number of lattice electrons or the chemical potential (Fermi level) of the electrons reaches such a low value that the mechanism of the reaction changes. Because of the dearth of electrons and the resulting drop in the Fermi level, chemisorption becomes the rate-controlling step. Equation (1), the chemisorption step, may be rewritten

$$N_2O_{(g)} \leftrightarrows N_2O^-_{(ads)} + \oplus_{(B)} \quad \text{(step consuming electrons, slow)}, \quad (6)$$

and

$$N_2O^-_{(ads)} \to O^-_{(ads)} + N_{2(g)} \quad \text{(step without electron consumption, fast)}. \quad (7)$$

Experimental results show that the forward rate of reaction is proportional to the first power of the partial pressure of $N_2O(p_{N_2O})$. Thus, we may write for the rate of decomposition of N_2O on a p-type semiconductor when the chemisorption step is controlling [Eq. (6)]:

$$-dn_{N_2O_{(g)}}/dt = k_1 p_{N_2O_{(g)}} - k_2 n_{N_2O^-_{(ads)}} n_{\oplus(B)}, \quad (8)$$

or, for an n-type catalyst:

$$-dn_{N_2O_{(g)}}/dt = k_1' p_{N_2O_{(g)}} n_{\ominus(B)} - k_2' n_{N_2O^-_{(ads)}}, \quad (9)$$

where $n_{\ominus(B)}$ is the concentration of free electrons in the boundary layer. When desorption is rate controlling, we easily find from Eq. (4) for a p-type catalyst that the rate expression becomes

$$dn_{O_{2(g)}}/dt = k_3 n_{O^-_{(ads)}} n_{\oplus(B)} p_{N_2O_{(g)}}, \quad (10)$$

and for an n-type,

$$dn_{O_{2(g)}}/dt = k_4 n_{O^-_{(ads)}} p_{N_2O_{(g)}} - k_5 n_{\ominus(B)}. \quad (11)$$

From Eq. (10), we see that the rate of the desorption process increases with increasing concentration of holes for a p-type catalyst, and from Eq. (11) that the rate of the desorption process decreases with increasing concentration of free electrons for an n-type catalyst. These relationships are confirmed by experiment. From Eq. (8), it is clear that the rate of chemisorption decreases with increasing concentration of positive holes. Thus, as the amount of added Li_2O is increased with consequent increase in positive hole concentration, the rate of chemisorption of N_2O continually decreases until it finally becomes rate controlling for the decomposition of N_2O.

We now turn to a more general case (*33*) and examine the complexities which can arise as a result of the multitude of possible single steps, each of which might be rate determining. Consider the catalyzed reaction

$$A + D \to AD, \quad (12)$$

where A can react as an electron acceptor (A^-) and D as an electron donor (D^+) or vice versa. Both A and D may be physically adsorbed $(A^\times, D^\times)$ and product AD is desorbed in a single step only from the physically adsorbed

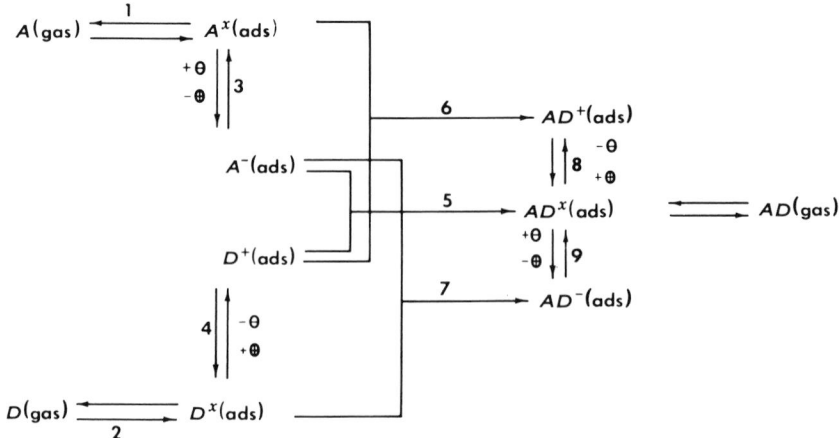

Fig. 15.3. Possible steps of a reaction A + D → AD catalyzed at the surface of an *n*- or *p*-type semiconductor: ⊖ is a free electron and ⊕ a free positive hole. [After Hauffe, in "Coloquio Sobre Qimica Fisica de procesos en superficies solidas (XXV Aniversario del C.S.I.C., Octubre 1964)." Libreria Cientifica Medinaceli, Madrid (1965).]

state (AD^x). The reaction scheme is shown in Fig. 15.3 where the steps which involve free electrons and holes are distinguishable from those which do not. Steps 5, 6, and 7 are not influenced directly by free electrons or holes, whereas steps 3, 4, 8, and 9 are. Therefore, if one of steps 5, 6, or 7 are rate controlling, the overall reaction is not dependent on the concentration of free electrons or holes. However, if one of steps 3, 4, 8, or 9 is rate controlling, then the overall reaction is directly dependent on the concentration of free electrons or holes. Such a scheme is oversimplified because it ignores the effect of lattice defects. We shall discuss the direct participation of lattice defects in catalytic reactions in Section 15.3.

Whether the rate-controlling step involves free electrons and holes or not, the rate of the electron or hole consuming step is usually not equal to the electron or hole supplying step. Consequently, a space-charge layer will build up. The chemisorption of one or both reactants creates a concentration gradient of electrons or holes and, at steady state, the diffusion of the current of electrons or holes will be just balanced by a current in the opposite direction caused by the gradient of the resulting electric field. How much of a layer is built up depends on the relative electron affinities and rates of chemisorption of A and D. In Fig. 15.4, two cases for an *n*-type semiconductor are illustrated schematically: Case (a) illustrates the situation where the rates of consumption and supply of electrons are equal and thus no space-charge layer is built up. Case (b) shows a space-charge layer with *exhaustion* of electrons when chemisorption of A is faster than the chemisorption of D or reaction step 7 in

Fig. 15.3 and with *enrichment* of electrons when the chemisorption of D is faster than the chemisorption of A or reaction step 6 in Fig. 15.3. In Fig. 15.4, E_A and E_D are the electron exchange levels of species A and D, respectively; E_L and E_V the electronic energies of the conduction and valence band edges; and E_F the Fermi level. The Fermi level determines the extent of chemisorption and the diffusion potential. The electronic exchange level of the initially nonchemisorbing molecules is moved away from the Fermi level by the chemisorption of the other molecule and thus the new position of the exchange level $E \pm V_D$ becomes more favorable for chemisorption. This enhancement of the chemisorption of one species by the chemisorption of the other occurs only when the two species are oppositely charged. A displacement should occur, according to this model, when the two species have the same charge; the gas with the higher electron exchange level E_D with respect to the Fermi level should displace the other gas.

Garrett (6) gives a more quantitative treatment of a reaction scheme similar to that in Fig. 15.3. He simplifies the scheme by ignoring the interaction of physically adsorbed species $(A^\times, D^\times)$ with ionized chemisorbed species

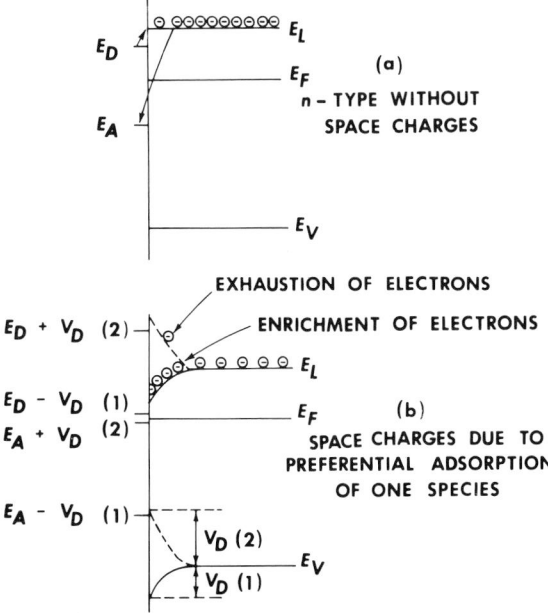

FIG. 15.4. Band model for an operating *n*-type catalyst without space charges (a) and with space charges if the donor or the acceptor reaction step is rate determining (b); E_A and E_D are the electronic exchange levels of the A and D molecules, respectively. [After Hauffe, in "Coloquio Sobre Qimica Fisica de procesos en superficies solidas (XXV Aniversario del C.S.I.C., Octubre 1964)." Libreria Cientifica Medinaceli Madrid (1965).]

(A^-, D^+). The product AD exists on the surface only as neutral molecules. Unfortunately, the treatment involves quite a number of physical parameters for semiconductor, bulk and surface properties, and catalytic activity, and there does not appear to be a single satisfactory case in the literature in which a sufficient number of different techniques have been applied to the same system under the same conditions to allow a quantitative check on the predictions of the theory. Until such information becomes available, the quantitative theory is no more useful than the purely descriptive models. It should be pointed out, however, that descriptive models, based on no more information than a set of reaction-rate measurements as a function of, for example, the bulk doping of a semiconductor, cannot provide unambiguous and unique answers about the correctness of the theory.

THEORIES OF THE RUSSIAN SCHOOL

Russian scientists, Wolkenstein and co-workers in particular, have extended their theory of chemisorption to catalysis. We saw in Section 8.3 that two types of chemisorption were defined: "strong" and "weak." "Strong" chemisorption involves free electrons or holes of the crystal lattice, and catalysis associated with this type of chemisorption will be our chief concern in this section. Catalysis associated with "weak" chemisorption will be discussed in the following section. We shall not go into great detail, because the same difficulty applies here as in the case of other charge-transfer theories, namely, the inability to claim uniqueness. We shall mention only a few special features.

Wolkenstein (7) considers acceptor and donor reactions. An acceptor reaction is a reaction that is accelerated by electrons, reaction step 7 in Fig. 15.3 for example. A donor reaction is one that is accelerated by positive holes, as in reaction step 6 in Fig. 15.3. An acceptor reaction proceeds faster the higher the Fermi level. This may be seen from Eq. (49), Chapter VIII which shows an increasing amount of negatively adsorbed species with increasing ϵ_s^+, the energy separation of the top of the valence band from the Fermi level. A donor reaction proceeds faster the lower the Fermi level as seen from Eq. (50), Chapter VIII. An example of an acceptor reaction is the dehydrogenation of ethyl alcohol, and an example of a donor reaction is the dehydration of ethyl alcohol.

As explained in Section 8.4, the participation of a free electron or hole in chemisorption leads to the transformation of a valence-saturated particle into an ion-radical and to the transformation of a radical into a valence-saturated charged particle. Thus "strong" chemisorption may give particles of radical or valence-saturated forms depending on the characteristics of the adsorbate. Wolkenstein considers that the radical or ion-radical forms will be most reactive, though nonradical mechanisms are not excluded. An exam-

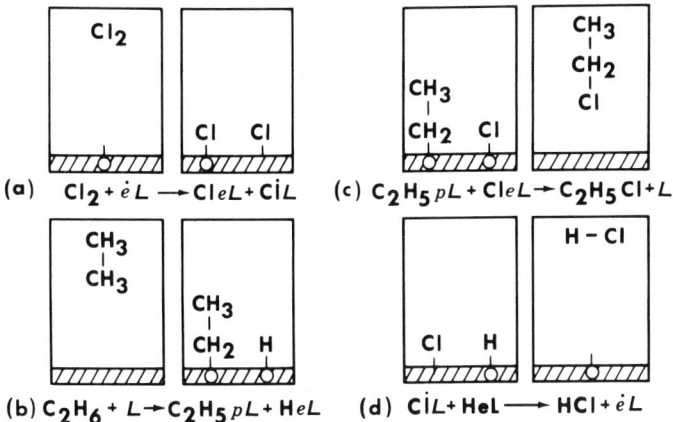

FIG. 15.5. Wolkenstein's radical mechanism (7) for the catalyzed reaction

$$C_2H_6 + Cl_2 \rightarrow C_2H_5Cl + HCl.$$

ple of a radical mechanism given by Wolkenstein is the chlorination of ethane:

$$C_2H_6 + Cl_2 \rightarrow C_2H_5Cl + HCl \qquad (13)$$

a reaction that takes place on $ZnCl_2$. Figure 15.5 shows a possible radical mechanism, where $\dot{e}L$ is a free electron of the lattice, $Cl\dot{e}L = Cl^-$, $Cl\dot{L}$ a chlorine atom adsorbed as a radical, C_2H_5pL an ethyl radical which has captured a free hole, and $H\dot{e}L$ a hydrogen atom which has captured a free electron. This reaction also illustrates a chain mechanism, since the free electron used in the reaction is regenerated. Mechanisms for other reactions, such as hydrogenation of ethylene or the addition of hydrogen halides to olefins, may be constructed.

WEAKNESSES IN THEORIES OF CATALYSIS INVOLVING COOPERATIVE INTERACTIONS

In the first place, such theories are limited by the lack of knowledge of surface states. Secondly, the concentration of free electrons and holes are taken into account, but their mobilities are ignored. Questions such as the degree of localization of electrons and holes could be extremely important in relation to catalytic problems. Thirdly, the mobilities of ions and interactions of defects are also neglected, a serious oversimplification at elevated temperatures. A fourth point which needs further investigation is the question of extent of d-orbital overlap or exchange interaction between nearest or next-nearest neighbors. Transition metal oxides show this interaction. By addition of diamagnetic ions such as Mg^{2+} to the lattice of a transition metal oxide, the exchange interaction is disturbed, and at high magnetic dilution may disappear. The effect has been strikingly shown in the systems CoO–MgO, Cr_2O_3–Al_2O_3 and UO_2–ThO_2 (*34*). The question arises as to whether such changes in the exchange interaction occur in the doping of semiconducting oxides, for example, the addition of Li_2O to NiO. If so, the simultaneous changes in conductive and magnetic properties add a note of confusion to the interpretation of catalytic phenomena in such systems. To test the point, Vrieland and Selwood (*35*) studied two systems for the decomposition of ammonia. The first system consisted of Eu_2O_3 and Gd_2O_3, whose properties are nearly identical except for a difference in semiconductivity by a factor of 19. The second system consisted of mixtures of MgO and MnO in which the exchange interaction diminishes significantly on increasing additions of MgO, but the semiconductivity remains essentially constant over the entire range of composition. The rate of decomposition of ammonia over Eu_2O_3 and Gd_2O_3 and the activation energy were found to be almost identical. In the MgO–MnO system, no perceptible changes in rate and activation energy for ammonia decomposition were found with changes in composition. These experiments imply that neither electronic nor magnetic factors were important for catalysis in these two examples. All that can be said at present is that many more experiments are needed to determine how universal these results are. A fifth point which needs more study is the effect of adsorbed residues of reactants or products on the catalyst surface, since these residues can influence the interpretation of semiconductivity-catalysis studies (*36*). Finally, not enough attention has been paid to localized interactions at semiconductor surfaces, a subject to which we now turn.

15.3   Localized Interactions and Catalysis

Wolkenstein was one of the first to clearly recognize that catalysis on semiconductors need not always be associated with the conducting free electrons and holes. Through his theory of " weak " chemisorption states, he emphasized the importance of catalysis by localized levels. However, progress, both experimental and theoretical, in elucidating the mechanisms of localized interactions has been slow for two general reasons. In the first place, the elegance and generality of the collective approach involving free electrons and holes was sufficiently great to command thorough investigation. And many papers have been written on the relationships between semiconductivity and catalysis in the last 15 years. For example, about a hundred papers have appeared on zinc oxide and nickel oxide alone. Zinc oxide was a good choice, and good correlations have been obtained in many instances. Nickel oxide, however, has introduced some serious problems, as we saw in connection with the oxidation of carbon monoxide in Section 15.2. The band theory and chemistry of lattice imperfections are not as well established for nickel oxide as for zinc oxide, and consequently the link between semiconductivity and catalysis is much less direct. The second reason for slow progress in the development of mechanisms of localized interactions is the tremendous difficulty in constructing theories and devising experiments to give quantitative answers which are at least reasonable approximations to the truth. Although localized models of covalent or ionic character have been proposed by Dowden (*37*) and others (*38*), the problems of estimating the relative contributions of Coulombic, polarization, and stabilization energies present a formidable barrier. Essentially, it is the many-electron quantum mechanical problem which is involved as opposed to the quantum statistical approach employed in collective electron treatments. There is a great need to design experiments and theories which will allow more quantitative discussion of chemisorption and catalysis on oxides from a localized point of view.

Before reviewing the rather meager direct evidence for the existence of localized interactions of importance in catalysis, we should describe in more detail what we mean by the term. In general, we refer to interactions which do not involve free electrons or holes of the solid. First, under localized interactions we include all interactions between adsorbed molecules and point defects of the catalyst surface, such as, anion and cation vacancies, interstitial anions and cations, F-centers, V-centers, ad-cations and ad-anions, and foreign impurities. Second, we include interactions associated with Wolkenstein's "weak" chemisorption, that is, the formation of a localized bond between a lattice ion, defect or normal, and an adsorbed molecule without the participation of free electrons of the conduction band or holes of the valence band. In such localized interaction, it is not necessary to stipulate that the bond be localized completely between a single lattice ion and the adsorbate; the bond may extend to near neighbors as quantum mechanical calculations imply. The detailed nature of the coordination between a metal ion and an adsorbate molecule will be discussed in Chapter XVI, for example, in connection with π-complexes. Third, we include interactions involving local symmetry, such as crystal- or ligand-field effects in chemisorption and catalysis. There are other localized interactions which may be involved in catalysis, such as chemisorption on foreign molecules strongly adsorbed on a surface, a phenomenon of considerable importance in catalysis on insulators to be discussed in Chapter XVI.

EXPERIMENTAL EVIDENCE FOR LOCALIZED INTERACTION

Simkovich and Wagner (*39*) have considered the decomposition of an ethyl halide

$$CH_3CH_2X \rightleftharpoons CH_2{=}CH_2 + HX \qquad (14)$$

on the surface of a metal halide at four different types of point defects: ad-cations, ad-anions, cation vacancies, and anion vacancies. Ad-ions are defined as surface cations or anions on top of the outermost complete lattice plane, and they are analogous to interstitial cations and anions in the bulk crystal. Figure 15.6a illustrates a possible mechanism at the plane surface of a metal halide crystal in which H and X atoms are formed followed by desorption of HX. Figure 15.6b–e show how the break-away of H and X might be facilitated by the presence of one of the four defects mentioned above.

We see in Fig. 15.6b a hydrogen atom and a halogen atom of ethyl halide

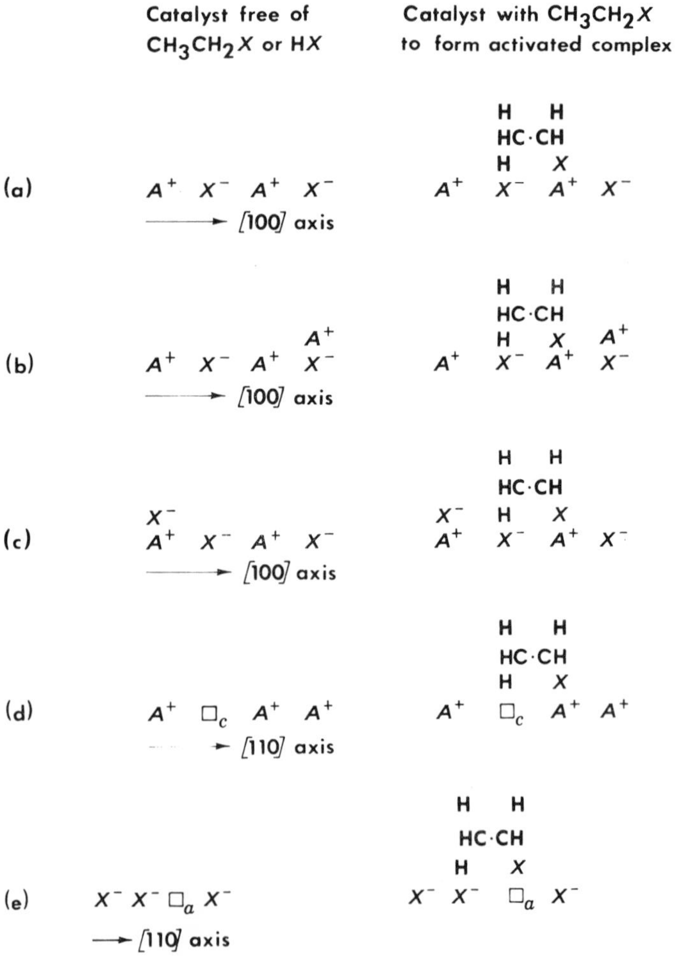

	Catalyst free of CH$_3$CH$_2$X or HX	Catalyst with CH$_3$CH$_2$X to form activated complex

(a)

A^+ X^- A^+ X^-
⟶ [100] axis

H   H
HC·CH
H   X
A^+ X^- A^+ X^-

(b)

            A^+
A^+ X^- A^+ X^-
⟶ [100] axis

H   H
HC·CH
H   X A^+
A^+ X^- A^+ X^-

(c)

X^-
A^+ X^- A^+ X^-
⟶ [100] axis

H   H
HC·CH
X^- H   X
A^+ X^- A^+ X^-

(d)

A^+ $\square_c$ A^+ A^+
⟶ [110] axis

H   H
HC·CH
H   X
A^+ $\square_c$ A^+ A^+

(e)

X^- X^- $\square_a$ X^-
⟶ [110] axis

H   H
HC·CH
H   X
X^- X^- $\square_a$ X^-

FIG. 15.6. Mechanisms for the reaction CH$_3$CH$_2$X → CH$_2$=CH$_2$ + HX on the [001] surface of a metal halide A^+X^- with rock salt structure for (a) plane surface, (b) surface with ad-cation, (c) surface with ad-anion, (d) surface with cation vacancy $\square_c$, (e) surface with anion vacancy $\square_a$ (39).

attached to an anion in the outermost plane and an ad-cation, respectively, corresponding to adsorption of HX on the catalyst. When HX is desorbed, the surface is restored to its initial state and catalytic reaction proceeds. An alternative, Fig. 15.6c is the adsorption of a hydrogen on an ad-anion and a halogen on a cation. If cation vacancies exist, a hydrogen atom may occupy a cation vacancy in the outermost plane and a halogen atom may simultaneously adsorb on top of a cation in the same plane (Fig. 15.6d). By diffusion, H and X may come together, combine and desorb, leaving the catalyst in its initial state. Alternatively, the halogen atom may occupy an anion vacancy and the hydrogen atom may be adsorbed on an anion as shown in Fig. 15.6è. The authors assume that in each case involving a defect the energy of the system is lower than the case without defects (Fig. 15.6a). Therefore, it might be expected that the activation energy for transfer of H and X from the organic compound to the surface may be lower in those cases where defects are involved. If the activation energy is lowered sufficiently to more than counter any compensation effect, then the reaction rate will increase for the mechanisms involving defects. To test their theory, the authors used AgCl doped with various percentages of $CdCl_2$ up to 5 % producing structures with cation vacancies and ad-anions, and $PbCl_2$ doped with KCl or $LaCl_3$ giving anion vacancies and ad-cations or cation vacancies and ad-anions. Tertiary butyl chloride was used as reactant instead of ethyl chloride because decomposition of tertiary butyl chloride occurs at lower temperature with less likelihood of side-reactions.

Doping AgCl with $CdCl_2$ gives a large effect; the reaction rate increases linearly with increasing ionic conductivity which is proportional to the concentration of cation vacancies in the bulk catalyst and thus presumed proportional to the surface vacancies. However, the result does not prove the mechanism shown in Fig. 15.6d because the concentration of ad-ions and Cd^{2+} ions on doped AgCl are proportional to the concentration of cation vacancies and therefore, also proportional to conductivity. Thus, the mechanisms in Fig. 15.6c and d are both consistent with the experimental results, as well as a mechanism involving catalysis by Cd^{2+} ions. In contrast, similar measurements using $PbCl_2$ doped with KCl or $LaCl_3$ showed no change in reaction rate with increased doping, although conductivities changed appreciably. The authors conclude that, in this case, point defects have no significant effect on catalytic activity of $PbCl_2$ for decomposition of tertiary butyl chloride. Although the specific nature of the defect mechanism on doped AgCl cannot be ascertained at present, the results have a heuristic value in the theory of point defect catalysis.

According to Benson et al. (40), the first unequivocal identification of active centers with the particular point defect consisting of positive holes trapped at surface levels (V-centers) was made by Kohn and by Boreskov et al. They

explained the low temperature activity of silica gel in the H-D exchange reaction resulting from gamma-ray irradiation. By ESR and optical spectroscopy, it was confirmed that the active centers were positive holes trapped at
surface oxygen ions near aluminum impurities in the silica gel. Benson *et al.*
(*40*) explain the unusual activity of copper–magnesia catalyst by the same
mechanism.

Cimino *et al.* (*41*) have obtained evidence of localized interaction for oxygen chemisorption and N_2O decomposition on NiO–MgO and related solid
solutions. By preparing solid solutions of formula $Ni_x Mg_{1-x}O$ at 1200°C
they were able to study the behavior of nickel ions in isolation. They came to
the conclusion that p-type conductivity with drift of charge across an array
of Ni^{2+}/Ni^{3+} ions is much less relevant for catalysis of N_2O decomposition
than the nature of the octahedral complex formed at the surface by chemisorbed oxygen resulting from the decomposition. We shall discuss this work
in more detail later in this section under *Local Symmetry.*

Thus, the experimental evidence for local interaction is meager, but
nevertheless sufficient to warrant more attention.

THEORIES OF THE RUSSIAN SCHOOL

Formal quantum mechanical calculations of chemisorption on atomic defects have been made by Bonch-Bruevich (*42*) and more recently by Kogan
and Sandomirsky (*43*). Since such calculations cannot be related quantitatively to specific real systems yet, we shall not discuss them, but shall instead
present some of the qualitative considerations of Wolkenstein (*7*) who describes chemisorption on an *F*-center as a preamble to catalysis. He considers
the case of a monovalent electropositive atom C on an *F*-center in a lattice
composed of ions M^+ and R^-. The *F*-center is a vacant R^- site with an
electron localized near it as shown in Fig. 15.7a and is denoted by the symbol
DL. Figure 15.7b shows the same *F*-center with its electron removed, and such
an ionized *F*-center is denoted by the symbol DpL. Two forms of chemisorption of C may occur. Fig. 15.7c shows the "strong" acceptor bond consisting
of two electrons which is formed when C interacts with an *F*-center. It is denoted by the symbol CDL. Figure 15.7d shows the "weak" bond consisting
of one electron which is formed when C interacts with an ionized *F*-center.
This form is denoted by the symbol CDpL. Thus, atom C interacts with a
metal ion M which is close to the vacancy of the *F*-center; the *F*-center electron is either localized on the ion M or it has been transferred to the conduction band. Figure 15.8 illustrates the energy-band scheme of the catalyst
surface with the y axis parallel to the surface. *F*-centers are labeled D for donor
levels, and the *F*-centers with chemisorbed C atoms are designated as donor
levels CD.

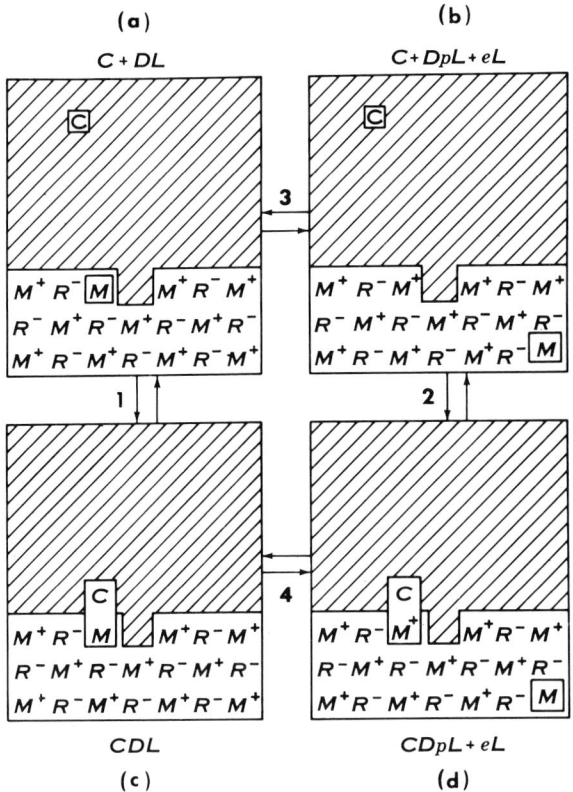

FIG. 15.7. Interaction of atom C with surface defects. (a) an *F*-center DL plus atom C, (b) an ionized *F*-center D*p*L plus atom C, (c) interaction of C with an *F*-center to form CDL, (d) interaction of C with an ionized *F*-center to form CD*p*L (7).

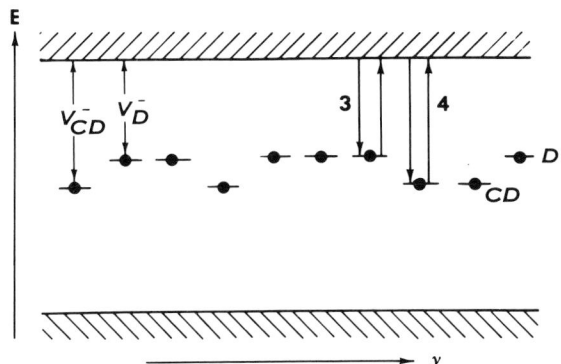

FIG. 15.8. Energy-band scheme for interaction of atom C with *F*-center (7).

The transformations in Fig. 15.7a–d may be written as follows with the exothermic direction of heats of reaction from left to right:

$$
\begin{array}{lll}
(1) & C + DL \leftrightarrows CDL, & q^- \\
(2) & C + DpL \leftrightarrows CDpL, & q^0 \\
(3) & DpL + eL \leftrightarrows DL, & V_D^- \\
(4) & CDpL + eL \leftrightarrows CDL, & V_{CD}^-
\end{array}
\tag{15}
$$

where q^-, q^0, V_D^-, and V_{CD}^- are the heats of the reactions. The reactions (3) and (4) of Eqs. (15) represent the electron transitions depicted in Fig. 15.8 by arrows (3) and (4). From Eq. (15), we find

$$q^- = q^0 + (V_{CD}^- - V_D^-) \tag{16}$$

from which it follows that

$$\text{if } q^- > q^0, \text{ then } V_{CD}^- > V_D^- \, ; \tag{17}$$

$$\text{if } q^- < q^0, \text{ then } V_{CD}^- < V_D^-. \tag{18}$$

Levels D in Fig. 15.8 are used up as adsorption proceeds and are transformed one-to-one into levels CD. Thus, the Fermi level is displaced upward if the inequality (17) holds, and downward for inequality (18). If V_D^- and V_{CD}^- are both large, that is, if levels D and CD lie far below the conduction band, e.g., as in alkali halide crystals, then transitions (3) and (4) may be ignored when thermal energy is the only ionizing agent. In this event, adsorption processes on F-centers and ionized F-centers are independent of each other, that is, there will be insufficient energy available at reasonable temperatures to ionize an F-center or CDL, an atom of C adsorbed on an F-center. Given a crystal containing F-centers and one of the same species containing an equal number of ionized F-centers, then the adsorptivity of the former will be

$$\exp[(q^- - q^0)/kT] = \exp[(V_{CD}^- - V_D^-)/kT] \tag{19}$$

times greater than the latter. Schematic representations of chemisorptions on other types of defects may be depicted and related to catalysis as in the example of Simkovich and Wagner (39) discussed above.

LOCAL SYMMETRY

Dowden and Wells (44) were the first to suggest relationships between local symmetry on solid surfaces and chemisorption and catalysis. They found that crystal field theory offered broad, qualitative explanations for many of the phenomena observed in heterogeneous catalysis. Following the lead of Dowden and Wells (44), Haber and Stone (45) studied crystal-field effects in the specific case of oxygen chemisorption, and later, Cimino et al. (46) extended the studies to include decomposition of N_2O. Before discussing these investigations, we shall give a brief background on field effects.

Bethe (*47*) first introduced crystal-field theory in 1929; valuable contributions were made by Van Vleck (*48*). In this theory, a molecule, neutral or charged, is considered as an aggregate of atoms or ions that interact with each other only electrostatically. Since it is assumed that there is no exchange of electrons, covalent bonding is excluded from consideration. A transition metal ion, for example, surrounded by a regular octahedron of negative ions or dipolar molecules arrayed with their negative ends pointing toward the central ion undergoes certain changes as a result of the field of negative charges. In particular, the five degenerate d-orbitals are split into more than one energy level, and the manner of splitting depends on the geometry of the surrounding charges. An octahedral environment splits the d-orbitals into two levels, the lower (t_{2g}) containing the d_{xy}, d_{xz}, and d_{yz} orbitals which are repulsed less by negative point charges on the x and y axes than are the $d_{x^2-y^2}$ and d_z^2 orbitals (e_g level). Figure 15.9 shows the d-orbital splitting for some important geo-

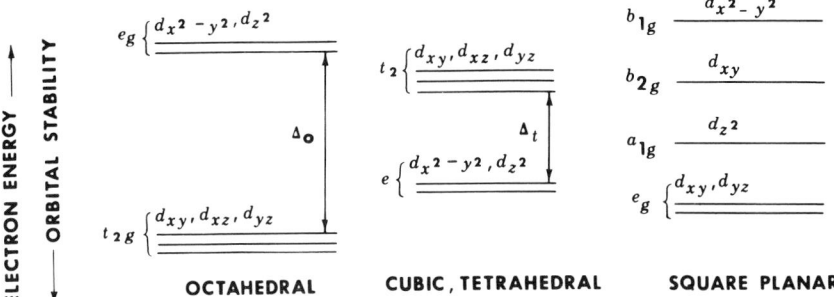

Fig. 15.9. Orbital splitting diagrams for four important types of complexes. Orbitals within each pair of braces are of identical stability.

metric arrangements. In the octahedral, tetrahedral, and cubic symmetries, the original fivefold degenerate orbitals are split into two levels whose energy separation is denoted by Δ. When Δ is small, electrons will fill the five orbitals singly before any pairing of electrons occurs, as in the case of the free ion. This is called the "weak field" approximation. When Δ is large, orbitals of the lower level are assumed to fill completely with a pair of electrons in each orbital before filling of the orbitals of the upper level commences. This is known as the "strong field" approximation. We shall consider only the weak approximation.

If the particles surrounding a transition metal ion are considered, in the light of the molecular-orbital theory, to interact covalently as well as electrostatically, then the effect of various geometries on the nature of the splitting of d-orbitals remains generally the same as in the crystal field theory, though the values of Δ calculated by each theory may vary considerably from one another. Crystal field theory is the cruder of the two theories, and, of course, such questions as the relative importance of σ and π bonding can only be treated by the molecular orbital theory (see Section 7.4). Orgel (49) uses the term "ligand-field theory" for a hybrid approach that encompasses parts of both theories selected for their value in the solution of particular problems.

An important quantity is the stabilization energy E_c referred to as the crystal field or ligand-field stabilization energy (CFSE or LFSE). This quantity is derived on the basis of the following rule: The stabilization energy is zero when all levels are equally occupied. Thus, in the case of octahedral geometry, if we assign an energy $-\frac{2}{5}\Delta_o$ for the t_{2g} level (Fig. 15.9) and $+\frac{3}{5}\Delta_o$ for the e_g level, the stabilization energy is $E_c = -\frac{2}{5}n_t\Delta_o + \frac{3}{5}n_e\Delta_o$, where n_t and n_e are the number of electrons in the t_{2g} and e_g levels, respectively. The rule is satisfied in the weak-field approximation for all possible cases of equal occupation, d^0, d^5, and d^{10}. This assignment is equivalent to choosing the zero of energy, that is, the energy of the d-orbitals of the free ion, as the weighted mean energy of the d-orbitals of the octahedral field. We show in Table 15.2

TABLE 15.2

STABILIZATION ENERGIES (E_c) IN OCTAHEDRAL AND TETRAHEDRAL FIELDS[a]

No. of Electrons	Stabilization energy[b], E_c		
	E_c, octahedral	E_c, tetrahedral	ΔE_c[c]
0, 5, 10	0	0	0
1, 6	$-\frac{2}{5}\Delta_o$	$-\frac{3}{5}\Delta_t$	$-\frac{1}{10}\Delta_o$
2, 7	$-\frac{4}{5}\Delta_o$	$-\frac{6}{5}\Delta_t$	$-\frac{2}{10}\Delta_o$
3, 8	$-\frac{6}{5}\Delta_o$	$-\frac{4}{5}\Delta_t$	$-\frac{8}{10}\Delta_o$
4, 9	$-\frac{3}{5}\Delta_o$	$-\frac{2}{5}\Delta_t$	$-\frac{4}{10}\Delta_o$

[a] Weak field approximation.
[b] Δ_t generally is $\leqslant \frac{1}{2}\Delta_o$ for a given ion with common ligands.
[c] E_c (octahedral) $- E_c$ (tetrahedral) with $\Delta_t = \frac{1}{2}\Delta_o$ illustrating the twin-peaked pattern.

the stabilization energies for octahedral and tetrahedral geometries with the number of electrons ranging from 0 to 10. In the tetrahedral configuration, the energies for e and t_2 are $-\frac{3}{5}\Delta_t$ and $+\frac{2}{5}\Delta_t$, respectively.

According to Dowden and Wells (44), rate-controlling steps in certain chemisorptions and catalytic reactions which involve a species so polarized as to restore near octahedral symmetry about a surface cation in an ionic solid contribute to higher energies of activation and lower activities at the cation electron configurations d^0, d^5, and d^{10}. These are the configurations which possess zero stabilization energy. In most transitions from some configuration to the octahedral configuration, the change in stabilization energy (ΔE_c) leads to an increase in stability of the complex ($\Delta E_c < 0$). Thus, there are often maxima in ΔE_c between d^0 and d^5 and between d^5 and d^{10}, as illustrated in Table 15.2 for the transition from tetrahedral to octahedral symmetry. A rough similarity to the twin-peaked ΔE_c pattern is found for the activity of hydrogen reactions as a function of the number of electrons in the d-band in the first transition series. And this suggests that ΔE_c may be a significant factor in the energetics of chemisorption and catalysis. Correlations also hold for oxygen chemisorption.

We now turn to the work of Haber and Stone (*45*) which illustrates the application of the general observations of Dowden and Wells (*44*) to the specific case of chemisorption of oxygen on nickel oxide. They consider chemisorption of oxygen ions (O^{2-}) on the three principal planes (100), (110), and (111) illustrated in Fig. 15.10a–c, respectively. The middle column represents the surface as it appears at the moment of cleavage; the left column, the bare

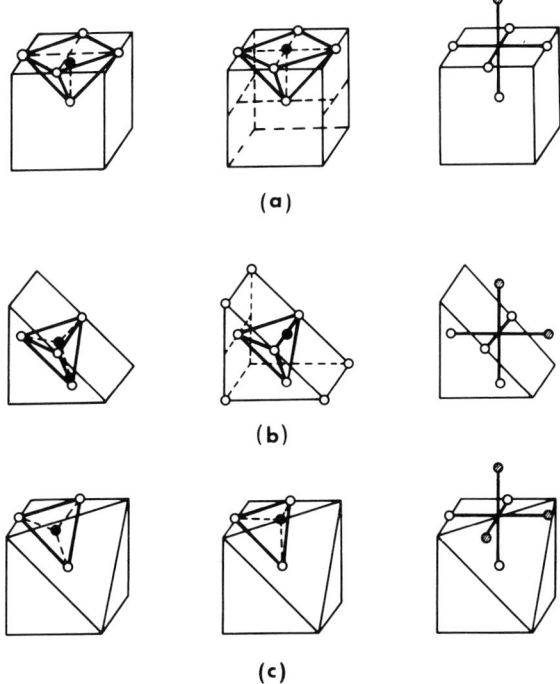

(a)

(b)

(c)

FIG. 15.10. Changes in nickel ion coordination during the chemisorption of oxygen on nickel oxide. (a), (100) plane; (b), (110) plane; (c), (111) plane. Diagrams in the middle column refer to a bare surface at the instant of cleavage; left column, bare surface, relaxed position; right column, after adsorption of oxygen, ● nickel ion; ○ lattice oxygen ions; ◐ adsorbed oxygen ions (*45*). [After Haber and Stone, *Trans. Faraday Soc.* **59**, 192 (1953). By permission of The Faraday Society.]

surface after relaxation; and the right column, after adsorption of oxygen. Sufficient adsorption on each face to give octahedral symmetry is assumed to take place. In the (100) face, a nickel ion has five nearest-neighbor oxygen ions distributed in the shape of a square pyramid. On cleavage, the surface may relax in such a way that the nickel ion enters the pyramid, as shown to the left in Fig. 15.10a. Actually, oxygen ions will move relative to the nickel ion, but for simplicity only movement of the nickel ion is depicted. Chemisorption of an oxygen ion completes the octahedron about the Ni^{2+} ion as shown to the right in Fig. 15.10a. The change in stabilization energy may be estimated by the change from square pyramidal to octahedral stabilization. Basolo and Pearson (50) give the stabilization energy for d^8 octahedral symmetry as $-12Dq$, where $10Dq = \Delta_o$ (see Table 15.2) is another nomenclature for the d-level separation energy in an octahedral field. For the stabilization energy in square pyramidal d^8 symmetry, they give $-10Dq$ ($= -\Delta_o$). The gain in stabilization energy is therefore $2Dq$ ($= \frac{1}{5}\Delta_o$) per atom adsorbed on the (100) plane. Analogous changes take place on the (110) and (111) planes as shown in Figs. 15.10b and c, respectively. The relaxed position for the (110) plane approximates tetrahedral symmetry for the cation, and upon adsorption of *two* oxygen ions octahedral symmetry is attained with a gain in stabilization energy of $8.4Dq$. On a freshly cleaved (111) plane, a nickel ion relaxes towards the center of the triangle of oxygen ions lying directly beneath it, giving trigonal symmetry. Octahedral symmetry is attained by adsorption of *three* oxygen ions, and the gain in stabilization energy is $1.1Dq$. Results are summarized in Table 15.3. It is evident that the theory predicts the greatest stabilization for the (110) face. At low overall coverage, we may expect the

TABLE 15.3

CHANGES IN CRYSTAL FIELD STABILIZATION ENERGY DURING
CHEMISORPTION OF OXYGEN ON NICKEL OXIDE[a,b]

Plane	Field	Stabiliza-tion[b] energy	Atoms adsorbed	Final field	Stabili-zation[b] energy	Gain in stabili-zation energy	Gain per adsorbed atom[c]
(100)	Square pyramid	$-10.0\ Dq$	1	Octa-hedral	$-12\ Dq$	$2.0\ Dq$	$2.0\ Dq$
(110)	Tetra-hedral	$-3.6\ Dq$	2	Octa-hedral	$-12\ Dq$	$8.4\ Dq$	$4.2\ Dq$
(111)	Trigonal	$-10.9\ Dq$	3	Octa-hedral	$-12\ Dq$	$1.1\ Dq$	$0.4\ Dq$

[a] See Haber and Stone (45).
[b] $Dq = 910\ cm^{-1} = 2.6$ kcal mole^{-1}.
[c] $10\ Dq = \Delta_o$ where Δ_o is the d-level separation energy in an octahedral field.

(110) planes to be preferentially covered at the expense of the (100) and (111) planes.

Haber and Stone (*45*) provide support for their theory through photo-desorption experiments. They found that adsorbed oxygen on nearly stoichiometric nickel oxide could be desorbed by illumination, the most effective region being the visible spectrum in the wavelength range 650–900 mμ. Using an Orgel diagram (*51*) for the Ni^{2+} ion in tetrahedral and octahedral fields, which gives the energy levels for these symmetries, Haber and Stone suggest that oxygen photodesorption takes place by a transition from the octahedrally coordinated nickel in its ground state $^3A_{2g}$ to the excited state $^3T_{1g}$—a transition whose absorption spectra occurs around 650 mμ. They consider the excited state $^3T_{1g}$ as an activated complex which is unstable with respect to the ground state of a tetrahedrally coordinated Ni^{2+} ion. Thus a surface Ni^{2+} ion which has chemisorbed two oxygen atoms to form an octahedral complex may lose the two atoms as an oxygen molecule and return to the tetrahedral state when irradiated with the wavelength corresponding to the transition $^3A_{2g} \rightarrow {}^3T_{1g}$. The model may be plausible, but Jongepier and Schuit (*52*) have serious doubts concerning the interpretation of the experimental results which they check. Among their most serious criticisms of the interpretations of Haber and Stone are: (1) no allowance is made for charge transfer and valence changes of the nickel ion or desorption of neutral oxygen, (2) it is not possible to have a continuous transition from 0_h to T_d symmetry while the $^3T_{1g}$ wave function conserves its symmetry. They believe that a more realistic interpretation of photodesorption must await a more comprehensive theoretical model based on molecular orbital treatment.

As mentioned above in this section, Cimino *et al.* (*41*) have studied the decomposition of N$_2$O in the light of localized interactions. Having shown that *p*-type semiconductivity of NiO, that is, the drift of charge across an array of Ni^{2+}/Ni^{3+} ions, is not relevant for catalytic decomposition of N$_2$O, since it occurs over extremely dilute solid solutions of NiO in MgO, they went on to speculate about the nature of the localized active centers. They found that in the temperature range -78–400°C a reversible adsorption of oxygen develops. This reversible adsorption is greater on a solid solution containing 0.01% Ni than on one containing 0.1%. Furthermore, the catalytic decomposition of N$_2$O parallels the amount of reversible adsorption, being greater over the catalyst of lower nickel content. Catalysts with nickel contents greater than 0.01% contain a progressively greater proportion of strongly adsorbed oxygen which cannot be pumped off by evacuation, and this is true up to pure NiO. Cimino *et al.* believe that the strongly adsorbed oxygen is present as O^{2-} and that it acts as a poison for the decomposition of N$_2$O. They suggest that the reversibly adsorbed oxygen, which correlates with catalytic activity, is present as O$^-$. It may be formed especially on (110) planes from decompos-

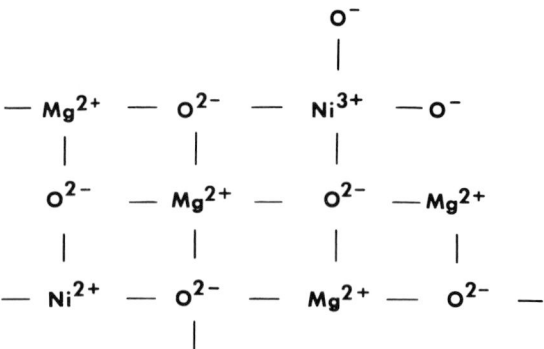

FIG. 15.11. Reversible adsorption of O^- from decomposition of N_2O. [After Cimino *et al.*, *Discussions Faraday Soc.* **41**, 350 (1966). By permission of The Faraday Society.]

ing N_2O molecules. Two separately chemisorbed O^- ions may then migrate a few lattice distances to form the adsorbed state shown in Fig. 15.11. Desorption of such oxygen is likely to be relatively easy and the decomposition of N_2O can continue. The same transition from tetrahedral to octahedral symmetry on the (110) face may be considered to occur on adsorption of two oxygen ions as was postulated for adsorption on pure NiO, except that now the central metal ion is trivalent instead of divalent and the oxygen ions are singly instead of doubly charged. Preferred adsorption at low coverage on the (110) face [or at steps on (100) faces] would be expected as before.

15.4   Surface States and Catalysis

In Section 8.3, we described briefly the work of Morrison on measurements of the surface states associated with oxygen chemisorbed on ZnO. Negatively charged, chemisorbed oxygen ions are formed by an electrical discharge through gaseous oxygen, and they retain their negative charge at room temperature. Thus, a negative layer is built up on the surface. As the temperature is raised, the surface charge decays as electrons gain the activation energy E required to raise them into the conduction band and so into the interior of the semiconductor. Morrison found as many as three separate activation energies, $E_1 = 0.6$, $E_2 = 0.8$, and $E_3 = 1.0$ eV. He associates these surface states with chemisorbed oxygen. To be more specific about surface states associated with chemisorbed molecules, we use the singly charged oxygen ion O^- for illustration. The empty orbital in the neutral oxygen atom O into which an electron enters to form the negative ion is the orbital of interest. As O^- approaches the surface of ZnO, the energy level of this orbital is modified, and it becomes the surface state associated with chemisorbed oxygen. In particular, this is called an acceptor surface state. When a neutral atom loses an electron and is adsorbed as a positive ion, the orbital from which the electron is lost represents a donor surface state. In addition to surface states associated with chemisorbed ions there are discrete defect states of the crystal surface itself. Defect surface states of the solid (also the conduction and valence bands), in the absence of external agencies, are the sole source or sink of electrons required in the formation of surface states associated with chemisorption.

Gray and Amigues (53) have also measured the surface states of ZnO by a direct and elegant method. They subjected ZnO under well-defined conditions of pretreatment and gas atmosphere to excitation at low temperature (liquid nitrogen) from a high-pressure mercury vapor discharge or from a voltage regulated strip-filament tungsten lamp and narrow band-pass interference filters. A stationary electronic state is developed with a specific occupancy of all impurity electronic levels which does not change significantly when the excitation is removed. Subsequently, slow uniform heating releases trapped electrons by thermal excitation at the temperature corresponding to the depth of the level. The thermally stimulated electron current is measured by an electrometer and millivolt recorder. Measurements have been made in in the presence of oxygen and in vacuum so that electronic surface states of the ZnO as well as surface states associated with adsorbed oxygen are obtained. The existence of levels in high-purity ZnO at 0.12–0.15 eV (the donor defect level of ZnO), 0.4–0.5, 0.6, 0.8, 1.0, and 1.1–1.2 eV have been established. The level at 0.4–0.5 eV is due to residual copper impurity (0.3 ppm). The level 0.8 eV is believed to be associated with O^- and the other levels in common with Morrison's measurements, 0.6 and 1.0 eV, are also believed to be associated with oxygen anions. Gray and Amigues found shallower levels not reported by Morrison, probably because they investigated thermal excitation starting as low as liquid nitrogen temperature, whereas Morrison started at ambient temperature. At this temperature, the shallower levels would already have been emptied. Figure 15.12 shows the disappearance of the electronic current peak due to the crystal defect donor level (0.12–0.15 eV) as its supply of electrons is transferred to chemisorbed oxygen states. Similar work has been reported by Gray and Cichowski (54) on TiO_2.

In addition to the experimental investigations just mentioned, the reader is referred to the theoretical papers of Mark (55, 56) on the electronic states of ionic lattices and of chemisorbed ions. All of this work, experimental and theoretical, is leading rapidly to a more quantitative understanding of catalytic reactions on semiconductor surfaces. Unfortunately, at the moment there are no published reports on specific relations between surface states of semiconductors and a catalytic reaction. But there is a significant paper (57) on the energy states associated with OH^- anions on the surface of the insulator MgO and their role as active centers in the catalysis of the H–D exchange reaction, which will be discussed in Chapter XVI.

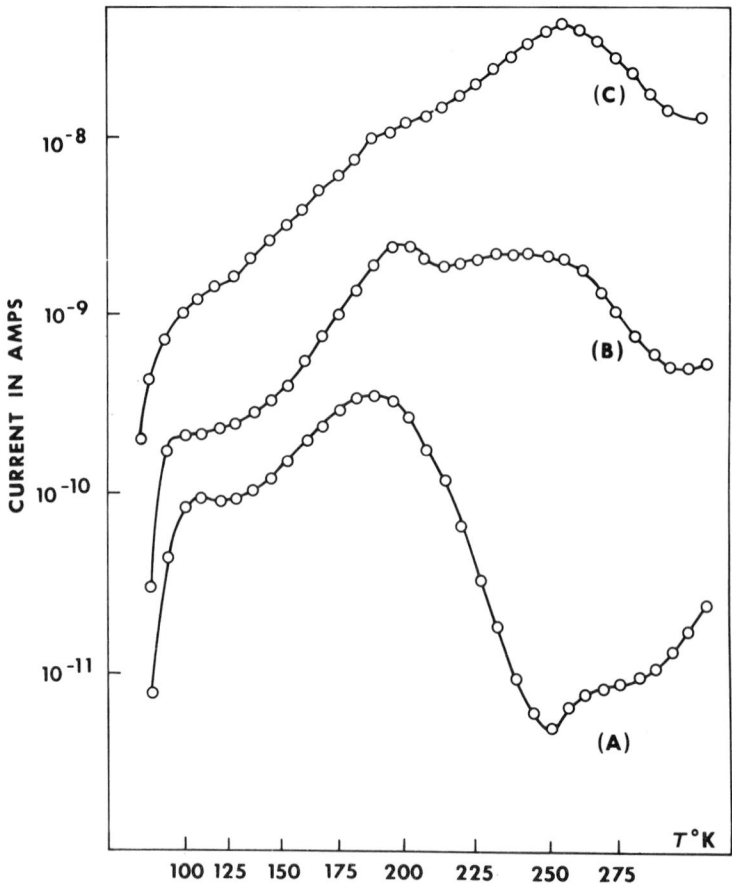

Fig. 15.12. Pure ZnO, influence of oxygen pressure on thermally stimulated currents: A, 150 mm O_2; B, 0.5 mm O_2; C, vacuum irradiated to saturation. [After Gray and Amigues, Am. Chem. Soc. Meeting San Francisco. Symp. Adsorption Catalysis Semiconducting Materials (1968). Courtesy T. J. Gray.]

REFERENCES

1. Wagner, C., and Hauffe, K., Z. Elektrochem. **44**, 172 (1938).
2. Garner, W. E., Gray, T. G., and Stone, F. S., Proc. Roy. Soc. (London) Ser. A **197**, 294 (1949).
3. Wolkenstein, F. F., Usp. Fiz. Nauk. **60**, 249 (1956).
4. Hauffe, K., "Semiconductor Surface Physics," (R. H. Kingston, ed.), p. 259, Univ. of Pennsylvania Press, Philadelphia, Pennsylvania, 1956.
5. Hauffe, K., and Schlosser, E. G., Z. Elektrokem. **61**, 506 (1957).

6. Garrett, C. G. B., *J. Chem. Phys.* **33**, 966 (1960).
7. Wolkenstein, F. F., *Advan. Catalysis* **12**, 189 (1960).
8. Wagner, C., and Hauffe, K., *J. Chem. Phys.* **18**, 69 (1950).
9. Dell, R. M., Stone, F. S., and Tiley, P. F., *Trans. Faraday Soc.* **49**, 201 (1953).
10. Amphlett, C. B., *Trans. Faraday Soc.* **50**, 273 (1954).
11. Engell, H. J., and Hauffe, K., *Z. Elektrochem.* **57**, 776 (1953).
12. Winter, E. R. S., *Discussions Faraday Soc.* **28**, 183 (1959).
13. Rheaume, L., and Parravano, G., *J. Phys. Chem.* **63**, 264 (1959).
14. Schwab, G. M., and Schultes, H., *Z. Physik. Chem.* **B9**, 265 (1930); Schwab, G. M., Staegar, R., and Baumbach, A. H., *Z. Physik. Chem.* **B21**, 65 (1933); Schwab, G. M., and Schultes, H., *Z. Physik. Chem.* **B25**, 411 (1934); Schwab, G. M., and Stalgar, R., *Z. Physik. Chem.* **B25**, 418 (1934).
15. Schmid, G., and Keller, N., *Naturwissenschaften* **37**, 43 (1950).
16. Saito, Y., Yoneda, Y., and Makishima, S., *Proc. 2nd Intern. Congr. Catalysis 1960,* p. 1937. Technik, Paris (1961).
17. Hauffe, K., Glang, R., and Engell, H. J., *Z. Physik Chem. (Leipzig)* **201**, 223 (1952).
18. Stone, F. S., "Chemistry of the Solid State" (W. E. Garner, ed.), p. 397. Butterworth, Scientific Publ. London, 1955.
19. Parravano, G., *J. Am. Chem. Soc.* **75**, 1452 (1952).
20. Schwab, G. M., and Block, J., *Z. Physik. Chem. (Frankfurt)* **1**, 42 (1954).
21. Keier, N. P., Roginskii, S. Z., and Sazonova, I. S., *Izv. Akad. Nauk SSSR Otdel. Fiz. Nauk* **21**, 183 (1957).
22. Dry, M. E., and Stone, F. S., *Discussions Faraday Soc.* **28**, 192 (1959).
23. Stone, F. S., *Advan. Catalysis* **13**, 35, 36 (1962).
24. Amigues, P., and Teichner, S. J., *Discussions Faraday Soc.* **41**, 362 (1966).
25. Molinari, E., and Parravano, G., *J. Am. Chem. Soc.* **75**, 5233, (1953).
26. Voltz, S. E., and Weller, S., *J. Am. Chem. Soc.* **75**, 5227 (1953).
27. Clark, A., *Ind. Eng. Chem.* **45**, 1476 (1953).
28. Baker, McD., M., and Jenkins, G. I., *Advan. Catalysis* **7**, 39 (1955).
29. Beeck, O., *Discussions Faraday Soc.* **8**, 34 (1950).
30. Krauss, W., *Z. Elektrochem.* **53**, 320 (1948).
31. Hart, A. B., and Ross, R. A., *J. Catalysis* **2**, 251 (1963).
32. Hauffe, K., *Advan. Catalysis* **7**, 237 (1955).
33. Hauffe, K., Publicado en "Coloquio Sobre Qimica Fisica de procesos en superficies solidas (XXV Aniversario del C. S. I. C., Octubre 1964)." Libreria Cientifica Medinaceli, Duque de Medinaceli 4. Madrid 14, Espana (1965).
34. Goodenough, J. B., "Magnetism and the Chemical Bond." Wiley (Interscience), New York, 1963.
35. Vrieland, V. G., and Selwood, P. W., *J. Catalysis* **3**, 539 (1964).
36. Bozon-Verduraz, F., and Teichner, S. J., *J. Catalysis* **11**, 7 (1968).
37. Dowden, D. A., "Chemisorption," (W. E. Garner, ed.), p. 13. Butterworth, London and Washington, D.C., 1957; Dowden, D. A., and Wells, D., *Actes 2e Congr. Catalyse,* p. 1499. Technip, Paris, 1961.
38. Stone, F. S., and Haber, J., *Trans. Faraday Soc.* **59**, 192 (1963).
39. Simkovich, G., and Wagner, C., *J. Catalysis* **1**, 521 (1962).
40. Benson, J. E., Walters, A. B., and Boudart, M., *Symp. Electron. Phenomena Chemisorption Catalysis Semiconduct. 4th Intern. Congr. Catalysis* Paper No. 7, 1968.
41. Cimino, A., Schiavello, M., and Stone, F. S., *Discussions Faraday Soc.* **41**, 350 (1966).
42. Bonch-Bruevich, V. L., *J. Phys. Chem. (USSR)* **27**, 662, 960 (1953).
43. Kogan, S. M., and Sandomirsky, V. B., *Compt. Rend. Acad. Sci USSR* **127**, 377 (1959); *Bull. Acad. Sci. USSR Ser. Chem.* 1681 (1959).

44. Dowden, D. A., and Wells, D., *Actes 2^e Congr. Intern. Catalyse* p. 1499. Technip, Paris (1961).
45. Haber, J., and Stone, F. S., *Trans. Faraday Soc.* **59**, 192 (1963).
46. Cimino, A., Schiavello, M., and Stone, F. S., *Discussions Faraday Soc.* **41**, 350 (1966).
47. Bethe, H., *Ann. Physik* **3**, 133 (1929).
48. Van Vleck, J. H., "Theory of Electric and Magnetic Susceptibilities." Oxford Univ. Press, London and New York, 1932.
49. Orgel, L. E., "An Introduction to Transition-Metal Chemistry, Ligand-Field Theory." Methuen, London, 1960.
50. Basolo, F., and Pearson, R. G., "Mechanisms of Inorganic Reactions." Wiley, New York, 1958.
51. Orgel, L. E., *J. Chem. Phys.* **23**, 1004, (1955).
52. Jongepier, R., and Schuit, G. C. A., *J. Catalysis* **3**, 464 (1964).
53. Gray, T. J., and Amigues, P., Am. Chem. Soc. Meeting San Francisco. *Symp. Adsorption Catalysis Semiconducting Materials* (1968).
54. Gray, T. J., and Cichowski, R. S., PhD. Thesis, New York State College of Ceramics at Alfred Univ. (1968).
55. Mark, P., *Catalysis Rev.* **1**, 165 (1967).
56. Mark, P., *J. Phys. Chem. Solids* **29**, 689 (1968).
57. Harkins, C. G., Shang, W. W., and Leland, T. W., *J. Phys. Chem.* **73**, 130 (1969).

XVI

THE NATURE OF SURFACE
SPECIES AND THEIR ROLE IN CATALYSIS

We shall now abandon completely the collective electron approach to catalysis and consider strictly localized interactions leading to a variety of chemisorbed species. These species may be tightly adsorbed residues that have nothing to do directly with the catalysis of reactions on the surface, though they probably influence the course of reaction through near-neighbor interactions or site-blocking. On the other hand, chemisorbed species may be involved directly in the reaction either as active centers or reaction intermediates. Such catalysis must be highly specific, and no general physical theories like those developed in the collective electron approach will apply. Generalized, quantitative theories would involve the principles of quantum chemistry. But surface problems are so complex that for the present only qualitative, often tenuous, guides can be developed. In the absence of general theories, we select a few important examples out of the many disparate ones which have been reported. We start with a discussion of active sites on insulator-type oxides in which free electrons and holes are practically nonexistent.

16.1 Chemisorbed Species as Active Sites

It is generally agreed that catalysts such as silica–alumina possess two kinds of acid sites: Lewis sites and Brönsted sites. Lewis sites are defined as aprotic and Brönsted sites as protonic. From this point, there is much diversity of opinion as to the nature and number of the two kinds of sites on a particular surface and how they are used in chemisorption and catalysis. We shall review the situation briefly.

LEWIS AND BRÖNSTED SITES ON SILICA–ALUMINA

There have been many schematic structures suggested for Brönsted and Lewis acid sites on silica–aluminas. Among the most recent and exhaustive studies are those of Leonard et al. (1) and Fripiat et al. (2) who employed the techniques of adsorption, and X-ray fluorescence and infrared spectroscopy. Their proposed structures are shown in Fig. 16.1. At low drying temperatures (about 100°C) and for alumina contents below 33%, structure A predominates. For alumina contents greater than 33%, structures A and C coexist in different regions of the solid. Structure A, characteristic of lower alumina content, is transformed by heating to remove water into structure E, a process representing the transformation of Brönsted sites into Lewis sites on aluminum atoms. The authors present evidence why the corresponding process of formation of Lewis sites on silicon atoms does not occur.

The C structure, in which alumina is octahedrally coordinated, predominates for silica–aluminas rich in aluminum. Upon removing water, structure D is formed, the precursor of structures F and G which contain Lewis sites. The transition of D to F and G is believed to be induced by interaction with basic adsorbates such as pyridine and ammonia. In structure F, silicon is a Lewis acid site, and the charge balance is internally compensated. In structure G, aluminum is a Lewis site. The relative production of F or G depends on the basic strength of the adsorbate. It is obvious from Fig. 16.1 that Brönsted sites are associated with hydroxyl groups formed from chemisorbed water. In the simplest picture of Brönsted sites, they consist of hydroxyl groups in which the hydrogen atom is polarized so that it possesses some protonic character. Not all hydroxyl groups form Brönsted sites.

Lewis and Brönsted sites may be distinguished by the differences in the spectra of chemisorbed bases such as pyridine (3) and ammonia (4–6). These substances are coordinately bonded on Lewis sites whereas they form pyridinium or ammonium ions, respectively on Brönsted sites. In addition, appreciable amounts of pyridine and ammonia are physically adsorbed, in which state they are hydrogen bonded to surface OH groups. By extended pumping at elevated temperature (150°C), all physically adsorbed species can be removed leaving only chemisorbed species on Lewis and Brönsted sites. Basila and Kantner (7) have determined the ratio of Lewis-to-Brönsted-acid sites on various silica–aluminas pretreated by heating to 500°C over a 3-hr period

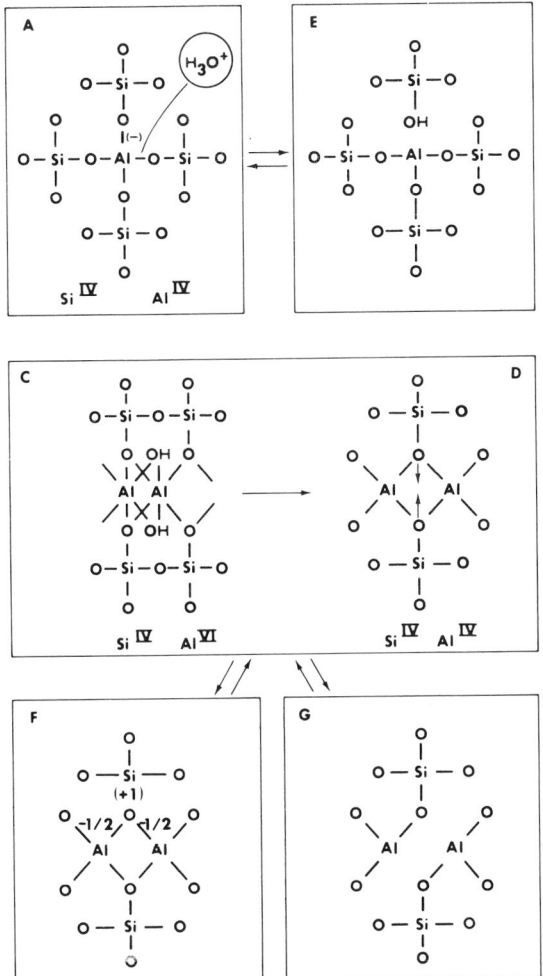

FIG. 16.1. Schematic structures of Lewis and Brönsted sites on silica–alumina. [After Fripiat *et al.*, *J. Phys. Chem.* **69**, 3274 (1965). By permission of The American Chemical Society.]

with continuous evacuation, calcining in pure oxygen for at least 4 hr at 500°C, and evacuating overnight at the same temperature. Table 16.1 shows a spread in the ratio from 3 to 6. Similar results are obtained with synthetic zeolites, but alumina reveals the presence only of Lewis sites.

Basila *et al.* (*8*) suggest that all of the primary acid sites on silica–alumina are Lewis sites centered on active surface aluminum atoms and that apparent Brönsted sites are produced by secondary interactions between a molecule

TABLE 16.1

Lewis–Brönsted Acid Site Distribution in Silica–Aluminas[a]

Catalyst	Al_2O_3, %	SA, m^2/gm	LPY/BPY[b]
American Cyanamid (Triple A)	25	430	6.0
Davison F-125	25	352	3.8
Houdry S-90	12	421	4.2
Houdry S-36	12	157	2.9

[a] See Basila and Kantner (7).
[b] Ratio of pyridine adsorbed on Lewis sites to that adsorbed on Brönsted sites.

chemisorbed on a Lewis site and a nearby surface hydroxyl group. Their conclusion is based on the following observations:

(1) Surface OH groups are attached predominantly to silicon atoms.
(2) Adsorption studies with both pyridine and ammonia indicate the presence of both Lewis and Brönsted acidity.
(3) Adsorbed H_2O transforms Lewis pyridine (LPY) to Brönsted pyradine (BPY). The interaction is reversible and relatively weak since the H_2O can be pumped off at 150°C.
(4) In the dual adsorption of H_2O and pyridine (PY) on silica–alumina, PY is the stronger base and remains on the acid site. The H_2O interacts with the surface OH as well as the chemisorbed PY.

The most striking and unexpected experimental observation is the transformation of LPY to BPY through relatively weak interaction with H_2O. The results indicate that H_2O changes the primary interaction of PY with Lewis acid sites to a secondary interaction involving H_2O and forming BPY. The addition of water not only increases the formation of BPY, but also increases the amount of hydrogen bonding of surface OH groups. It appears that the added water interacts simultaneously with chemisorbed PY and surface OH groups, but does not displace PY from its initial Lewis acid site. Brönsted acidity thus is thought to arise by the secondary interaction which involves transfer of a proton. It is for this reason that the authors suggest that all primary acid sites on silica–alumina are Lewis sites. The strength of interaction between a chemisorbed molecule on a Lewis site and a nearby surface OH group, and thus the ratio of LPY to BPY, depends on the interaction distance, the nature of the chemisorbed molecule, and the degree of activation of the molecule by the acid site. Similar results are obtained with ammonia. (9).

The nature of the chemisorbed molecule is an extremely important factor. An olefin, for example, requires a proton to form a classical carbonium ion, whereas a paraffin can form a classical carbonium ion by hydride ion extraction and therefore does not necessarily need a proton. Leftin and Hall (*10*) have proposed that Lewis sites extract hydride ions from triphenylmethane to form the carbonium ion:

$$\varphi_3 \, CH + Surface \rightarrow \varphi_3 \, C^+ + H^-. \tag{1}$$

By analogy, they extend their mechanism to include simpler paraffins such as isobutane. Arguments (*11*) are briefly as follows: If hydride ions are formed, they apparently remain with the catalyst. They are not transferred to a carbonium ion formed by reaction of catalyst protons with impurity olefins, nor is H_2 generated by reaction with those protons. Attempts were made to eliminate the possibility of oxidation of H^- to OH^- by treating the catalyst with hydrogen at 500°C to remove any chemisorbed oxygen, but carbonium ion formation was not suppressed. In fact, it was found that chemisorption of triphenylmethane on silica–alumina is enhanced by oxygen gas only when the system is irradiated. Because of the relative scarcity of catalyst protons, Leftin and Hall (*10*) do not believe that catalyst oxygen is necessarily involved in forming triphenylcarbinol intermediate which could then react with catalyst protons as follows:

$$\varphi_3 \, CH + O \rightarrow \varphi_3 \, C^+OH^-$$
$$\varphi_3 \, C^+OH^- + H^+ \rightarrow H_2O + \varphi_3 \, C^+. \tag{2}$$

They admit that the fate of H^- is not known exactly, and that until it is, the role of catalyst oxygen cannot be accurately assessed. However, since carbonium ions can be formed in the dark on reduced catalysts sealed in vacuo while at 550°C, they prefer the mechanism of Eq. (1) with hydride ions held by Lewis sites rather than the mechanism of Eq. (2) in which carbonium ions are formed by reaction with Brönsted sites. In a later paper, Porter and Hall (*12*) reaffirm their preference for the formation of carbonium ions from triphenylmethane by interaction with non-Brönsted sites but in a modified form. They propose the abstraction of hydride ions by cation–anion vacancy pairs instead of conventional Lewis sites.

Hirschler (*13*) and Hirschler and Hudson (*14*) favor carbonium ion formation from triphenylmethane, not by Lewis sites or cation–anion vacancy pairs, but by Brönsted sites. Their reasons are (1) about ten molecules of NH_3 or *n*-butyl amine per Lewis site are required to poison the reaction; (2) protonic acids, such as sulfuric acid, supported on silica also catalyze this reaction; (3) 90% or more of the chemisorbed ion is recovered as triphenylcarbinol, not triphenylmethane; (4) the rate of formation of chemisorbed ions is strongly accelerated by light. Neither set of arguments are unequivocal, and it would be extremely difficult, if not impossible, to find sufficient experimental evidence at the present time for an unequivocal argument. Indeed, it seems reasonable that both mechanisms could occur to a significant extent depending on local conditions.

ACTIVE CENTERS FOR BUTENE-1 ISOMERIZATION ON SILICA–ALUMINA

A very complex picture has been built up for the isomerization of butene-1 on silica–alumina (15–22) and there are still a lot of unsettled questions. When butene-1 is passed in helium at 25°C over silica–alumina, an appreciable quantity remains behind until apparent saturation of the high-energy sites is complete. The butene retained covers no more than 10% of the surface sites. In order to throw light on the nature of the adsorbed butene complex, an ammonia-blocking technique has been employed by Clark and Finch (22). The technique consists in saturating the surface of the catalyst with ammonia at various temperatures; stripping with helium at adsorption temperature; noting the quantity of strongly adsorbed, retained ammonia, and then determining as before the amount of butene taken up by these ammonia-saturated catalysts. When the catalyst is saturated with ammonia at 25°C, no butene is retained by it at 25°C. Similarly for all temperatures of ammonia saturation

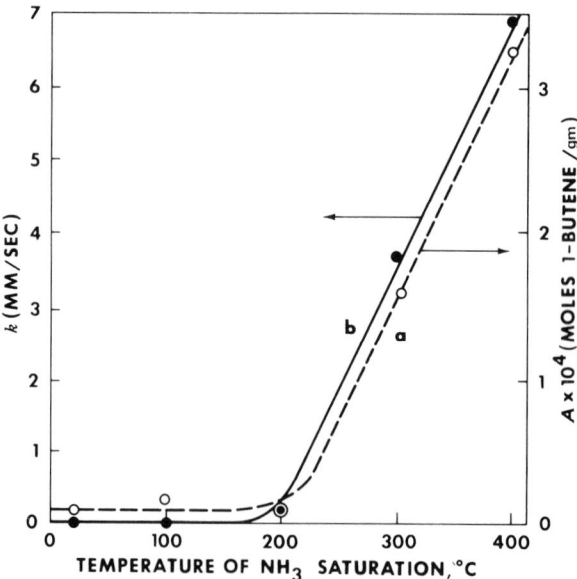

FIG. 16.2. Butene-1 on NH_3-saturated silica-alumina: (a) fixed adsorption of butene, (b) velocity constant of butene-1 isomerization. [After Clark and Finch, *4th Intern. Congr. Catalysis*, Preprint 75, Moscow (1968).]

below 200°C, no butene is retained at 25°C. However, as the temperature of ammonia saturation is increased above 200°C, the amount of butene retained by the catalyst at 25°C increases proportionately. These results imply that there is a rather sharp boundary between those sites whose adsorption energies are so great with respect to kT that molecules adsorbed on them stay indefinitely and those whose adsorption energies are so low with respect to kT that molecules are easily stripped from them. The higher the temperature, the higher the lower boundary of the energy required for fixed adsorption. The ammonia adsorption temperature, 200°C, marks the point at which ammonia just covers all the sites involved in the fixed adsorption of butene at room temperature. The results are shown in Fig. 16.2a.

When the number of butene molecules adsorbed on the catalyst at room temperature and a partial pressure of 30 mm is divided by the number of ammonia molecules adsorbed at 200°C, a measure of the number of butene molecules per occupied site is obtained. The number is approximately 4, that is, butene appears to be adsorbed on the average as tetramer. This figure is approximately confirmed by determining the average molecular weight of the polymer obtained by stripping polymer from the catalyst with cyclohexane. The calculation assumes one molecule of ammonia per occupied site, for which there is good evidence. Changes in partial pressure of olefin do not alter the temperature (200°C) at which the catalyst must be saturated with ammonia to just prevent all fixed adsorption of butene. However, the average degree of polymerization does decrease with decreasing pressure. The results suggest that some chain-terminating mechanism for polymerization is operating on the surface of the catalyst, possibly at constant rate, while the chain growth mechanism decreases with butene partial pressure.

Another question that concerns us is—what are the sites for isomerization of butene-1 to butene-2 on silica–alumina? Ammonia-blocking experiments (22) have shed some light on this question also. The experiments show that when ammonia is adsorbed at temperatures below 200°C, no isomerization of butene takes place at room temperature. But when ammonia is adsorbed at temperatures above 200°C, isomerization does occur, and the higher the temperature of application of ammonia, the greater the amount of isomerization. The results are shown in Fig. 16.2b and imply that isomerization may be associated with strongly adsorbed butene polymer. When this polymer is absent, no isomerization occurs; when it is present, isomerization does occur, and the more that is present, the more isomerization takes place. Of course, ammonia may block sites other than those holding butene polymer, certain OH groups, for example which could be the seat of catalysis.

These results are augmented by deuterium tracer studies (16, 18, 22). The products obtained by passing pulses of butene-1 over exhaustively deuterated silica–alumina (1.2×10^{14} OD groups/cm^2) contain less than 0.2% of the deuterium available on the catalyst. But nearly all mechanisms (*17*) proposed for the isomerization of butene-1 to butene-2 involve proton transfer. Thus, when an exhaustively deuterated catalyst is contacted with butene-1, the initial butene-2 product should contain one deuterium atom per molecule, if the OD groups are all involved in the isomerization process. The small exchange observed suggests that only a small fraction of the hydrogen or deuterium atoms are directly involved as isomerization sites, otherwise strong H–D exchange would have occurred. The maximum number of deuteroxyl groups exchanged with the butene is 1–$3 \times 10^{11}/cm^2$. Considered as an estimate of the concentration of active sites, this figure is much less than the usual estimates of Brönsted (*14*) or Lewis (*10*) acidity, and, of course, is valid only so far as the assumed mechanism of isomerization is valid. Thus, the almost negligible amount of H–D exchange may indicate that butene-1 isomerizes either by repeated use of a very small number of Brönsted sites or by use, in some manner, of the chemisorbed butene polymer. Gerberich *et al.* (*17*) point out the possibility of randomizing the hydrogen of chemisorbed butene polymer with catalyst D. If this hydrogen (8×10^{14} H/cm^2) is randomized with all the catalyst D (1.2×10^{14} D/cm^2), then the overall concentration would be about 13% D and the initial butene-2 product would be expected to contain about this amount, which is far higher than the experimental values. If, however, only a small portion of the catalyst D is randomized with the butene H, the experimental results can be explained. For example, if $8 \times 10^{14}/cm^2$ butene H is randomized with $1.5 \times 10^{12}/cm^2$ catalyst D, the overall concentration would be about 0.2% and no more than 0.2% C_4H_7D would be expected in the initial butene-2 product. In contrast, butene-2 product obtained by passing butene-1 over silica–alumina having perdeuterobutene polymeric complex on its surface is highly deuterated. The initial product contains 36 times more deuterium than does the initial product from the exhaustively deuterated catalyst, while in both cases the reactant butene-1 contains a negligible amount. If isomerization is assumed to take place on Brönsted sites, one would not expect deuterium in the initial butene-2 product unless randomizing of the catalyst H and butene-D had occurred. Randomizing to the same extent that fits the experimental results for the reverse case, that is, $8 \times 10^{14}/cm^2$ butene D and $1.5 \times 10^{12}/cm^2$ catalyst H, should give an initial butene-2 product with essentially one D per molecule, whereas the actual product contains only about 0.3. It is possible that only a portion of the butene D is available for randomizing with catalyst H.

If we suppose that isomerization is associated with chemisorbed butene polymer, several detailed mechanisms may be considered. The butene polymer chemisorbed on Lewis sites may act as a proton donor as suggested by Ozaki and Kimura (*16*), or butene may react directly with hydroxyl groups forming induced Brönsted sites, or, finally, butene polymer chemisorbed on Lewis sites may use a small portion of catalyst OH, transferring a proton from it to form the active polymer site, corresponding to the mechanism of Basila *et al.* (*8*) discussed above. Since it has been shown that isomerization correlates directly with catalyst OH content, the last two mechanisms are favored. It is not possible in our present state of knowledge to choose between them.

SURFACE STATES OF MgO AND H_2–D_2 EXCHANGE ACTIVITY

As a final example of chemisorbed species in the role of actual sites, we describe the work of Harkins *et al.* (*23*) on the catalytic activity of MgO in relation to its electron energy states. They studied the catalytic activity of MgO for the H_2–D_2 exchange reaction as a function of surface radiation with controlled frequencies of ultraviolet from a monochromator. Relative enhancement of the rate constant was obtained at well-defined peaks, 5.7, 4.9, and 4.0 eV, as shown in Fig. 16.3, where k and k_0 are the rate constants with and without radiation, respectively. Results of previous studies by Lunsford (*24*) and Lunsford and Leland (*25*) reveal two distinct mechanisms involved in

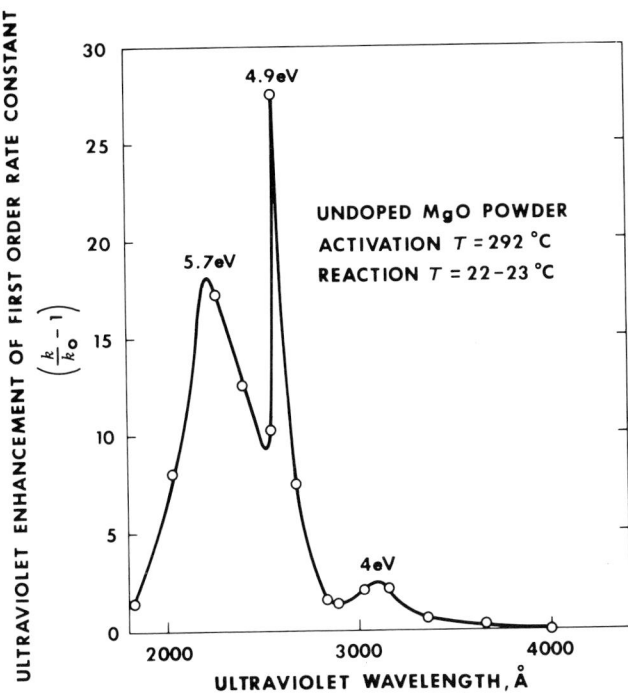

FIG. 16.3. Ultraviolet enhancement of first-order rate constant for H_2–D_2 exchange as a function of wavelength. [After Harkins *et al.*, *J. Phys. Chem.* **73**, 130 (1969). By permission of The American Chemical Society.]

the exchange depending on whether the catalyst has been activated at 500°C or at 300°C. Catalyst activated at the lower temperature retains considerably more hydroxyl groups and catalytic activity is shown to be associated with these groups. Furthermore, the catalytic activity of samples pretreated at the lower temperature is sensitive to ultraviolet radiation, whereas the activity of samples pretreated at the higher temperature is not. We shall restrict our attention to catalysts treated at the lower temperature, for they illustrate activity associated with chemisorbed species.

In order to gain evidence for a hydrogen-containing active site, deuterium was placed in contact with a fresh catalyst sample not previously exposed to hydrogen, and the rate of exchange with the hydrogen species (OH groups) on the catalyst surface was determined. The reactor was then evacuated and the catalyst sample irradiated with 4.9 eV light for about 30 min. After completion of the radiation, D_2 was introduced again to the catalyst and the rate of exchange with any residual hydrogen determined. This second exchange proceeded faster than the original thermally activated exchange which took place in the presence of a larger reservoir of surface hydrogen. Thus, any H_2–D_2 exchange reaction will be associated with a hydrogen-containing site which is sensitive to ultraviolet radiation. Logically, it is assumed that these hydrogen-containing sites are OH^- anions.

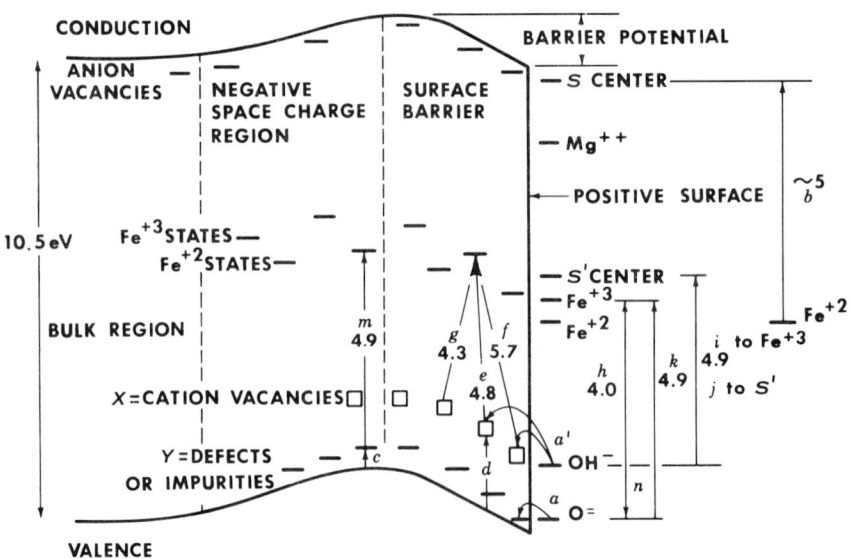

SURFACE REGION TRANSITIONS IN MgO

Fig. 16.4. Surface-region transition in MgO. [After Harkins *et al.*, *J. Phys. Chem.* **73**, 130 (1969). By permission of The American Chemical Society.]

On the basis of their own work (*23–25*) and the work of others on the photoconductivity (*26–28*), intrinsic optical absorption bands (*29–32*), impurity optical absorption bands (*27, 33–38*), and electron spin resonance studies (*37, 39–43*), Leland *et al.* have constructed a diagram of energy transitions in the surface region of MgO. In Fig. 16.4, the various states and the transitions between them are reproduced from their paper (*23*). There are cation vacancy states (X) which are called V_1 centers when they capture a positive hole. A positive hole is produced by promotion of an electron to some level, such as an impurity level; for example, transitions e, f, and g which represent the promotion of an electron to an iron impurity level. There is also the S center, a surface anion vacancy (O^{2-} vacancy) with a trapped electron, and the S' center in which a surface cation-anion vacancy pair has trapped an electron. Finally, the diagram shows the OH^- and O^{2-} levels. Leland *et al.* have ruled out certain levels and energy transitions as responsible for catalytic activity. For example, V_1 centers, surface cation impurities, S centers, and S' centers are not considered to be active catalytic sites and transition (b) is an unlikely site generator. They have selected others consistent with their data. The photo-enhanced activity is postulated to be associated with a hole trapped at the OH^- surface site which is equivalent to a bound OH^0 radical. The various transitions responsible for OH^0 radical formation are summarized in Table 16.2.

TABLE 16.2

ENERGY TRANSITIONS ASSOCIATED WITH H_2–D_2 EXCHANGE ON MgO[a]

Energy, eV	Transition	Effect
~4.9	k	O^- produced by excitation of electron from O^{2-} to a surface impurity which then forms OH^0 by transition n and fluorescence. $OH^- + O^- \rightarrow OH^0 + O^{2-}$.
~4.9	h	OH^0 produced on a surface by excitation of an electron from OH^- to a surface impurity, e.g., Fe^{3+}.
~4.9	i, j	OH^0 produced on surface by excitation to form an S' center or to an impurity below the surface.
~4.9	e	Positive hole formed by promotion of an electron (e), capture of the hole by a cation vacancy to form a V_1 center near the surface, transfer of an electron from OH^- to the V_1 center to form OH^0 (a').
5.7	f	A V_1 center formed near the surface (f) which then receives an electron from OH^- forming OH^0 (a').
4.0	h	OH^0 formed directly by excitation of an electron from OH^- to a surface impurity.

[a] See Harkins *et al.* (*23*).

Photo-enhancement of the H_2–D_2 exchange is greatest for 4.9 eV radiation, as seen in Fig. 16.3, for this energy produces OH^0 by more paths than any of the other energies as Table 16.2 shows. The course of the exchange reaction is depicted as follows:

$$H_2 + OH^0 \leftrightarrows HOH + H,$$

$$D_2 + OH^0 \leftrightarrows DOH + D,$$

$$D + HOH \leftrightarrows OH^0 + HD,$$ (3)

$$H + DOH \leftrightarrows OH^0 + HD.$$

The entire scheme is plausible, but it should be pointed out that the OH^0 group which is proposed as the active site has never been detected in ESR studies.

16.2 π-Complexes

Investigations of catalytic reactions on metal surfaces have emphasized the electronic state of the metal and less attention has been given to the nature of the bonds around the atoms in adsorbed species. Since more is known about the bonding orbitals of an atom or molecule than those of a surface, it is surprising that only recently attempts have been made to contruct models of adsorption based on the bonding orbitals of the atoms in the adsorbate. Such studies have led to detailed pictures of surface species which play the role of intermediates in chemical reactions. The main evidence for these intermediates comes from interpretation of hydrogen exchange reactions. We shall begin with a brief description of the classical exchange mechanisms, and follow with a review of the more recent π-complex mechanisms.

CLASSICAL EXCHANGE MECHANISMS

These may be divided into two classes, dissociative and associative mechanisms. In dissociative theories, originally proposed by Farkas and Farkas (44–46), hydrocarbon molecules are chemisorbed at a carbon atom after rupture of a carbon–hydrogen bond. In associative mechanisms, proposed by Horiuti and Polanyi (47), unsaturated hydrocarbons are chemisorbed by the opening of a double bond. Chemisorption and exchange of saturated molecules will be considered later in this section. The two mechanisms for exchange, dissociative and associative are shown, Eqs. (4)–(8). A point in favor of the associative mechanism is the fact that exchange reactions with unsaturated hydrocarbons are very much faster than those with saturated hydrocarbons. This observation is difficult to explain for the dissociative mechanism, since π electrons are not involved in that mechanism and carbon–

Classical dissociative mechanism

$$D_2 \;+\; 2\,M \;\rightleftharpoons\; 2 \;\underset{M}{\overset{D}{|}} \tag{4}$$

(5)

(6)

Classical associative mechanism

(7)

(8)

hydrogen bond strengths are approximately the same for saturates and unsaturates. On the other hand, the fact that exchange reactions take place between normal and deuterated hydrocarbons, e.g., C_2H_4–C_2D_4, is a strong argument in favor of the dissociative mechanisms. Another argument in favor of the dissociative mechanism concerns the loss of resonance energy attending the associative adsorption of aromatics. The seriousness of this effect is illustrated by hydrogenation studies which show that the first pair of hydrogen atoms add to the ring endothermically. In general, the dissociative mechanism is favored over the associative mechanism.

Yet there are strong arguments against either classical mechanism. Data on the relative adsorption strengths of aromatic molecules on metals (*48*), determined by studies of competitive adsorption in exchange or hydrogenation reactions, indicate considerable differences within various aromatic series. These differences are difficult to interpret by the classical associative or dissociative mechanisms, for classical mechanisms fail to explain why different aromatic hydrocarbons possessing nearly the same carbon–hydrogen bond strengths should form carbon–metal chemisorptive bonds of markedly different strengths. Difficulties of interpretation in terms of the associative mechanisms are not as severe, since π-electron energies differ greatly for members of various aromatic series. However, as mentioned above, this mechanism is considered relatively unimportant. In view of the difficulty with classical mechanisms, a new approach was felt necessary.

π-COMPLEX EXCHANGE MECHANISMS WITH AROMATICS

In the new approach, chemisorption of aromatics is assumed to proceed with the plane of the ring parallel to the surface by formation of a π-complex, (*48–52*) denoted in Eq. (9). The π-bonded aromatic molecule is similar to a sandwich compound, such as chromium dibenzene. The π-electrons of the

$$\langle\bigcirc\rangle \;+\; M \;\rightleftarrows\; \underset{\underset{M}{\mid}}{\langle\bigodot\rangle} \tag{9}$$

$$(\pi\text{-Bonded})$$

aromatic ring are delocalized forming a collective π orbital which interacts with appropriate empty *d*-orbitals of the transition metal atom. Back-bonding between filled metal *d*-orbitals and the empty antibonding π orbitals of the aromatic may also occur. Garnett and Sollich (*49*) propose two mechanisms analogous to those in the classical case. The associative π complex substitution mechanism is pictured in Eq. (10). It is similar to a conventional aromatic

$$\underset{\underset{M}{\mid}}{\langle\bigodot\rangle} \;+\; \underset{M}{\overset{D}{\mid}} \;\longrightarrow\; \left[\underset{\underset{M}{\mid}}{\langle\bigodot\rangle}\overset{H}{\underset{D}{\diagup}}\right] \;\longrightarrow\; \overset{H}{\underset{M}{\mid}} \;+\; \underset{\underset{M}{\mid}}{\langle\bigcirc\rangle}{-}D \tag{10}$$

substitution reaction except that the aromatic is π-bonded to the metal. In this mechanism, the π-bonded aromatic nucleus is attacked by a chemisorbed deuterium atom resulting in the partial localization of a π electron in the transition complex. Note the similarity of the transition state to the "half-hydrogenated" intermediate of the classical associative mechanism, Eq. (8).

Garnett and Sollich visualize the dissociative π-complex substitution mechanism as in Eq. (11) and (12). The π-bonded aromatic reacts with a *metal*

Horizontally
π-bonded

Inclined 45°

Vertically
(90°) σ-bonded

$$(11)$$

$$(12)$$

radical in a substitution reaction equation (11). During the process, the molecule is assumed to rotate through 90° resulting in an edge-on σ-bonded adsorbate. The transition state from π-to σ-bonding is postulated to form an angle of about 45° with the catalyst surface. The exchange is completed by a reverse process with D, Eq. (12). Garnett and Sollich present evidence why the dissociative π-complex substitution mechanism is of major importance in platinum-catalyzed exchange reactions.

π-COMPLEX EXCHANGE MECHANISMS WITH SATURATED COMPOUNDS

Even stronger evidence for π-complex exchange mechanisms has been obtained by Kemball, Burwell and their colleagues (*53–63*) for paraffins and cycloparaffins. Early results of Anderson and Kemball (*55*) gave the observed values listed in Table 16.3 for initial distribution of deuterium exchanged with ethane. They explained the results by postulating the interconversion of adsorbed ethyl radicals and ethylene molecules formed by the stepwise dissociative chemisorption of ethane according to the following scheme, where X represents an H or D:

$$X_2C\!-\!CX_2 + 2 \underset{M}{\overset{X}{\underset{\text{(iv)}}{\big|}}} \overset{\text{(iii)}}{\underset{\text{(iv)}}{\rightleftharpoons}} X_2C\!-\!CX_3 + \underset{M}{\overset{X}{\underset{\text{(i)}}{\big|}}} \overset{\text{(ii)}}{\underset{\text{(i)}}{\rightleftharpoons}} C_2X_6 \,(g).$$

$$(13)$$

This interconversion, steps (iii) and (iv), is commonly referred to as the $\alpha\beta$ process. On the basis of Mechanism (*13*), it is possible to calculate a distribution for comparison with the observed distribution. Let P denote the ratio of the chance of an adsorbed ethyl radical forming an adsorbed ethylene molecule by step (iii) to the chance of its desorption by step (ii) to form gaseous ethane. Then the chances of these two events are $P/(1 + P)$ and $1/(1 + P)$, respectively. Expressions for the initial distribution can be obtained mathematically in terms of the single parameter P. Selecting values of P, distribu-

TABLE 16.3

INITIAL DISTRIBUTIONS FROM EXCHANGE OF ETHENE[a]

Metal		Percentage					
		d_1	d_2	d_3	d_4	d_5	d_6
Mo	observed	82	14	3.0	0.7		
	calculated	80.0	17.2	2.5	0.3		
	($P = 0.25$)						
Pd	observed	5	6	8	11	19	51
	calculated	5.3	7.3	8.7	11.2	16.9	50.6
	($P = 18$)						

[a] See Rooney (*64*).

tions are calculated, as shown in Table 16.3, which match the observed distributions. The results give evidence that the deuterium exchange occurs by the $\alpha\beta$ process. $\alpha\beta$-diadsorption of ethylene, as depicted in Eq. (13), would appear to proceed most readily by cis-elimination of H atoms, and to be followed by cis-addition of D; and, therefore, rapid rotation of the methyl group must occur if all H atoms are replaced by D atoms in one sojourn of the molecule on the catalyst surface. It was generally accepted that both the ethyl radical and ethylene were σ-bonded to the metal.

The classical $\alpha\beta$ mechanism soon ran into difficulties. For instance, the exchange of propane, *n*-pentane, and *n*-hexane with deuterium on palladium films (*57*) gave initial distributions where the perdeutero-isomer amounted to more than 80% of the total products under the same conditions where ethane exchange gave only about 50% of the perdeutero-isomer. If an $\alpha\beta$ process is operating, the possibility of triadsorbed, tetra-adsorbed, and other multiadsorbed species must be considered in order to explain the initial formation of products with a high deuterium content. A generalized $\alpha\beta$ process could be visualized:

$$\alpha \leftrightarrows \alpha\beta \leftrightarrows \alpha\beta\gamma \leftrightarrows \cdots \alpha\beta\gamma \cdots. \tag{14}$$

However, such tri- and tetra- and other multiadsorbed species become highly improbable, if only σ-bonded species are considered. The geometric restrictions imposed on a paraffinic molecule σ-bonded through sp^3-hybrid, tetrahedrally oriented, carbon-atom orbitals are too severe. Another difficulty faced by the classical $\alpha\beta$ mechanism is the explanation of the initial exchange of all ten hydrogen atoms in cyclopentane on palladium. By the classical $\alpha\beta$ mechanism, only five hydrogen atoms should exchange initially, and then the molecule would have to desorb, "turn over," and re-adsorb in order to exchange the five atoms on the other side of the planar ring.

To circumvent these and other difficulties, the classical $\alpha\beta$ mechanism was discarded, and instead multicentered bonding to the same metal atom was considered. It was proposed that intermediates undergo stepwise interconversion as σ- and π-bonded ligands to one atomic center in a metal surface. The interconversion of σ-bonded alkyl and π-bonded olefin is still restricted to cis elimination and addition. Furthermore, Burwell and Briggs (*62*) have shown that a *gem* dimethyl group in a hydrocarbon chain blocks the propagation of an exchange process. For example, the highest deutero-isomer obtained initially from 3,3 dimethylhexane was the d_7-isomer. Also the protons in such methyl groups are relatively inactive for exchange. We illustrate the stepwise interconversions of σ- and π-bonded ligands by the reactions of cyclohexane and its unsaturated derivatives, Eq. (15).

To explain the initial exchange of all ten hydrogen atoms in cyclopentane, it is postulated that arene-type complexes are responsible, and that during the formation of π-bonded cyclopentenyl (and possibly of cyclopentadiene and cyclopentadienyl) hydrogen atoms can be removed and replaced on both sides of the ring. In Eq. (16), we illustrate with the interconversion of cyclopentene and cyclopentenyl intermediates. In the transition state, the C–H bond

(15)

$$(16)$$

above the plane of the ring at the allylic carbon becomes labile and there is a change toward sp^2-hybridization. The H atom is eliminated by reaction with a surface D atom and a π-allylic complex is formed. The D′ atom originally below the plane of the ring is now in the plane; but, on addition of a D atom below the ring, the D′ atom is finally located above the ring. A D atom may add above or below the ring, which, in the case of dialkyl substituted π-allylic derivatives of cyclic paraffins, can lead to cis or trans isomers from a common intermediate. It should be pointed out that π-allylic species can form only when there are more than two adjacent carbon atoms with replaceable hydrogen atoms. With only two adjacent carbon atoms having replaceable hydrogens, for example, two adjacent carbon atoms each blocked by an adjacent *gem* dimethyl structure, π-allylic species cannot form, so that initial exchange on the side of the ring only would be expected.

In order to check the ideas related to π-complexes, exchange reactions were carried out on a series of polymethyl-cyclopentanes (*59, 60, 65*) over several metal films. We shall consider first the results over palladium, keeping in mind the rules developed above, namely:

(1) By the $\alpha\beta$ mechanism, initial exchange can occur only on one side of the ring.
(2) The π-complex mechanism cannot occur unless there are three or more adjacent carbon atoms with replaceable hydrogen atoms.
(3) By the π-complex mechanism, initial exchange can occur on both sides of the ring.
(4) In both mechanisms, only *cis* elimination or addition is allowed, and *gem* dimethyl groups do not exchange appreciably.

Table 16.4 shows the initial distributions from exchange on palladium of the structures depicted in Fig. 16.5. In Fig. 16.5, dotted bonds are oriented away from the surface and full bonds toward the surface. An $\alpha\beta$ mechanism will

TABLE 16.4

INITIAL DISTRIBUTIONS FROM EXCHANGE ON PALLADIUM[a]

Compound T °C	I 75	II 80	III 40	IV 80
d_1	43	54.7	<2.0	8.1
d_2	57	1.5		1.8
d_3	<1	<1.0		1.1
d_4		<1.0		1.8
d_5		4.4		6.6
d_6		39.4		0.2
d_7		<1.0		0.4
d_8				0.5
d_9				0.8
d_{10}				1.1
d_{11}			7.5	9.6
d_{12}			92.5	66.9
d_{13}			<0.5	1.1

[a] See Rooney (64).

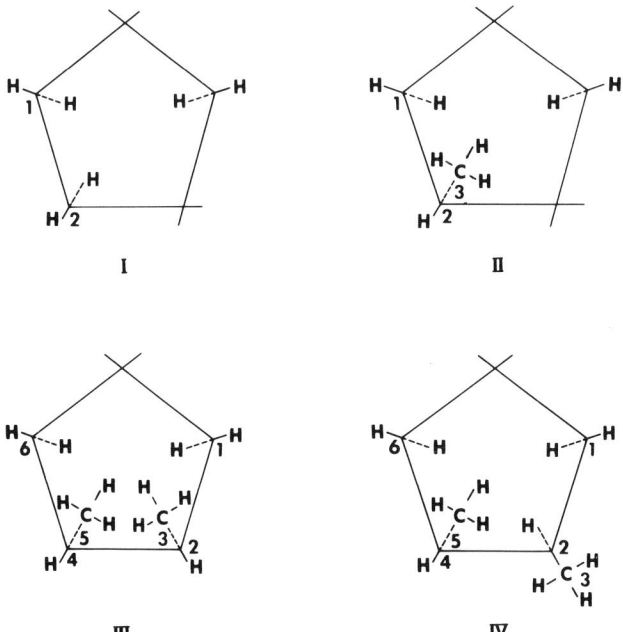

FIG. 16.5. Polymethylcylopentanes.

allow initial exchanges within one set or the other, not both, as in the π-complex mechanism, In structure I, 1,1,3,3-tetramethycyclopentane, only two adjacent carbon atoms have replaceable hydrogens (1 and 2), so a π-allylic complex cannot form, and consequently the cis exchange of two hydrogen atoms initially is the maximum allowed. The observed result in Table 16.4 agrees with this prediction.

In structure II, 1,1,3,3,4-pentamethylcyclopentane, a maximum of five H atoms (1, 2, 3) may be exchanged by the $\alpha\beta$ process. Since there are three adjacent reactive carbon atoms, a π-allylic complex may form and a sixth hydrogen located on carbon atom 1 and oriented away from the surface should also exchange initially. Table 16.4 shows a maximum at d_6.

In structure III, cis-1,1,3,4-tetramethylcyclopentane, a maximum of ten hydrogen atoms are expected to change initially by the $\alpha\beta$ process. In addition, the two hydrogens on carbon atoms 1 and 6, oriented away from the surface, will exchange by the π-complex mechanism, making a total of twelve exchangeable hydrogens. Note the maximum at d_{12} in Table 16.4. A striking confirmation of the π-complex mechanism is offered by the corresponding trans isomer, Structure IV, in which only five H atoms can be exchanged by the $\alpha\beta$ mechanism, either those on carbon atoms 1, 2, 3, or 4, 5, 6. If the exchange processes are to propagate from one group to the other, a π-allylic complex must be formed either by elimination of a hydrogen atom on carbon 2 or carbon 4 whichever is oriented away from the surface. Since π-allylic complexes are readily formed, a maximum of twelve exchangeable hydrogens is expected, which checks with the observed result in Table 16.4.

Exchange reactions for several of these compounds were also carried out on films of platinum, nickel, and rhodium (64). The order of importance of π-allylic complexes in these reactions was found to be Pd $\gg$ Pt > Ni > Rh. We make a few comparisons between palladium and rhodium which appears to operate to the least extent by the π-complex mechanism. All four H atoms on carbon atoms (1) and (2) Fig. 16.5, structure I, are now replaced initially and maxima occur at both d_2 and d_4. The suggestion is made that in this case the "turnover" mechanism responsible for d_4 involves $\alpha\alpha$-diadsorbed species, that is, two σ-bonds to a single carbon atom. Support for this mechanism in the case of rhodium stems from the fact that this metal is most active in catalyzing the $\alpha\alpha$ process in the exchange of methane. In compounds II and III, only five and ten H atoms are readily replaced indicating that in the case of rhodium only a rapid $\alpha\beta$-exchange process occurs without participation of π-allylic complexes.

16.3   Cyclic Intermediates

An interesting and significant reaction reported recently by Banks and Bailey (66) is the olefin disproportionation reaction, also referred to as olefin dismutation (*67*) and olefin metathesis (*68*). Two identical olefin molecules, $R_1CH_2CH{=}CHR_2$ can react as follows:

$$R_1CH_2CH{=}CHR_2 \atop R_1CH_2CH{=}CHR_2 \; \leftrightarrows \; {R_1CH_2CH \atop R_1CH_2CH}{\|} + {HCR_2 \atop HCR_2}{\|}, \qquad (17)$$

where R_1 and R_2 may be alkyl groups or hydrogen atoms. An olefin symmetrical about the double bond reproduces itself. Two dissimilar olefins, ethylene not excluded, can also react. Diolefins, acetylenes, and cyclic olefins may also react, but we shall restrict our attention to acyclic olefins, which suffices to illustrate the general mechanism. Catalysts for the reaction are supported transition metal oxides such as molybdenum or rhenium oxide on alumina, tungsten oxide on silica, and various homogeneous organo-metallics containing the above-mentioned transition metals. A particular catalyst, derived from WCl_6, C_2H_5OH, and $C_2H_5AlCl_2$ (*68*), is reported to be active only for internal olefins.

The nature of the products obtained from these reactions lends support to a four-center mechanism, involving a quasi-cyclobutane intermediate formed by proper alignment of the carbon atoms at the double bonds of the two reacting olefins. We illustrate with butene-1:

$$\begin{matrix} C{=}C{-}C{-}C \\ C{=}C{-}C{-}C \end{matrix} \; \leftrightarrows \; \begin{matrix} C{\cdots}C{-}C{-}C \\ C{\cdots}C{-}C{-}C \end{matrix} \; \leftrightarrows \; {C \atop C}{\|} + {C{-}C{-}C \atop C{-}C{-}C}{\|}. \qquad (18)$$

Further reactions may follow by isomerization of butene-1 to butene-2 and disproportionation of the two isomers:

$$\begin{matrix} C{=}C{-}C{-}C \\ C{-}C{=}C{-}C \end{matrix} \; \leftrightarrows \; \begin{matrix} C{\cdots}C{-}C{-}C \\ C{-}C{\cdots}C{-}C \end{matrix} \; \leftrightarrows \; {C \atop C{-}C}{\|} + {C{-}C{-}C \atop C{-}C}{\|}. \qquad (19)$$

New olefins are not produced by interaction of two butene-2 molecules.

As an alternative to the transalkylidenation mechanism described above, Calderon *et al.* (*68*) have suggested the possibility, for internal olefins, of a transalkylation mechanism:

$$\begin{matrix} R_1CH{=}CH{-}R_2 \\ R_1{-}CH{=}CHR_2 \end{matrix} \; \leftrightarrows \; {R_1CH{=}CH \atop R_1}{|} + {R_2 \atop CH{=}CHR_2}{|} \qquad (20)$$

where R_1 and R_2 are alkyl groups. Although the same product olefins are predicted by both mechanisms, their experiments with deuterated olefins indicated that the distribution of deuterium in the products is consistent only with the transalkylidenation mechanism, Eqs. (18) and (19). For example, the interaction of 2-butene with 2-butene-d_8 gave $C_4H_4D_4$ as the only new product in the absence of isomerization of 2-olefin to 1-olefin. The products anticipated by each mechanism are listed in Table 16.5, and it is obvious that the transalkylation mechanism is eliminated.

TABLE 16.5

ANTICIPATED PRODUCTS FROM 2-BUTENE WITH 2-BUTENE-d_8

Mechanism	Anticipated products	Mass
Transalkylation	$CH_3CH{=}CHCH_3$	56
Eq. (20)	$CH_3CH{=}CDCH_3$	58
	$CH_3CH{=}CHCD_3$	59
	$CH_3CD{=}CDCD_3$	61
	$CD_3CH{=}CHCD_3$	62
	$CD_3CD{=}CDCD_3$	64
Transalkylidenation	$CH_3CH{=}CHCH_3$	56
Eq. (17)	$CH_3CH{=}CDCD_3$	60
	$CD_3CD{=}CDCD_3$	64

Further evidence in support of the four-center, quasi-cyclobutane intermediate was obtained by Clark (*69*) through application of the radioactive carbon isotope ^{14}C. Substitution of ^{14}C for ^{12}C in the 1-position of propylene [asterisk in Eq. (21)] results in ^{14}C being present in the product only in the ethylene, and not in the butene, after reaction over molybdena–alumina catalyst.

$$\begin{matrix} C{-}C{=}C^* \\ C{-}C{=}C^* \end{matrix} \leftrightarrows \begin{matrix} C{-}C{\cdots}C^* \\ C{-}C{\cdots}C^* \end{matrix} \leftrightarrows \begin{matrix} C{-}C \\ C{-}C \end{matrix} + \begin{matrix} C^* \\ C^* \end{matrix} . \tag{21}$$

To carry out the experiment successfully, it was necessary to avoid isomerization of the propylene-1-^{14}C to propylene-3-^{14}C. For this reason, the temperature was kept as low as possible (80°C), consistent with sufficient reaction products for accurate detection. The results support the postulated mechanism. As the temperature of the reaction is increased above 80°C, disproportionation of propylene-1-^{14}C yields products with an increasing amount of radioactivity in the butene as a result of isomerization to propylene-3-^{14}C, until, at 160°C, nearly as much ^{14}C is present in the butene as in the ethylene. Other experiments have been reported (*69, 70*) using propylene-2-^{14}C. In this case, as expected, practically all of the radioactivity in the products is concentrated in the butenes. And this is true over the temperature range 80–160°C, because isomerization in this case does not change the nature of the radioactive carbon atom in the molecule.

$$\begin{matrix} C{-}C^*{=}C \\ C{-}C^*{=}C \end{matrix} \leftrightarrows \begin{matrix} C{-}C^*{\cdots}C \\ C{-}C^*{\cdots}C \end{matrix} \leftrightarrows \begin{matrix} C{-}C^* \\ C{-}C^* \end{matrix} + \begin{matrix} C \\ C \end{matrix} . \tag{22}$$

There are many other reactions in which the evidence for a four-center mechanism is favorable. Among them are the dehydrogenation of 1-butene over chromia–alumina catalyst (*71*) and the polymerization of propylene over nickel oxide–silica alumina catalyst (72). Although the evidence for a cyclic intermediate is strong in all of these reactions, further work is necessary to provide a more detailed picture.

16.4 Stereospecificity

Of the many examples of stereospecificity, the most important ones are vinyl polymers made from monomers of the type $CH_2{=}CHR$. On polymerization of such a monomer, asymmetric carbon atoms are generated along the chain backbone. Each carbon atom bearing the R substituent can take up either a d- or l-configuration. In most polymerization processes, d- and l-configurations are random and therefore the polymer does not exhibit any optical activity in the absence of activity of the R substituent. There are, however, certain catalysts such as the Ziegler–Natta and Phillips catalysts in which the d- and l-configurations are not random, so that in a regular sequence such as head-to-tail monomer units different arrangements are possible. If all the R groups are on the same side of the plane, as in Fig. 16.6a, a regular sequence of head-to-tail monomer units has every other carbon atom of the same conformation, and the polymer has an isotactic structure (73). Optical activity of the polymer is not observed in practice because its mirror image is produced in an equal amount. If the R groups are alternately above and below the plane of the zigzag backbone as in Fig. 16.6b, the polymer is said to

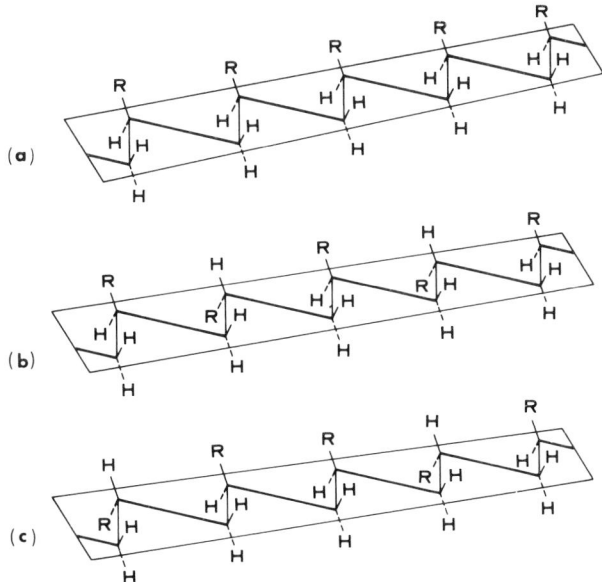

FIG. 16.6. (a) Isotactic polymer, (b) syndiotactic polymer, (c) atactic polymer.

have a syndiotactic structure. Finally, a polymer of random structure with no regular sequence is referred to as atactic polymer, Fig. 16.6c. Spatial arrangements are not exactly as depicted in the planar zigzag conformation. Often the steric interaction between R groups is sufficient to force the chain out of its planar arrangement. By suitable rotations around the carbon–carbon single bonds, the interactions are relieved, and the R substituents serve as a barrier to free rotation. Chains in the solid state tend to fit into a spiral spatial arrangement in which a full 360° spiral might require three or more structural recurring units in a vinyl polymer.

The most detailed theory proposed to explain stereospecificity in polymers produced from monomers of the type $CH_2{=}CHR$ is that of Cossee and Arlman which describes the mechanism for production of isotactic polypropylene using a Ziegler–Natta catalyst system, α-$TiCl_3$ and $Al(C_2H_5)_3$. We shall review their theory to illustrate the complex localized interactions of catalyst surface structures and olefins which are considered to lead to the growth of isotactic polymers.

Several investigators have put forth schemes for the mechanism of the reaction (74–82), and, with one exception, (74) they agree that polymer growth occurs by insertion of a monomer molecule between a metal atom and a carbon atom of an organometallic compound. Disagreement still exists concerning the nature of the active centers. Most theories require two metal ions bound in a complex, one to hold the monomer group, the other to carry the alkyl group. The theory of Cossee and Arlman postulates the use of only one metal ion. There is still discussion as to whether the growing polymer chain is bonded to the transition metal or to the nontransition metal. Nevertheless, most recent investigations support the view that the growth reaction occurs at the transition metal ion. In accordance with Natta's evidence (77, 83), it is believed that the reaction requires a solid catalyst. The various modifications of $TiCl_3$ (α, β, and γ), when used as catalysts in the polymerization of propylene, give products which differ in percentage of isotacticity, the α modification producing the highest percentage. With the exception of the treatment of Cossee and Arlman, little attention has been given to theoretical concepts such as ligand-field theory and molecular orbital methods, which allow a more detailed picture of the role of the transition element.

In their treatment, Cossee and Arlman start with the bonding of the olefin monomer. They proceed to postulate the ligand-field characteristics of a titanium ion which are conducive to the polymer growth process and show that these characteristics are most likely on a particular crystal face of the solid complex. Finally, they discuss the conditions for stereoregularity.

In postulating the complex formation between a transition metal ion and an olefin, Cossee (84) adopts the ideas of Chatt and Duncanson (85) which

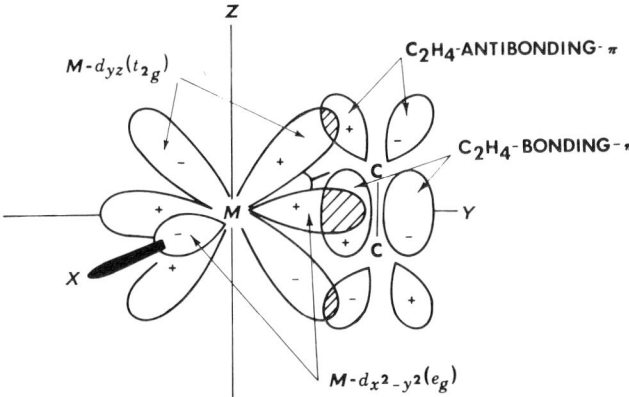

FIG. 16.7. Schematic picture showing spatial arrangement of the relevant orbitals in a π-bond between a transition metal and C_2H_4 *(84)*.

involve the formation of a π-complex similar to those discussed in Section 16.2. Figure 16.7 shows the bond of σ symmetry between a metal d-orbital, $d_{x^2-y^2}$, and the olefin π-orbital and the bond of π symmetry between a metal d-orbital, d_{yz}, and the antibonding π orbitals of the olefin. The structures of many such stable compounds are now considered well established, and would appear to be appropriate for olefin complexes with the surfaces of transition metal catalysts.

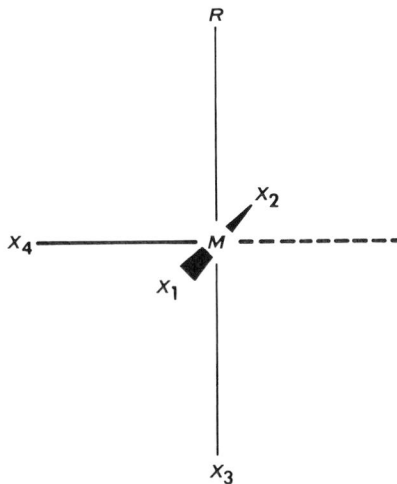

FIG. 16.8. Configuration supposed to be the active center in a Ziegler–Natta catalyst. M: transition metal; R: alkyl (growing polymer chain); X_1–X_4 are anions *(84)*.

As mentioned above, the propagation step in these polymerizations is strongly believed to take place by insertion of an olefin molecule between a metal atom and an alkyl group:

$$M-R + CH_2=CH_2 \rightarrow M-CH_2-CH_2-R. \tag{23}$$

Thus, the transition metal ion must have an alkyl ligand and the possibility to coordinate an olefin molecule. The simplest configuration meeting these requirements is shown in Fig. 16.8, where the metal ion is octahedrally co-ordinated to form halogen ions (X_1, X_2, X_3, X_4), an alkyl group R and a vacant position capable of bonding an olefin. When an olefin like ethylene is coordinated into such a configuration, the metal d_{yz} orbital (see Fig. 16.7), originally having an energy comparable to that of the d_{xy} and d_{xz} orbitals, now combines with the π antibonding orbital of ethylene and forms an orbital of considerably lower energy. The energy gap between this orbital and the orbital representing the bond between the alkyl group and the transition metal, M$-$R, is significantly reduced. Since the promotion of an electron from the M$-$R-bonding orbital by thermal excitation is now easier, the MR bond will be more susceptible to breaking into radicals. The reaction mechanism may thus be written as in Eq. (24).

From the configuration about an active center pictured in Fig. 16.8, it is evident that the availability of metal ions with vacant positions of ligand co-ordination is a critical factor for polymer chain growth. Arlman (86) presents arguments indicating that the surfaces of transition metal chlorides (e.g., TiCl$_3$) carry chlorine vacancies which satisfy this requirement for polymer

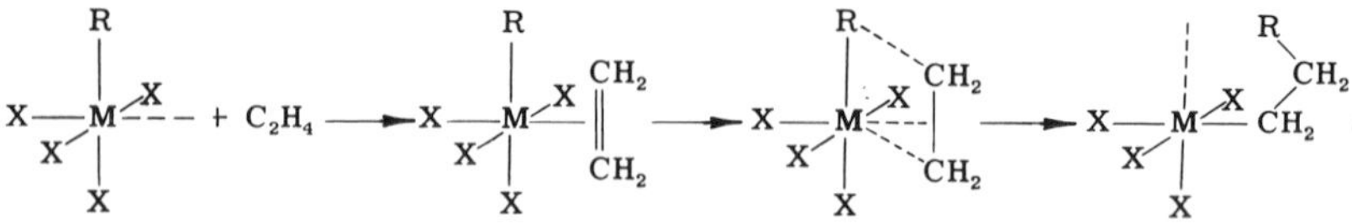

$$\tag{24}$$

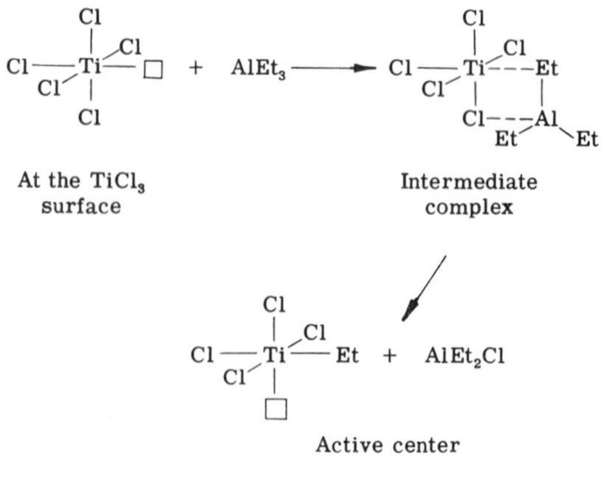

At the TiCl$_3$
surface

Intermediate
complex

$$\tag{25}$$

Active center

chain growth. For a layer lattice structure, he shows that the number of surface vacancies, i.e., incompletely coordinated titanium ions at the surface, is a
function of crystal size. These vacancies do not occur at the usually best
developed crystal face of these layer-lattice structures, e.g., (0001), but rather
at other faces which form the sides of crystal platelets. An example is the
$10\overline{1}0$ face, which Arlman works out in detail.

Starting with an octahedrally coordinated titanium ion on the surface of
$TiCl_3$ containing five chlorine ligands and one vacancy, Arlman and Cossee

(a)

(b)

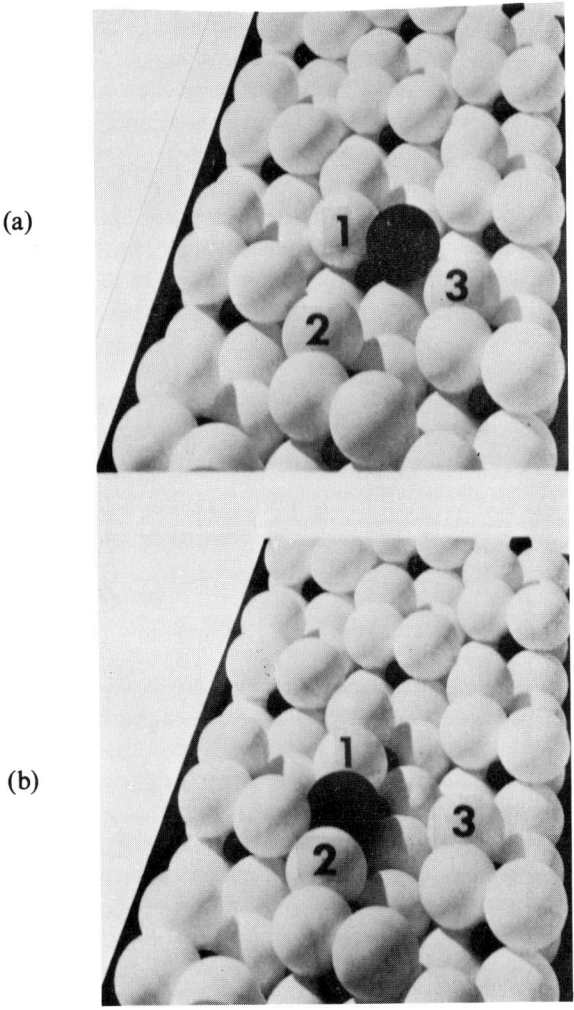

Fig. 16.9. Nonequivalence of the sites of alkyl group and chlorine vacancy in model of
$(10\overline{1}0)$. Black sphere represents an alkyl group.

(87) visualize the active center to be formed as in Eq. (25). One of the five chlorine ions around an exposed titanium ion is presumed to be loosely bound, since it is attached directly to only one titanium ion. It is this chlorine ion which will be removed in reaction. The remaining four chlorine ions are attached directly to two titanium ions and thus firmly bound. Only the sites of the loosely bound chlorine ion and of the vacancy are available for the alkyl group and the monomer entering into the propagation step. As demonstrated in Fig. 16.9, these two sites are not equivalent—an important factor which will be utilized below in explaining the stereospecific nature of the polymerization. In Fig. 16.9a, the black sphere (alkyl group) has four while that in Fig. 16.9b has seven of its twelve surrounding sites occupied. We see that in Fig. 16.9b, the alkyl group lies in the plane of chlorine ions 1, 2, and 3. Therefore, the most stable arrangement of an active center is a square base with three chlorine ions and the alkyl group at its corners and the titanium ion in the middle. The square is anchored by the fourth chlorine ion to the interior of the crystal, and the vacancy extends upward out of the crystal. The plane of the square base lies at an angle of 54°44′ with the (0001) face and thus in most cases oblique to other crystal faces. Consequently, two of the chlorine ions of the square base lie deeper in the crystal and are relatively blocked compared to the other chlorine ion of the base lying closer to the surface and thus exposed to a certain extent.

We now consider the requirements for stereospecificity in the polymerization of propylene. Arlman and Cossee (87) postulate that reaction can proceed according to mechanism (24) only when the C=C double bond is parallel to the titanium–alkyl bond (M—R). There are then four alternative positions for the propylene molecule of which two are ruled out immediately since they would require space for the methyl group over the blocked chlorine ions, which is unavailable. Referring to Fig. 16.10, in which one looks down on the square base around an exposed titanium ion, one can readily verify

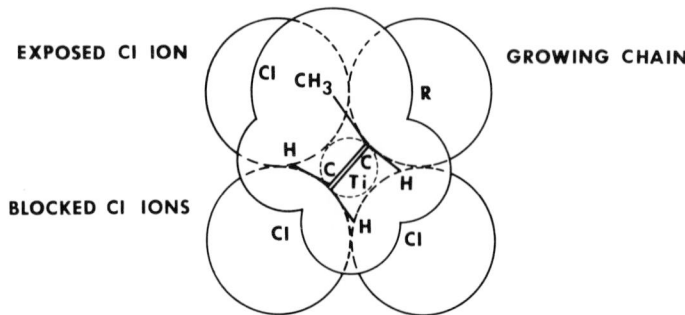

FIG. 16.10. Reactive position of propene on an active center. The propene molecule is projected perpendicularly onto the TiCl₃R plane.

that the plane perpendicular to this square base and going through the titanium–methyl bond is not a plane of symmetry because the exposed and corresponding blocked chlorine ions have different environments. Thus, the two remaining orientations are not equivalent. One partially covers a blocked chlorine ion, which is nearly impossible because of the steric hindrance of the surrounding chlorine ions. In the other orientation, the methyl group protrudes out of the crystal and is consequently the preferred one.

Thus, one of the requirements for stereoregular polymerization is satisfied, namely, the single orientation of the monomer on the active center. For the specific case of isotactic polymerization, there is a second requirement. Equation (24) shows that the alkyl group and the vacancy change sites after incorporation of monomer. These two sites were shown above not to be equivalent and we may therefore assume that one of them is preferred, so that after one propagation the alkyl group will move back to its original position. Thus, the process may be continued by a series of identical steps, which is the second requirement for isotactic polymerization. We see that the production of isotactic polypropylene follows logically from the particular surface structure of $TiCl_3$ crystals and the environment of the penta-coordinated titanium ions.

The production of syndiotactic polymers may also be fitted into the picture (*87*). When the rate of movement of the alkyl group back to its original position is decreased and the coverage of active sites increased, a new monomer molecule may react before the alkyl group has migrated. The configuration around the new asymmetric carbon will be the mirror image of the previous one which is the requirement for the production of syndiotactic polymers.

In a more recent publication (*88*), Cossee *et al.* have carried out quantitative molecular orbital calculations for the propagation step which give a fairly detailed picture of this catalytic reaction.

REFERENCES

1. Leonard, A., Suzuki, S., Fripiat, J. J., and De Kimpe, C., *J. Phys. Chem.* **68**, 2608 (1964).
2. Fripiat, J. J., Léonard, A., and Uytterhoeven, J. B., *J. Phys. Chem.* **69**, 3274 (1965).
3. Parry, E. P., *J. Catalysis* **2**, 371 (1963).
4. Mapes, J. E., and Eischens, R. P., *J. Phys. Chem.* **58**, 1059 (1954).
5. Eischens, R. P., and Pliskin, W. A., *Advan. Catalysis* **10**, 1 (1958).
6. Nicholson, D. E. *Nature* **186**, 630 (1960).
7. Basila, M. R., and Kantner, T. R., *J. Phys. Chem.* **70**, 1681 (1966).
8. Basila, M. R., Kantner, T. R., and Rhee, K. H., *J. Phys. Chem.* **68**, 3197 (1964).
9. Basila, M. R., and Kantner, T. R., *J. Phys. Chem.* **71**, 467 (1967).
10. Leftin, H. P., and Hall, W. K., *Actes Congr. Intern. Catalyse 2ᵉ Paris*, Vol. 1, p. 1353, Technip, Paris, 1961.

11. Hall, W. K., and Porter, R. P., *J. Catalysis* **5**, 544 (1966).
12. Porter, R. P., and Hall, W. K., *J. Catalysis* **5**, 366 (1966).
13. Hirschler, A. E., *J. Catalysis* **5**, 390 (1966).
14. Hirschler, A. E., and Hudson, J. O., *J. Catalysis* **3**, 239 (1964).
15. Peri, J. B., *Proc. 3rd Intern. Congr. Catalysis*, Vol I, 72, Amsterdam (1964).
16. Ozaki, A., and Kimura, K., *J. Catalysis* **3**, 395 (1964).
17. Gerberich, H. R., Larson, J. G., and Hall, W. K., *J. Catalysis* **4**, 523 (1965).
18. Hightower, J. W., and Hall, W. K., *Symp. Mech. Heterogeneous Catalysis*, Div. Petroleum Chem. Inc., Am. Chem. Soc. New York Meeting, September 11–16 (1966).
19. Gerberich, H. R., and Hall, W. K., *J. Catalysis* **5**, 99 (1966).
20. Hightower, J. W., Gerberich, H. R., and Hall, W. K., *J. Catalysis* **7**, 57 (1967).
21. Hightower, J. W., and Hall, W. K., *J. Phys. Chem.* **71**, 1014 (1967).
22. Clark, A., and Finch, J. N., *4th Intern. Congr. Catalysis* Preprint 75, Moscow (1968).
23. Harkins, C. G., Shang, W. W., and Leland, T. W., *J. Phys. Chem.* **73**, 130 (1969).
24. Lunsford, J. H., PhD Thesis, Rice Univ. Houston, Texas (1961).
25. Lunsford, J. H., and Leland, T. W., *J. Phys. Chem.* **66**, 2591 (1962).
26. Day, H. R., *Phys. Rev.* **91**, 822 (1953).
27. Peria, W. T., *Phys. Rev.* **112**, 423 (1958).
28. Yamaka, E., *Phys. Rev.* **96**, 293 (1954).
29. Johnson, P. D., *Phys. Rev.* **94**, 845 (1954).
30. Nelson, J. R., *Phys. Rev.* **99**, 1902 (1955).
31. Cohen, M. L., Lin, P. J., Roessler, D. M., and Walker, W. C., *Phys. Rev.* **155**, 992 (1967).
32. Roessler, D. M., and Walker, W. C., *Phys. Rev.* **159**, 733 (1967).
33. Clark, F. P., *Phil. Mag. Ser. 8*, **2**, 607 (1957).
34. Soshea, R. W., Dekker, A. J., and Sturtz, J. P., *J. Phys. Chem. Solids*, **5**, 23 (1958).
35. Low, W., and Weger, M., *Phys. Rev.* **118**, 1130 (1960).
36. Wong, J. Y., *Phys. Rev.* **168**, 337 (1968).
37. Armstrong, W. M., Chaklader, A. C. D., and Evans, D. G., *Trans. Brit. Ceram. Soc.* **61**, 246 (1962).
38. Reiling, G. H., and Hensley, E. B., *Phys. Rev.* **112**, 1106 (1958).
39. Lunsford, J. H., *J. Phys. Chem.* **68**, 2312 (1964).
40. Wertz, J. E., Auzins, P., Griffiths, J. H. E., and Orton, J. W., *Trans. Faraday Soc.* **28**, 136 (1959).
41. Wertz, J. E., Auzins, P., Griffiths, J. H. E., and Orton, J. W., *Trans Faraday. Soc.* **26**, 66 (1958).
42. Wertz, J. E., Orton, J. W., and Auzins, P., *J. Appl. Phys.* **33**, 322 (1962).
43. Nelson, R. L., Tench, A. J., and Harmsworth, B. J., *Trans. Faraday Soc.* **63**, 1427 (1967).
44. Farkas, A., and Farkas, L., *Proc. Roy. Soc.* (*London*), *Ser. A* **144**, 467, 481 (1934).
45. Farkas, A., and Farkas, L., *Trans. Faraday Soc.* **35**, 906 (1939).
46. Farkas, A., and Farkas, L., *Trans. Faraday Soc.* **36**, 522 (1940).
47. Horiuti, J., and Polanyi, M., *Nature* **132**, 819, 931 (1933).
48. Garnett, J. L., and Sollich-Baumgartner, W. A., *Advan. Catalysis* **16**, 95 (1966).
49. Garnett, J. L., and Sollich, W. A., *J. Catalysis* **2**, 339 (1963).
50. Garnett, J. L., and Sollich, W. A., *J. Catalysis* **2**, 350 (1963).
51. Garnett, J. L., and Sollich, W. A., *Australian J. Chem.* **14**, 441 (1961).
52. Garnett, J. L., and Sollich, W. A., *Australian J. Chem.* **15**, 56 (1962).
53. Kemball, C., *Advan. Catalysis* **11**, 223 (1959).
54. Kemball, C., *Proc. Chem. Soc.* 264 (1960).
55. Anderson, J. R., and Kemball, C., *Proc. Roy. Soc.* (*London*), *Ser. A* **226**, 472 (1954).

56. Kemball, C., *Proc. Roy. Soc. (London), Ser. A* **217**, 376 (1953).
57. Gault, F. G., and Kemball, C., *Trans. Faraday Soc.* **57**, 1781 (1961).
58. Erkelens, J., Galwey, A. K., and Kemball, C., *Proc. Roy. Soc. (London), Ser. A* **260**, 273 (1961).
59. Rooney, J. J., Gault, F. G., and Kemball, C., *Proc. Chem. Soc.* 407 (1960).
60. Gault, F. G., Rooney, J. J., and Kemball, C., *J. Catalysis* **1**, 255 (1962).
61. Burwell, R. L., Shim, B. K. S., and Rowlinson, H. C., *J. Am. Chem. Soc.* **79**, 5142 (1957).
62. Burwell, R. L., and Briggs, W. S., *J. Am. Chem. Soc.* **74**, 5096 (1956).
63. Meyer, E. F., and Burwell, R. L., *J. Am. Chem. Soc.* **85**, 2881 (1963).
64. Rooney, J. J., *Chem. Brit.* **2**, 242 (1966).
65. Rooney, J. J., *J. Catalysis* **2**, 53 (1963).
66. Banks, R. L., and Bailey, G. C., *Ind. Eng. Chem. Prod. Res. Develop.* **3**, 170 (1963).
67. Bradshaw, C. P. C., Howman, E. J., and Turner, L., *J. Catalysis* **7**, 269 (1967).
68. Calderon, N., Ofstead, E. A., Ward, J. P., Judy, W. A., and Scott, K. W., *J. Am. Chem. Soc.* **90**, 4133 (1968).
69. Clark, A., Am. Chem. Soc. Meeting, San Francisco, Phys. Chem. Div., April (1968).
70. Mol, J. C., Moulijn, J. A., and Boelhouwer, C., *Chem. Commun.* 633 (1968).
71. Okamoto, Y., Happel, J., and Kōyamo, H., *Bull. Chem. Soc. Japan* **40**, 2333, (1967).
72. Imai, H., Hasegawa, T., and Uchida, H., *Bull. Chem. Soc. Japan* **41**, 45 (1968).
73. Natta, G., and Danusso, F., *J. Polymer Sci.* **34**, 3 (1959).
74. Nenitzescu, C. O., Huch, C., and Huch, A., *Angew. Chem.* **68**, 438 (1956).
75. Eirich, F., and Mark, H., *J. Colloid Sci.* **11**, 748 (1956).
76. Bier, G., *Kunstoffe* **48**, 354 (1958).
77. Natta, G., *J. Inorg. Nucl. Chem.* **8**, 589 (1958).
78. Uelzmann, H., *J. Polymer Sci.* **32**, 457 (1958).
79. Patat, F., and Sinn, H., *Angew. Chem.* **70**, 496 (1958).
80. Gumboldt, A., and Schmidt, H., *Chem. Ztg.* **83**, 636 (1959).
81. Boor, J., Jr., *J. Polymer Sci.* **C1**, 257 (1963).
82. De Bruyn, P. H., *Chem. Weekblad.* **56**, 161 (1960).
83. Natta, G., *J. Polymer Sci.* **34**, 21 (1959).
84. Cossee, P., *J. Catalysis* **3**, 80 (1964).
85. Chatt, J., and Duncanson, L. A., *J. Chem. Soc.* 2939 (1953).
86. Arlman, E. J., *J. Catalysis* **3**, 89 (1964).
87. Arlman, E. J., and Cossee, P., *J. Catalysis* **3**, 99 (1964).
88. Cossee, P., Ros, P., and Schachtschneider, J. H., *4th Intern. Congr. Catalysis, Moscow,* Paper 14 (1968).

AUTHOR INDEX

Numbers in parentheses are reference numbers and indicate that an author's work is referred to although his name is not cited in the text. Numbers in italics show the page on which the complete reference is listed.

SUBJECT INDEX

DATE DUE

LIB 200 - 8 - TR